天津市蓟州区元古宇地层与沉积野外考察指导书

金振奎 朱小二 王金艺 王昕尧 任奕霖 王 凌 等编著

石油工业出版社

内容提要

本书通过对天津市蓟州区元古宇剖面进行详细的踏勘实测、系统描述、采样和室内分析，精选了13条野外地质考察线路和60个考察点，包括大量线路图、信手剖面图、地层柱状图、野外露头和镜下显微照片等，并配以精练的文字，让地质考察者能够宏观把握、细微深入地了解蓟州区元古宇地质特征。

本书可供地质相关专业学生、科研人员和地质爱好者参考阅读。

图书在版编目（CIP）数据

天津市蓟州区元古宇地层与沉积野外考察指导书 / 金振奎等编著 . —北京：石油工业出版社，2020.10

ISBN 978-7-5183-4007-1

Ⅰ . ①天… Ⅱ . ①金… Ⅲ . ①元古宙 – 区域地质 – 地质调查 – 天津 Ⅳ . ① P562.21

中国版本图书馆 CIP 数据核字（2020）第 078122 号

出版发行：石油工业出版社

（北京安定门外安华里 2 区 1 号　100011）

网　　址：www.petropub.com

编辑部：（010）64523544　　图书营销中心：（010）64523633

经　　销：全国新华书店

印　　刷：北京中石油彩色印刷有限责任公司

2020 年 10 月第 1 版　2020 年 10 月第 1 次印刷

889 毫米 ×1194 毫米　开本：1/16　印张：17.75

字数：460 千字

定价：140.00 元

（如出现印装质量问题，我社图书营销中心负责调换）

版权所有，翻印必究

《天津市蓟州区元古宇地层与沉积野外考察指导书》

编写人员

金振奎　朱小二　王金艺　王昕尧

任奕霖　王　凌　郭芪恒　李　阳

史书婷　李　硕　袁　坤　黎　瑞

闫　伟

PREFACE 前 言

　　天津市蓟州区北部山区广泛分布的元古宇露头剖面以其地层出露良好、连续、地质现象丰富、交通方便等得天独厚的特色而闻名于世，被李四光教授称道为"在欧亚大陆同时代地层中，蓟县剖面之佳，恐无出其右者"。这里是一座世界少有的、地质现象十分丰富多样的天然地质博物馆。同时，长期的国家地质公园建设修建了考察专用路，在各考察点设立了介绍地质现象的标牌。可以说，这里是地质类、地质工程类大学生和研究生实习、学者专家考察、地质科普的绝佳之地。

　　该区元古宇剖面是距今1650—780Ma地球演化史的最真实记录，隐含着丰富的前寒武纪古地理、古气候、古生物和古构造等方面的宝贵地质信息，在区域地层对比、生命形成与演化、地球演化、古地理和古气候演化等基础科学问题的研究方面具有重大价值，犹如一座有待发掘的自然宝库。其沉积地层从陆相到海相，从碎屑岩到碳酸盐岩，从浅水到深水应有尽有，岩石类型、沉积构造类型、沉积相类型丰富多样。不仅可以看到滨岸砂岩到深水陆棚页岩的连续过渡，还可以看到深水到浅水碳酸盐岩的连续过渡、碳酸盐岩与碎屑岩混合沉积、类型丰富的叠层石、生物丘、海绿石、"鸡爪石"、"虾米石"、气孔发育的火山岩、火山角砾岩、特征的米级沉积旋回以及笔者新发现的冰碛岩。元古宇之下的太古宇变质岩类型也是丰富多彩。

　　该区元古宇总厚度9547m，考察路线总长度约35km。正常情况下，考察完整个剖面需2～3天。

　　在过去多年间，笔者先后十余次带领研究生和油田专家对蓟州区元

古宇剖面进行地质考察。然而，目前关于蓟州区元古宇剖面还没有一本详尽的考察指导类书籍，难以满足考察者在该区域进行地质考察的过程中随时参考、查阅，并高效、迅速掌握相关地质资料的需求。正是基于这方面的考虑，笔者于2019年4月又带领9位研究生重新系统实测、描述了该剖面，并采集大量样品进行分析化验，收集了大量资料，在此基础上编写了该指导书，为未来考察此剖面的国内外学者提供相关资料和参考。

本指导书共分4章，系统地介绍了天津市蓟州区地理与交通概况（第一章），元古宇地层划分历史、地层序列特征及大地构造演化（第二章），13条野外地质考察线路的考察内容及各组沉积相分析（第三章），最后简单介绍了野外地质考察的基本技能和工作方法（第四章）。为了方便考察者使用，本书囊括了大量的地质图件、地层柱状图、信手剖面图及详尽的点位信息、野外露头和镜下显微照片等，并配以精练的文字，让考察者能够宏观把握、细微深入地了解所观察的地质现象。此外，本书还附有一些野外工作中常用的工具图件，从而极大地增强了书本的实用性。

为了方便国际学术交流，本书还包括了详细英文摘要，把核心内容（主要是第三章）都翻译成了英文。

本书由金振奎和其研究生共同完成。金振奎负责本书的整体内容设计和统稿。第一章和第二章由朱小二、黎瑞、袁坤、李硕和闫伟编写；第三章由金振奎、朱小二、王金艺、王昕尧、任奕霖、王凌、郭芪恒、

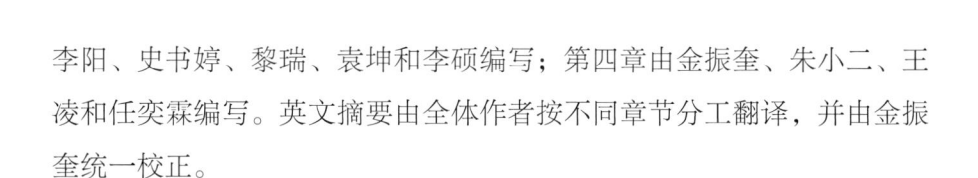

李阳、史书婷、黎瑞、袁坤和李硕编写；第四章由金振奎、朱小二、王凌和任奕霖编写。英文摘要由全体作者按不同章节分工翻译，并由金振奎统一校正。

在以往对该区元古宇的考察中，笔者曾与中国石油大学（北京）的鲍志东教授、季汉成教授和梁婷副教授及其研究生杨飞、张靓等一同考察研讨；与华北油田分公司原总经理（现为大港油田分公司总经理）赵贤正教授级高工以及金凤鸣教授级高工、杨得相教授级高工、韩春元高工等一同考察研讨，均获益匪浅，在此表示衷心感谢！本指导书在编写过程中参考和引用了大量已发表或未正式发表的国内外著作、文献、考察指南和手册等，以及一些专业网站，在此不再一一列举，一并表示诚挚感谢！

金振奎
中国石油大学（北京）地球科学学院教授
2020 年 6 月

目 录 CONTENTS

第一章　天津市蓟州区地理概况 ·· 1
　　第一节　地理位置和交通状况 ·· 1
　　第二节　地貌特征 ·· 2
　　第三节　气候特征 ·· 3

第二章　地层及大地构造演化 ·· 5
　　第一节　地层研究和划分历史 ·· 5
　　第二节　地层划分及特征 ·· 7
　　第三节　大地构造演化 ··· 18
　　第四节　资源概况 ··· 20

第三章　考察路线描述及沉积相分析 ······································· 22
　　第一节　概述 ··· 22
　　第二节　下营镇常州村—青山岭村北长城系常州沟组（Chc）
　　　　　　考察路线 ··· 24
　　第三节　下营镇青山岭村北—刘庄子村长城系串岭沟组（Chch）
　　　　　　考察路线 ··· 32
　　第四节　下营镇团山子村西长城系团山子组（Cht）考察路线 ··· 37
　　第五节　下营镇大红峪沟长城系大红峪组（Chd）考察路线 ····· 43
　　第六节　下营镇大红峪沟—罗庄子镇翟庄村北蓟县系
　　　　　　高于庄组（Jxg）考察路线 ····································· 47

第七节　罗庄子镇翟庄村北—青山村—花果峪村蓟县系
　　　　杨庄组（Jxy）考察路线 ·········· 57

第八节　罗庄子镇青山村—磨盘峪村—二十里铺村蓟县系
　　　　雾迷山组（Jxw）考察路线 ·········· 65

第九节　渔阳镇小岭子村蓟县系洪水庄组（Jxh）考察路线 ······ 71

第十节　渔阳镇小岭子村—罗庄子镇铁岭子村南蓟县系铁岭组（Jxt）
　　　　考察路线 ·········· 74

第十一节　罗庄子镇铁岭子村南—下庄子村待建系下马岭组（?x）
　　　　　考察路线 ·········· 80

第十二节　渔阳镇后寺沟青白口系龙山组（Qbl）考察路线 ······ 83

第十三节　渔阳镇西井峪村北青白口系景儿峪组（Qbj）
　　　　　考察路线 ·········· 88

第十四节　渔阳镇府君山公园西侧南华系西井峪组（Nhx）
　　　　　考察线路 ·········· 91

第十五节　沉积相分析 ·········· 96

第四章　野外地质工作基本技能和方法 ·········· 115

第一节　野外地质工作的基本原则 ·········· 115

第二节　野外地质工作常用物品及其使用方法 ·········· 115

第三节　野外地质工作的基本步骤及内容 ·········· 123

第四节　常见岩石和矿物的野外鉴定 ·········· 133

参考文献 ·········· 152

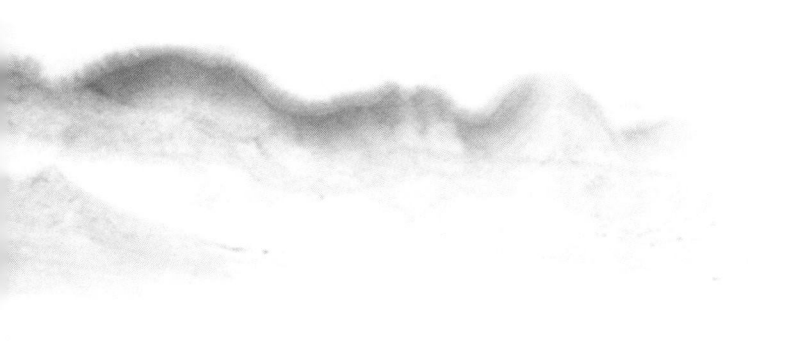

Abstract　Field Guide of Stratigraphy and Sedimentary Facies of the Proterozoic in Jizhou District, Tianjin ·················· 157
　　SECTION Ⅰ　Introduction ·················· 157
　　SECTION Ⅱ　Investigation Route of the Changzhougou Formation (Ch*c*) of Changcheng System from Changzhou Village to north of Qingshanling Village, Xiaying Town ·················· 161
　　SECTION Ⅲ　Investigation Route of the Chuanlinggou Formation (Ch*ch*) of Changcheng System from north of Qingshanling Village to Liuzhuangzi Village, Xiaying Town ·················· 170
　　SECTION Ⅳ　Investigation Route of the Tuanshanzi Formation (Ch*t*) of Changcheng System west of Tuanshanzi Village, Xiaying Town ·················· 175
　　SECTION Ⅴ　Investigate Route of the Dahongyu Formation (Ch*d*) of Changcheng System in Dahongyugou Valley, Xiaying Town ·················· 183
　　SECTION Ⅵ　Investigation Route of the Gaoyuzhuang Formation (Jx*g*) of Jixian System from Dahongyugou Valley to Zhaizhuang Village, Xiaying Town ·················· 187
　　SECTION Ⅶ　Investigation Route of the Yangzhuang Formation (Jx*y*) of Jixian System from Zhaizhuang Village to North Qingshan Village and Huaguoyu Village, Luozhuangzi Town ··· 196
　　SECTION Ⅷ　Investigation Route of the Wumishan Formation (Jx*w*) of Jixian System along Qingshan Village–Mopanyu Village–Ershilipu Village, Luozhuangzi Town ·················· 205
　　SECTION Ⅸ　Investigation Route of the Hongshuizhuang Formation (Jx*h*) of Jixian System in Xiaolingzi Village, Yuyang Town ···211

SECTION Ⅹ　　Investigation Route of the Tieling Formation （J$x t$） of Jixian System from Xiaolingzi Village in Yuyang Town to Tielingzi Village in Luozhuangzi Town ………… 215

SECTION Ⅺ　　Investigation Route of the Xiamaling Formation （?x） of System to be estabilished from the South of Tielingzi Village to Xiazhuangzi Village，Luozhuangzi Town ………… 222

SECTION Ⅻ　　Investigation Route of the Longshan Formation （Qbl） of Qingbaikou System in Housigou Valley in Yuyang Town ………… 226

SECTION ⅩⅢ　Investigation Route of the Jingeryu Formation （Qbj） of Qingbaikou System north of Xijingyu Village， Yuyang Town ………… 232

SECTION ⅩⅣ　Investigation Route of the Xijingyu Formation （Nhx） of Nanhua System west of Fujunshan Park in Yuyang Town ………… 236

SECTION ⅩⅤ　Sedimentary facies of each formation ………… 242

附录

实测地质剖面记录表 ………… 267

倾角换算表 ………… 268

剖面垂直比例尺放大后，岩层倾角大小歪曲结果表 ………… 269

含量估算模版 ………… 270

INTERNATIONAL CHRONOSTRATIGRAPHIC CHART ………… 271

国际年代地层表 ………… 272

第一章　天津市蓟州区地理概况

第一节　地理位置和交通状况

天津蓟县国家地质公园位于天津市北部的蓟州区，古称"渔阳郡"，西距北京市区约88km，南距天津市区115km，东距唐山115km，北距承德市200km，是京秦发展轴和津承发展轴的重要节点，地理区位优越，交通发达。大秦、京秦和津蓟铁路及京哈、津围、邦喜、宝平、平蓟等公路和县、乡公路在区内交织纵横，具备完善、便捷的交通体系（图1-1）。

图1-1　天津市蓟州区地理位置和交通状况

Fig.1-1　Geographical location and traffic condition of Jizhou District, Tianjin

蓟县元古宇国家自然保护区是天津蓟县国家地质公园的主要组成部分，是我国第一个国家级地质类自然保护区，地处蓟州区的北部山区，地理坐标介于北纬40°16′—40°21′和东经117°16′—117°30′之间，属于燕山山脉东段南麓，构造上为马兰峪背斜的西南翼。保护区北起长城脚下的常州村，南至蓟州北部府君山，南北长约24km，平均宽约350m，核心保护区面积达9km²（图1-2）。

保护区内元古宇剖面出露齐全、构造简单、现象丰富，且基础设施（道路、地质现象解说牌等）完备，是地质科研、旅游观光的极佳之地。

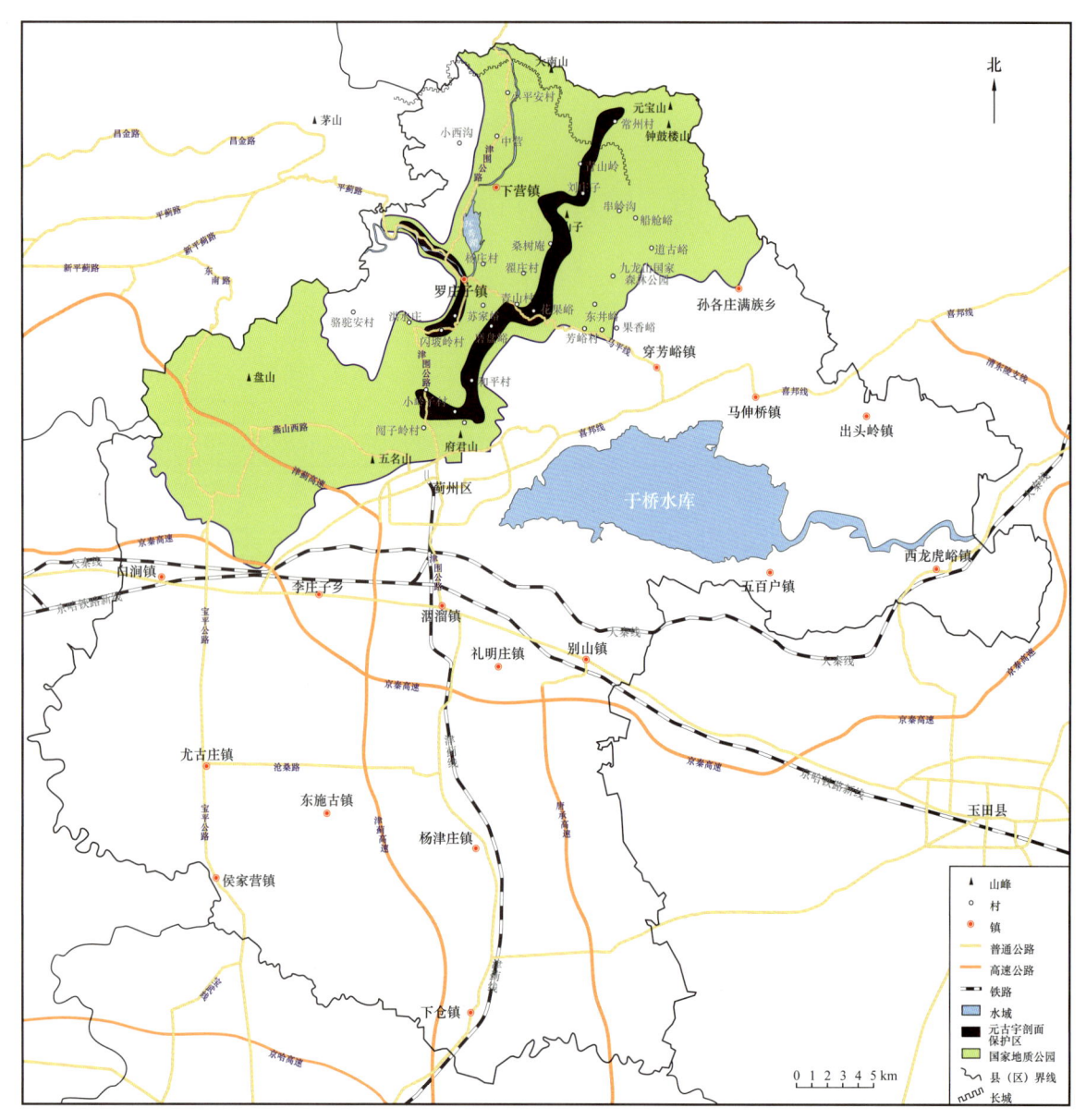

图 1-2 天津市蓟州区国家地质公园及元古宇剖面保护区位置

Fig.1-2 Locations of national geological park and natural reserve of Proterozoic section in Jixian District, Tianjin

第二节 地貌特征

蓟州区位于燕山山脉与华北平原的连接带上，古老的地质构造活动（如印支运动、燕山运动和喜马拉雅运动等）奠定了现今地貌发育的基础。其中中生代燕山运动使得区内发生强烈的断裂、褶皱、隆起和岩浆活动，对现今地貌形态起到主要控制作用，形成北部地区纬向的山脉，以及南部因断裂下沉而堆积的平原。

蓟州区地势北高南低，北部山地高耸，南部平原低平，地形高差悬殊，对比明显（图1-3）。

区内北缘与兴隆县交界处的九山顶，海拔 1078.5m，为蓟州区的最高峰，而最低处为南部的马槽洼，海拔仅 1.8m，海拔高差达 1076.7m。区内地貌类型主要表现为山区、平原和洼地，其中山区面积 840.5km²，平原面积 504.72km²，洼地面积 245.2km²。元古宇剖面保护区主要位于北部山区，地貌类型可以细分为中山、低山、丘陵、宽谷、盆地和坎谷等，中山和低山主要位于长城沿线及北部的石英砂岩分布区，山体高耸，峭壁悬崖；丘陵主要位于蓟州区城区以北的白云岩、石灰岩、泥岩和页岩分布区，海拔在 300~500m 之间，山体宽缓、慢坡缓岗。

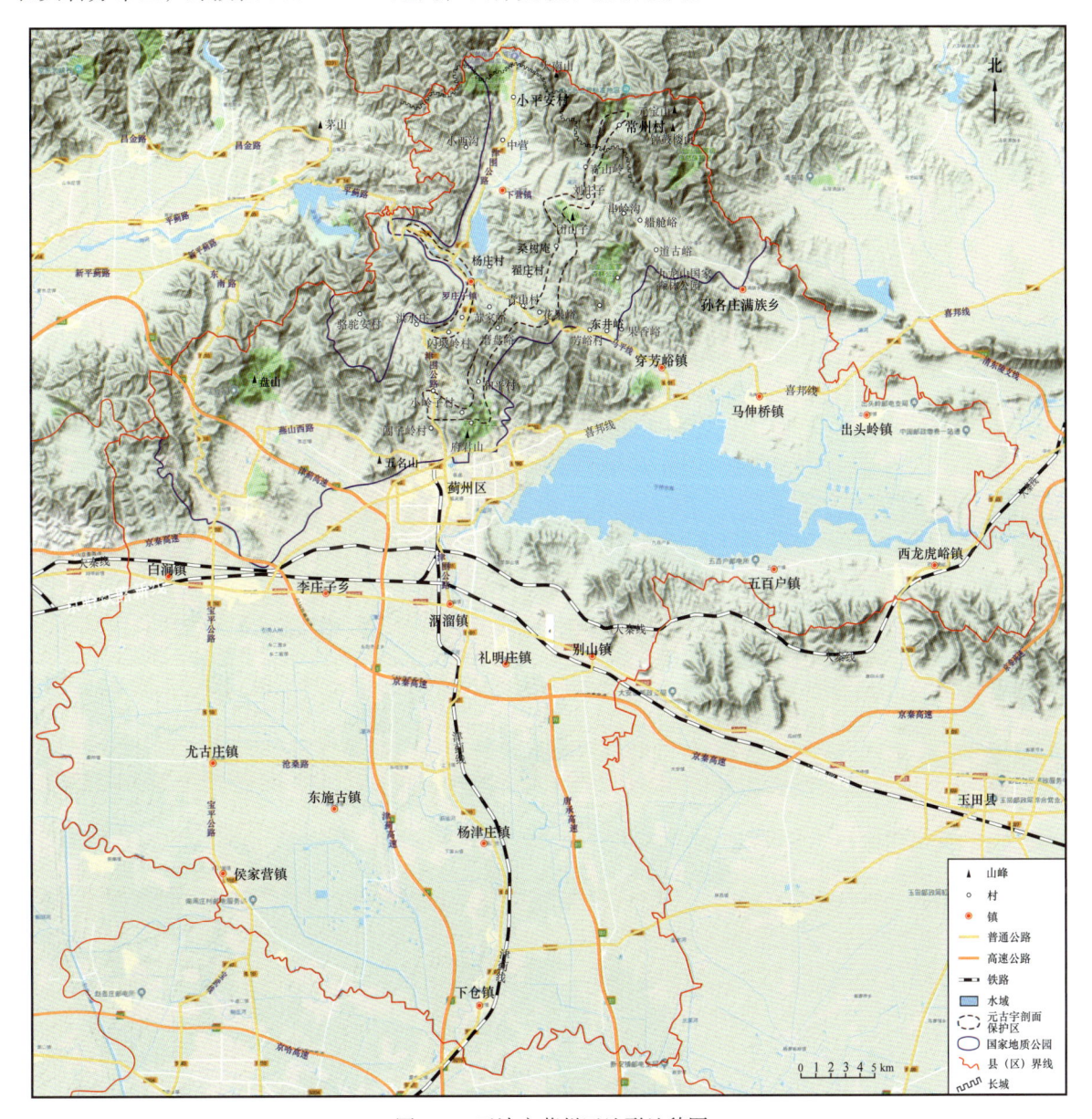

图 1-3 天津市蓟州区地形地貌图

Fig.1-3 Topographic map of Jizhou District, Tianjin

第三节 气候特征

蓟州区属于暖温带半湿润大陆性季风型气候，四季分明，雨热同期，阳光充足。区内年平均气温 11.5℃，南部洼区平均温度高于北部山区，低于山前平原区。这里季风气候鲜明，风向季节更替明显。因受蒙古—西伯利亚高气压和北太平洋副热带高压交替影响，冬季盛行西北风，夏季盛行东南风，年均风速为 2.2m/s 左右，最大风速达 25m/s。蓟州区为华北地区的多雨地带之一，年平均降水量达 678.6mm，但降水分配极度不均。夏季从北太平洋吹来的东南季风暖湿气流，受到北部高耸的燕山山脉的阻挡、抬升，在 7—9 月多形成集中降水，降水占全年总降水量的 76% 以上；而冬季多冰雪寒冷，大地封冻。冬雪一般从 11 月下旬起，到翌年 3 月中旬止，无霜期约 195 天。

　　蓟州区每年 3—6 月份，乍寒还暖，草木未丰，多晴少雨，是全年中野外考察的最佳时期；10—12 月，草木萧疏，风清气爽，也是野外考察的良好时期。不过由于剖面上植被不发育，全年皆可观察，但冬季寒冷。去考察之前，需提前了解气温和降雪情况。如果降雪量大，剖面会被积雪覆盖。

第二章　地层及大地构造演化

第一节　地层研究和划分历史

天津市蓟州区在元古宙处于燕辽裂陷带内裂陷最为强烈的地带，其间沉积了岩性复杂的巨厚地层序列，因地层厚度大、发育齐全，而成为中国北方元古宇的标准剖面所在地。该区元古宇剖面总厚度约9547m，以碳酸盐岩为主（约占73.5%），部分碎屑岩（砂砾岩约占17.6%，泥页岩约占7.6%），以及少量火山岩、火山碎屑岩和冰碛岩（约占1.3%）。

该套地层序列的研究起始于高振西、熊永先和高平于1931年建立的蓟县震旦系剖面（Kao et al.，1934），并在其后的几十年里，随着研究程度日益加深和研究手段的不断革新，该套地层序列的研究不断取得重大突破和进展，并对其地层层序和划分进行了多次的调整，大致经历了"震旦系""震旦亚界""中、上元古界"和"中、新元古界"及《中国地层表（2014）》等不同阶段（朱士兴等，2016）。

一、"震旦系"阶段

"震旦"一词最先由李希霍芬于1882年首次用于地层学领域（Richthofen，1882），并由Grabau（1922）做出了明确的定义，即指一套"不整合在深变质五台系之上和平行不整合在含化石的下寒武统馒头组页岩之下的未变质或轻微变质地层组成的岩系"。高振西等（1934）将蓟县的震旦系划分为三群十层（表2-1A），自下而上分为下震旦统南口群（包括长城石英岩、串岭沟页岩、大红峪石英岩和喷出熔岩、高于庄灰岩）、中震旦统蓟县群（包括杨庄页岩、雾迷山灰岩、洪水庄页岩、铁岭灰岩）和上震旦统青白口群（包括下马岭页岩和景儿峪灰岩），各群之间以假整合接触（表2-1A）。此后，孙云铸（1957）发现景儿峪灰岩上部地层属于下寒武统，且与"震旦系"之间呈不整合接触；王曰伦（1960）将该套下寒武统地层命名为"府君山组"，并于1963年发现三叶虫化石，从而确定了景儿峪灰岩的顶界；申庆荣和廖大（1958）提出高于庄灰岩与杨庄页岩间为整合接触，而高于庄灰岩与大红峪层之间为平行不整合，提议将高于庄灰岩划为蓟县群；王曰伦（1963）提出中国南北方的"震旦系"并非平行关系，而是上下关系，中国北方的"震旦系"位于南方的之下。

二、"震旦亚界"阶段

1975年关于前寒武纪的全国性座谈会（北京）提出将以三峡剖面所代表的南方"震旦系"置于以蓟县剖面为代表的北方"震旦系"之上，并统称为"震旦亚界"。同年，华北区前寒武纪地层专题会议（太原）决定将景儿峪组下部的陆源碎屑岩单独分为一组，并采用郝贻纯（1954）命名的"龙山组"一名（乔秀夫，1976）。后因与南方"龙山群"重复，北京市地质研究院将其改称"长龙山组"。1982年邢裕盛等提出将其替代为"骆驼岭组"（邢裕盛等，1989）。近年来学界使用"龙山

表 2-1 燕山地区元古宇沉积地层划分方案沿革表（据朱士兴等，2016 修改）

Table 2-1 Evolution table of stratigraphic division of Proterozoic in Yanshan area (modified from Zhu et al., 2016)

*注：*1. 据李怀坤等，2009；*2. 据高林志等，2007；*3. 据李怀坤等，2011；*4. 据苏文博等，2010；*5. 据李怀坤等，1995；*6. 据颉颃松年等，1991；*7. 据李怀坤等，1995；*8. 据高维华等，2008；*9. 据和政军等，2011。

组"较多，本书采用这一名称。此后，国内兴起了前寒武纪地层研究的新高潮，并系统地进行了岩石学、古生物学和地球化学等多学科的研究，给出了华北元古宇的年龄框架（表2-1B）。

三、"中、上元古界"和"中、新元古界"阶段

为了与国际地科联发布的前寒武系划分方案（表2-1F）相统一，以及鉴于"震旦亚界"的不合理性，我国于1982年再次召开前寒武纪座谈会，决定废弃"震旦亚界"一词，将"震旦系"限用于南方，并确定北方原"震旦系"的三系十二组的新划分方案，且均归属于中、上元古界（表2-1C）。1998年，全国地层委员会在关于推荐《中国地质年代表》的通告中将元古宙（宇）细分为古元古代（界）、中元古代（界）和新元古代（界），并将长城纪（系）、蓟县纪（系）归入中元古代（界），而青白口纪（系）归入新元古代（界），并正式成为我国中、新元古代（界）的地质年代（地层）单位（表2-1D）。

四、《中国地层表（2014）》阶段

最近十余年，由于测年新方法的突破，在下马岭组中的斑脱岩（凝灰岩）及侵入岩中获取了一系列的最新测年数据（Gao et al., 2007; 2008; 2009; Su et al., 2008; 李怀坤等，2009; 苏文博等，2010; 苏犁等，2016），从而将下马岭组的沉积时限从原来的1000—900Ma下拉至1400—1320Ma，使得原有的地层格架必须做出明显的调整。因而，全国地层委员会于2014年正式公布了《中国地层表（2014）》。在最新的地层表中将下马岭组从新元古界青白口系底部下移至中元古界的中部（称为"待建系"），并承认下马岭组与青白口系之间存在长期缺失（约420Ma），同时将长城系和蓟县系的界线下移至大红峪组和高于庄组之间，且为与国际地层表相接轨，将长城系（纪）归入古元古界（代）（表2-1E）。

综上所述，随着研究的不断深入，学者对华北元古宙地层的认识和划分，虽然仍存在着不同的认识，但地层划分的基本框架逐渐趋于统一，因而为了避免地层划分上的争议，本书主要参照全国地层委员会2014年推荐的最新的地层表（表2-1E）对天津市蓟州区的元古宇进行论述。

第二节 地层划分及特征

天津市蓟州区元古宇剖面整体位于马兰峪背斜的西南翼，由北向南元古宙地层由老至新依次出露。除局部区域因少量断层切割而缺失、重复外，地层整体连续、齐全，保存完好。该区整个元古宇厚9545m，由老至新可划分为古元古界长城系的常州沟组（Chc，厚859m）、串岭沟组（Chch，厚889m）、团山子组（Cht，厚518m）、大红峪组（Chd，厚408m）；中元古界蓟县系高于庄组（Jxg，厚1700m）、杨庄组（Jxy，厚773m）、雾迷山组（Jxw，厚3416m）、洪水庄组（Jxh，厚131m）和铁岭组（Jxt，厚303m）；待建系的下马岭组（?x，厚165m）；新元古界青白口系的龙山组（Qbl，厚118m）和景儿峪组（Qbj，厚110m）（图2-1—图2-3和表2-1）。此外，本次研究新建了一个组，即"西井峪组"（Qbx，厚155m），位于景儿峪组之上、寒武系府君山组之下，并认为属于新元古界。

一、古元古界长城系（Ch）

古元古界长城系（Ch）主要分布在天津市蓟州区下营镇一带，总体厚度达2674m。该系下部以砂岩为主，中部主要为页岩夹粉砂岩，中上部以含铁白云岩夹泥岩、砂岩为主，上部主要为石英砂岩、夹富钾基性火山岩和燧石质白云岩；自下而上分为常州沟组（Chc）、串岭沟组（Chch）、团山子组（Cht）和大红峪组（Chd）等四个组。该系底部角度不整合接触于太古宇遵化杂岩群的石榴角闪斜长片麻岩之上，顶部与蓟县系呈平行不整合接触。

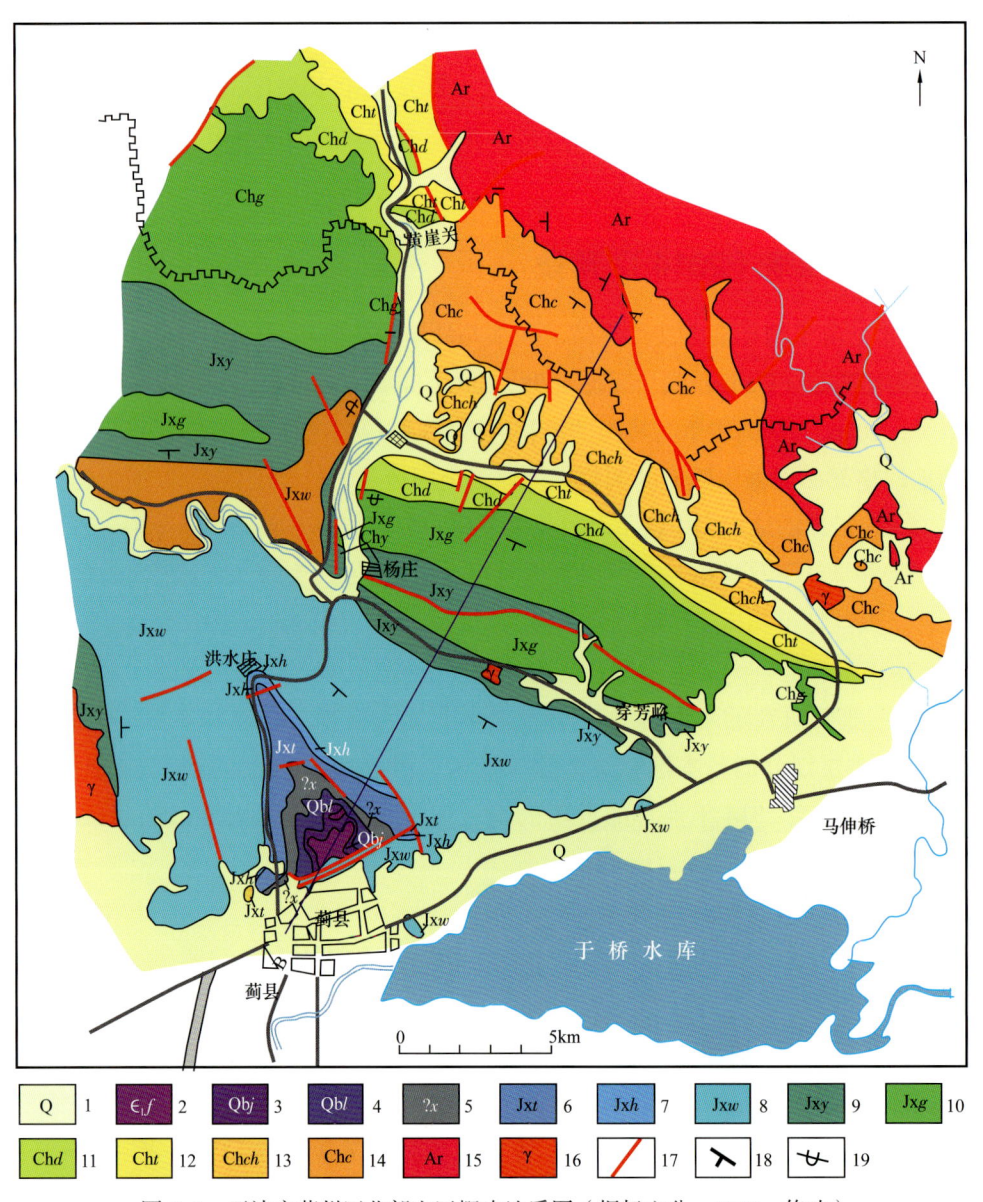

图 2-1　天津市蓟州区北部山区概略地质图（据杨立公，2010，修改）

Fig.2-1　Geological sketch map of northern mountain area of Jizhou District, Tianjin (modified from Yang, 2010)

1—第四系；2—下寒武统府君山组；3—新元古界青白口系景儿峪组；4—新元古界青白口系龙山组；5—中元古界待建系下马岭组；6—中元古界蓟县系铁岭组；7—中元古界蓟县系洪水庄组；8—中元古界蓟县系雾迷山组；9—中元古界蓟县系杨庄组；10—中元古界蓟县系高于庄组；11—古元古界长城系大红峪组；12—古元古界长城系团山子组；13—古元古界长城系串岭沟组；14—古元古界长城系常州沟组；15—太古宇遵化岩群；16—花岗岩；17—断层；18—地层产状；19—倒转地层

第二章 地层及大地构造演化

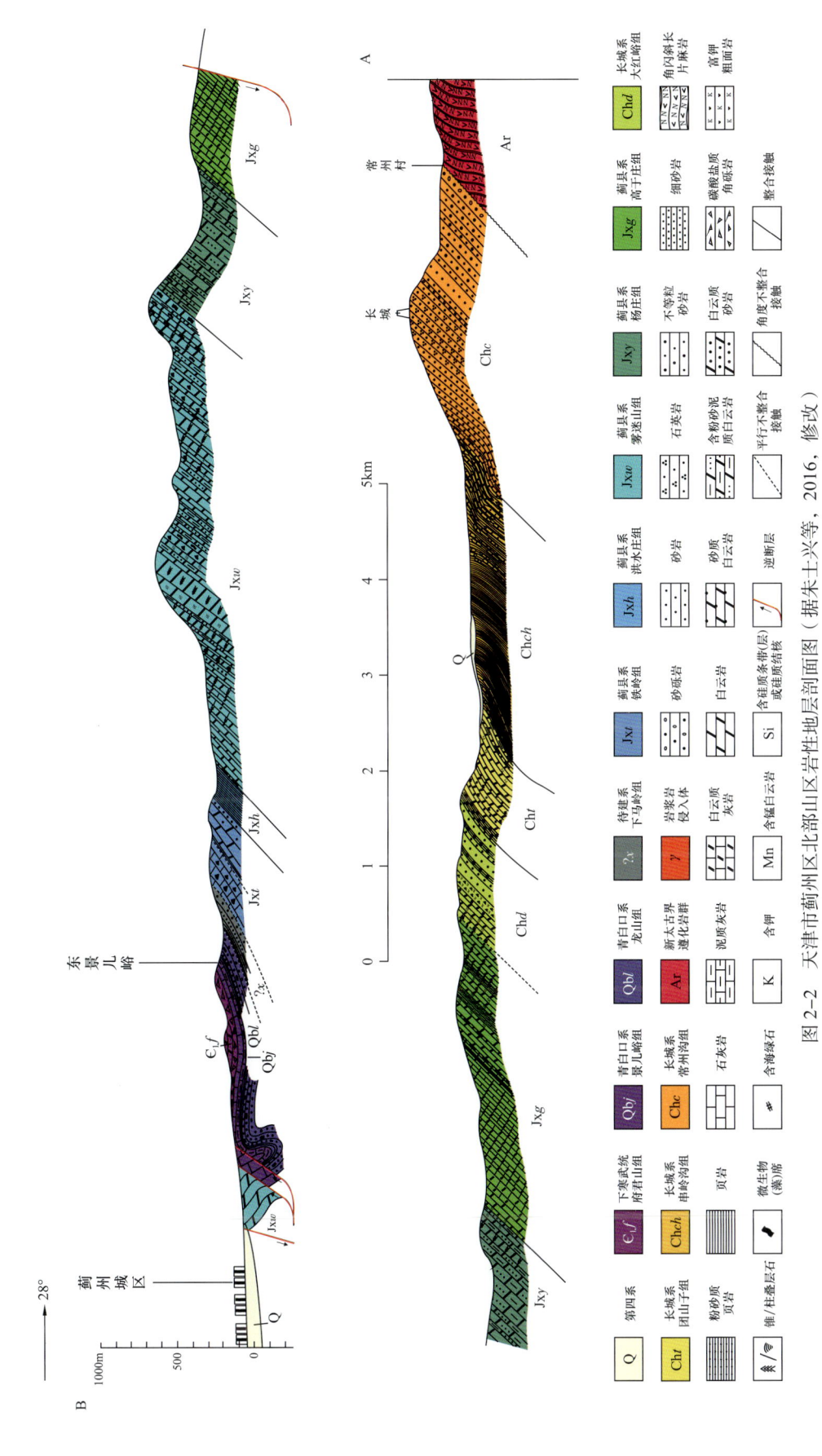

图 2-2 天津市蓟州区北部山区岩性地层剖面图（据朱士兴等，2016，修改）

Fig.2-2 Stratigraphic section of northern mountain area of Jizhou District, Tianjin (modified from Zhu et al., 2016)

- 9 -

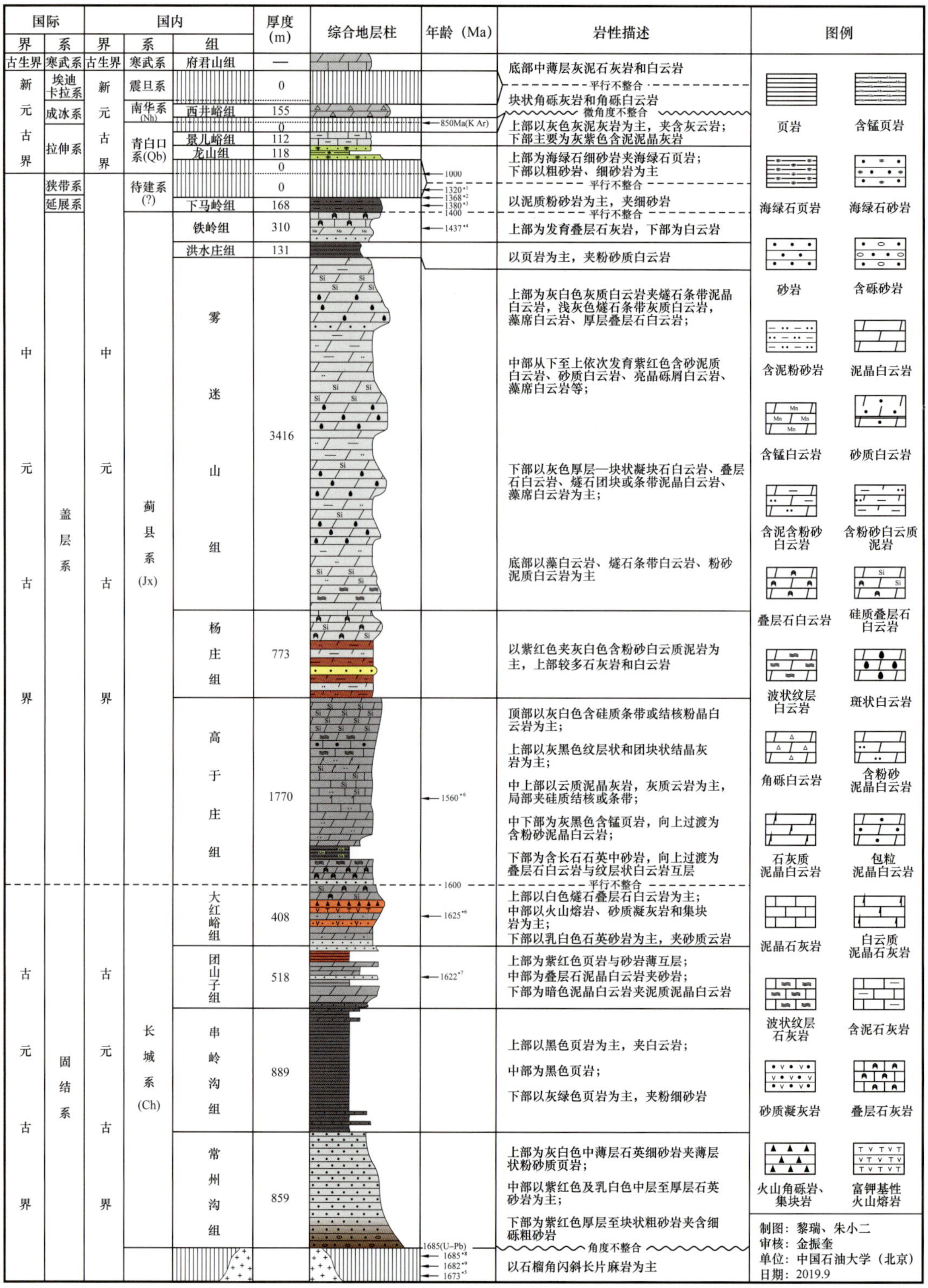

国际			国内		厚度(m)	综合地层柱	年龄(Ma)	岩性描述	图例			
界	系	界	系	组								
古生界	寒武系	古生界	寒武系	府君山组	—			底部中薄层灰泥质石灰岩和白云岩	页岩		含锰页岩	
新元古界	埃迪卡拉系	新元古界		震旦系	0		平行不整合	块状角砾岩和角砾白云岩				
	成冰系		南华系(Nh)	西井峪组	155		微角度不整合 850Ma(K Ar)	上部以灰白色灰泥质灰岩为主，夹灰岩；下部主要为灰绿色泥晶灰岩				
	拉伸系		青白口系(Qb)	景儿峪组	112		←1000	上部为灰绿色细砂岩夹海绿石页岩；下部以粗砂岩、细砂岩为主				
				龙山组	118				海绿石页岩		海绿石砂岩	
	狭带系延展系		待建系(?)	下马岭组	0 168		平行不整合 ←1320*³ ←1368*² ←1380*³ ←1400	以泥质粉砂岩为主，夹细砂岩				
				铁岭组	310		平行不整合 ←1437*⁴	上部为发育叠层石灰岩，下部为白云岩	砂岩		含砾砂岩	
				洪水庄组	131			以页岩为主，夹粉砂质白云岩				
中元古界	盖层系	中元古界	蓟县系(Jx)	雾迷山组	3416			上部为灰白色灰质白云岩夹燧石条带泥晶白云岩，浅灰色燧石条带泥质白云岩，藻席白云岩、厚层叠层白云岩；	含泥粉砂岩		泥晶白云岩	
								中部从下至上依次发育紫红色含砂泥质白云岩、砂质白云岩、亮晶砾屑白云岩、藻席白云岩等；	含锰白云岩		砂质白云岩	
								下部为灰色厚层—块状凝晶石白云岩、叠层石白云岩、燧石团块或条带泥晶白云岩、藻席白云岩为主；	含泥含粉砂白云岩		含粉砂泥质白云岩	
								底部为藻白云岩、燧石条带白云岩、粉砂泥质白云岩为主				
				杨庄组	773			以紫红色夹灰白色含粉砂灰质泥岩为主，上部较多石灰岩和白云岩	叠层石白云岩		硅质叠层石白云岩	
				高于庄组	1770		←1560*⁵	顶部以灰白色含硅质条带或结核粉晶白云岩为主；	波状纹层白云岩		斑状白云岩	
								上部以灰黑色纹层状和团块状结晶灰岩为主；				
								中上部以云质泥晶灰岩，灰质云岩为主，局部夹硅质结核或条带；	角砾白云岩		含粉砂泥晶白云岩	
								中下部为灰黑色含锰页岩，向上过渡为含粉砂泥晶白云岩；				
								下部为含长石石英中砂岩，向上过渡为叠层石白云岩与白云岩互层	石灰质泥晶白云岩		包粒泥晶白云岩	
							←1600 平行不整合					
古元古界	固结系	古元古界	长城系(Ch)	大红峪组	408		←1625*⁶	上部以灰白色燧石叠层石白云岩为主；中部以火山熔岩、砂质凝灰岩和集块岩为主；下部以乳白色石英砂岩为主，夹砂质云岩	泥晶石灰岩		白云质泥晶石灰岩	
				团山子组	518		←1622*⁷	上部为紫红色页岩与砂岩薄互层；中部叠层泥晶云岩夹砂岩；下部为暗色泥质白云岩夹泥质泥晶灰岩	波状纹层石灰岩		含泥石灰岩	
				串岭沟组	889			上部以黑色页岩为主，夹白云岩；中部为黑色页岩；下部以灰绿色页岩为主，夹粉细砂岩	砂质凝灰岩		叠层石灰岩	
				常州沟组	859		←1685(U-Pb) ←1685*⁸ ←1682*⁹ ←1673*⁹	上部为灰白色中薄层石英细砂岩夹薄层状粉砂质泥岩；	火山角砾岩、集块岩		富钾基性火山熔岩	
								中部为紫红色及乳白色中层至厚层石英砂岩为主；				
								下部为紫红色厚层至块状粗砂岩夹含细砾粗砂岩	制图：黎瑞、朱小二 审核：金振奎 单位：中国石油大学（北京） 日期：2019.9			
							角度不整合	以石榴角闪斜长片麻岩为主				

*注：*1：据李怀坤等，2009；*2：据高林志等，2007；*3：据su et al.，2007；*4：据苏文博等，2010；*5：据李怀坤等，2011；*6：据陆松年，1991；*7：据李怀坤等，1995；*8：据高维等，2008；*9：据和政等，2011。

图2-3 天津市蓟州区元古宇综合柱状图

Fig.2-3 Stratigraphic column of Proterozoic in Jizhou District，Tianjin

长城系的时限过去一直被认定为1800—1600Ma之间，但是近十年来获取的一系列定年数据对其时限进行了重新厘定。例如部分学者通过对北京密云地区侵入太古宇变质岩基底并被常州沟组不整合覆盖的花岗斑岩岩脉进行锆石定年，获取的岩脉侵入时间均集中在1685—1670Ma之间（高维等，2008；何政军等，2011；李怀坤等，2011），从而将较此年龄新的、不整合于其上的长城系常州沟组的沉积年龄限定在晚于1650Ma；同时部分学者也在长城系顶部大红峪组中部的火山熔岩中获取了一些最新的锆石年龄，皆分布在1625—1622Ma之间（陆松年和李惠民，1991；Lu et al., 2008；Gao et al., 2009），因而将较该年龄稍新的长城系顶界沉积时间限定在1620Ma之前；因此燕山地区长城系的沉积时限基本可以限定在1650—1620Ma之间（朱士兴等，2016），这与过去一直认定的长城系1800—1600Ma的时限存在巨大的差异，仅相当于国际地层表古元古界固结系（1800—1600Ma）的上部地层。

1. 常州沟组（Chc）

常州沟组角度不整合于太古宇遵化杂岩群的石榴角闪斜长片麻岩之上，是天津市蓟州区最古老的沉积地层，对应于高振西等（1934）划分的"南口群长城石英岩"。后因申庆荣和廖大从（1958）将"南口统"改为"长城统"，为避免重复，1958年陈荣辉等将"长城石英岩"改称"黄崖关组"（陈荣辉和陆宗斌，1963）。后因黄崖关处的碎屑岩并非属于"黄崖关组"，而长城石英岩在常州村一带出露良好，因而俞建章等（1964）建议改称为"常州村组"。但常州村的居民房屋多分散在太古宇变质岩之上，所以1964年召开的"蓟县震旦系现场学术讨论会议"提议改为常州沟组，并一直沿用至今。

常州沟组的正层型位于天津市蓟州区下营镇常州村至青山岭村北一带，为一套以砂岩为主的沉积岩系，总厚度约859m。由于该组岩石坚硬，不易风化，所以其分布区大多形成巍峨的山峰和峭壁，且区内的长城相当一部分建造于该组形成的山脊之上。

根据岩性变化和旋回，可细分为三段：下部常一段（Chc_1）主要为紫红色厚层至块状的砂质细砾岩和含细砾粗砂岩等，板状、楔状及槽状交错层理都很发育；中部常二段（Chc_2）为乳白色夹淡紫红色中层至厚层石英细砂岩，槽状及冲洗交错层理发育，在地貌上形成陡崖；上部常三段（Chc_3）的下部为中到厚层灰白色石英细砂岩夹灰绿色泥质薄层（单层厚度<1cm），向上逐渐过渡为中层灰白色石英细砂岩夹灰绿色薄层粉砂质页岩。该组自下而上，由氧化色变为还原色，粒度变细，层变薄，泥质夹层变多。

2. 串岭沟组（Chch）

串岭沟组连续过渡地整合接触于常州沟组之上，对应于高振西等（1934）划分的"串岭沟页岩"的下部碎屑岩系，并经第一次全国地层会议（1959）改称串岭沟组，并一直沿用至今。

串岭沟组主要出露在天津市蓟州区下营镇青山岭村北至刘庄子一带，主要以粉砂质伊利石页岩为主，夹少量粉砂岩及白云岩，厚度约889m，与下伏常州沟组呈渐变接触关系。根据岩性组合特征，可自下而上分为三段：串一段（Chch_1）以灰绿色、黄绿色粉砂岩和粉砂质伊利石页岩互层为主，向上粉砂岩出现频率逐渐降低；串二段（Chch_2）主要为灰黑色、黄绿色伊利石页岩，局部含粉砂；串三段（Chch_3）为黑色伊利石页岩，上部见辉绿岩岩墙。由于串岭沟组岩性较软，易于风化，沿走向形成山谷，夹于常州沟组山脉与团山子组—大红峪组山脉之间。

3. 团山子组（Cht）

团山子组底部以含铁白云岩和粉砂质含铁白云岩的出现与下伏串岭沟组分界，两者之间呈渐变整合接触。该组对应于高振西等（1934）划分的"串岭沟组"的上部碳酸盐岩系，与在河北省庞家堡矿区命名的"庞家堡灰岩"相似，因在天津市蓟州区下营镇团山子村一带出露良好，故建议改称团山子组，并被1964年召开的"蓟县震旦系现场学术讨论会议"所采纳，一直沿用至今。

团山子组为一套以碳酸盐岩为主，夹碎屑岩的沉积，总厚度约518m。根据岩性组合变化，在团山子村一带自下而上可划分为四段：

团一段（Cht_1）为深灰色中—薄层泥质泥晶白云岩和粉砂质泥晶白云岩与深灰色白云质页岩互层。

团二段（Cht_2）以灰色中—厚层泥晶白云岩夹薄层泥质泥晶白云岩为主。由于白云岩含铁量较高，其风化面上多呈黄褐色。

团三段（Cht_3）下部主要为深灰色薄层泥晶白云岩（风化后黄褐色）夹灰色白云质石英细砂岩，中部为深灰色页岩与细砂岩薄互层，夹数层具气孔构造的玄武岩（单层厚度几十厘米），上部为中—薄层泥晶白云岩夹中—薄层白云质石英砂岩。白云岩中层状和波状叠层石、泥裂常见；砂岩中交错层理和波痕常见。

团四段（Cht_4）为薄层紫红色粉砂质页岩夹细砂岩，砂岩层内常见小型交错层理；向上，细砂岩层变多、变厚，并过渡为大红峪组的灰白色厚层石英细砂岩。该组的分布区多呈低山地貌特征，但显著比串岭沟组地势高。

4. 大红峪组（Chd）

大红峪组与下伏团山子组呈渐变整合接触，对应于高振西等（1934）命名的"大红峪石英岩及安山熔岩"，申庆荣和廖大从（1958）称之为"大红峪层"，1959年第一次全国地层会议推荐使用"大红峪组"，并一直沿用至今。

大红峪组在天津市蓟州区大红峪沟一带出露完整，为一套火山—沉积岩系，总厚度约408m。根据岩性组合特征，自下而上可细分为三段：下部的大一段（Chd_1）以中—厚层乳白色石英砂岩为主，夹紫红色粉砂岩、含燧石砂质白云岩或白云质砂岩以及翠绿色凝灰岩。石英砂岩波痕和交错层理发育；中部的大二段（Chd_2）以富钾火山熔岩、火山角砾岩、集块岩为主，夹少量石英砂岩和凝灰岩；上部的大三段（Chd_3）以灰白色含燧石白云岩为主，含大量锥状叠层石（天津地质矿产研究所等，1979）。由于大红峪组岩石较为坚硬，其分布区多形成中等高度的山岭地貌。

二、中元古界蓟县系（Jx）

中元古界蓟县系主要分布在天津市蓟州区罗庄子镇和穿芳峪乡一带，是一套碳酸盐岩夹碎屑岩沉积，总厚度约6158m。根据岩性组合特征，该系自下而上划分为高于庄组（Jxg）、杨庄组（Jxy）、雾迷山组（Jxw）、洪水庄组（Jxh）和铁岭组（Jxt）五个组。

蓟县系（Jx）与下伏长城系（Ch）和上覆待建系（?）均呈区域性平行不整合接触，而在蓟县系内部地层是连续的，不存在沉积缺失。蓟县系底部高于庄组与下伏大红峪组之间虽然存在区域性的不整合接触关系，但在盆地中央仍有整合接触的地方，说明两者之间的沉积间断时间不会太长（朱士兴等，2016），因此可以用大红峪组的锆石定年数据将蓟县系的底界年龄限定为小于1620Ma；同时，李怀坤等（2010）在燕山西段高于庄组内凝灰岩中获取的SHRIMP U-Pb年龄

1559Ma±12Ma 和 LA-MC-ICPMS U-Pb 年龄 1560Ma±5Ma，蓟县系底界应该较此年龄数据老。结合上述测年数据的限定，同时考虑到 1600Ma 在国际地层表中是一个非常重要的地层界线，也是古元古代和中元古代的年代分界，可将蓟县系的底界（即长城系和蓟县系的分界）限定在 1600Ma，这样既可以得到现有数据的支撑，也有利于国际间的地层对比。同样地，根据近年来在蓟县系顶部铁岭组钾质斑脱岩中获取的年龄数据 1437Ma±21Ma（苏文博等，2010），以及上覆待建系下马岭组斑脱岩和侵入岩中获取的年龄数据（Gao *et al.*，2007；2008；2009；Su *et al.*，2008；李怀坤等，2009；苏文博等，2010；苏犁等，2016），可将蓟县系的顶界年龄限定在 1400Ma 左右。综上所述，蓟县系的沉积时限为 1600—1400Ma，与国际地层表中元古界盖层系完全对应。

1. 高于庄组（Jxg）

高于庄组最早由高振西等（1934）根据天津市蓟州区北的高各庄村和于各庄村所创名的"高于庄灰岩"，经 1959 年召开的"第一次全国地层会议"改称为"高于庄组"，并一直沿用至今。

高于庄组在天津市蓟州区下营镇大红峪沟至罗庄子镇翟庄村北一带出露良好，是位于大红峪组之上、杨庄组之下以碳酸盐岩占绝对优势的一套地层，总厚度约 1770m，与下伏长城系大红峪组呈平行不整合接触，与上覆杨庄组呈整合接触。陈晋镳等（1980）将高于庄组划分为四个亚组和10 个岩性段。四个亚组皆按照临近地名进行命名，自下而上依次为：官地亚组、桑树庵亚组、张家峪亚组和环秀寺亚组。

本书根据岩性组合的差异，将该组自下而上细分为 6 个岩性段：

高一段（相当于官地亚组）为灰色中—厚层含燧石条带和结核叠层石泥晶白云岩夹灰绿色薄层页岩，底部有一套厚约 3m 的石英砂岩，小型波痕发育。

高二段（相当于桑树庵亚组）为灰褐色含锰粉砂质页岩与含锰粉砂质白云岩、含锰粉砂岩互层，中下部含有"蓟县式锰（硼）矿"。

高三段（相当于张家峪亚组下部）为灰色中—厚层泥晶白云岩，夹少量灰绿色页岩。白云岩含铁，风化后呈褐色。

高四段（相当于张家峪亚组上部）为灰色中层灰质白云岩、白云质灰岩夹薄层泥质灰质白云岩、白云质灰岩。底部有一层几十厘米厚的瘤状灰岩。顶部有"臼齿灰岩"或称"鸡爪灰岩"。

高五段（相当于环秀寺亚组下部）下部为纹层状沥青质细晶白云岩和细晶灰岩，上部为核形石状粗晶白云岩（又称"虾米石"）与纹层状中细晶白云岩、泥晶白云岩互层。

高六段（相当于环秀寺亚组上部）为灰色中厚层含燧石结核和条带的泥晶白云岩。

2. 杨庄组（Jxy）

杨庄组与下伏的高于庄组除了在局部地区（如河北滦县等）呈平行不整合接触，在整个燕山地区主要为整合接触。杨庄组对应于高振西等（1934）命名的"杨庄页岩"，经由 1959 年召开的"第一次全国地层会议"改称为杨庄组，并一直沿用至今。

杨庄组在天津市蓟州区罗庄子镇杨庄村一带广泛出露，岩性主要以醒目的紫红色与灰白色相间（即红白相间）的泥岩为特征，像"五花肉"。泥岩常含粉砂和泥晶白云石。该组上部夹块状藻白云岩和石灰岩以及叠层石白云岩，总厚度约 773m。按照岩性组合，自下而上可以划分为三段：

杨一段（Jxy_1）为"红白相间"的泥岩，泥岩中常含粉砂和泥晶白云石，厚 209m。

杨二段（Jxy$_2$）为"红夹白"泥岩，泥岩中常含粉砂和泥晶白云石。上部见一层几十厘米厚的粗砂岩，厚434m。

杨三段（Jxy$_3$）为"红白相间"的泥岩夹灰色块状藻白云岩、石灰岩以及叠层石白云岩，厚130m。

3. 雾迷山组（Jxw）

雾迷山组整合接触于下伏的杨庄组之上，对应于高振西等（1934）命名的"雾迷山灰岩"，其中雾迷山是五名山的谐音，后经1959年召开的"第一次全国地层会议"改称雾迷山组，并一直沿用至今。天津综合地质大队于1962年曾将该组自下而上划分为罗庄层、二十里铺层、闪坡岭层，其后河北区测队又划分为四个段。陈晋镳等（1980）将其自下而上划分为罗庄亚组（雾一段和雾二段）、磨盘峪亚组（雾三段和雾四段）、二十里堡亚组（雾五段和雾六段）和闪坡岭亚组（雾七段和雾八段）等四个亚组和八个岩性段。

本书采用雾迷山组的"四分"方案，即划分为4个亚组（或4个段）。

雾迷山组在天津市蓟州区罗庄子镇磨盘峪村北山顶至磨盘峪南山（考察路线之一），及二十里铺起沿公路至洪水庄东南一带（考察路线之二）出露较好，主要为一套碳酸盐岩夹少量白云质页岩的地层，总厚度为3416m。

雾迷山组最显著的特征是米级沉积旋回发育。每个旋回厚度为数米，有两种类型：A型自下而上依次为绿灰色白云质页岩—灰色凝块石细晶白云岩—含燧石层状叠层石泥粉晶白云岩；B型自下而上依次为灰色凝块石细晶白云岩—含燧石层状叠层石泥粉晶白云岩。整个雾迷山组总体上就是由大量这样的旋回叠置而成。

前人虽然将雾迷山组划分为多个段，但各段的差异并不十分明显。各段的划分主要基于绿灰色白云质页岩的多少和燧石结核和条带的多少，此外还考虑了特殊岩性，如紫红色白云质泥岩、白云质砂岩等。

1）罗庄亚组（雾一段）

该亚组厚860m，以灰色凝块石细晶白云岩—含燧石层状叠层石泥粉晶白云岩旋回层为主，夹一些绿灰色白云质页岩—灰色凝块石细晶白云岩—含燧石层状叠层石泥粉晶白云岩旋回层。总体上，页岩夹层较少。该亚组顶部发育白云质角砾岩。

2）磨盘峪亚组（雾二段）

该亚组厚766m。与罗庄亚组相比，磨盘峪亚组以绿灰色白云质页岩—灰色凝块石细晶白云岩—含燧石层状叠层石泥粉晶白云岩旋回层发育为特征，页岩夹层较多。顶部夹红色白云质泥岩层。叠层石以锥—柱状为主。

3）二十里堡亚组（雾三段）

该亚组厚963m，底部为红色含砂泥质白云岩、亮晶砾屑白云岩；主体以灰色凝块石细晶白云岩—含燧石层状叠层石泥粉晶白云岩旋回层为主。

4）闪坡岭亚组（雾四段）

该亚组厚827m，底部为厚1m左右的灰白色白云质石英砂岩；主体以灰色凝块石细晶白云岩—含燧石层状叠层石泥粉晶白云岩旋回层为主，顶部夹翠绿色海绿石页岩。

4. 洪水庄组（Jxh）

洪水庄组与下伏雾迷山组呈整合接触，对应于高振西等（1934）命名的"洪水庄页岩"，经由1959年召开的"第一次全国地层会议"改称洪水庄组，并一直沿用至今。

洪水庄组在天津市蓟州区罗庄子镇洪水庄村及其东南一带出露良好，厚131m，自下而上可分洪一段和洪二段。洪一段为中层白云岩夹灰绿色、翠绿色和灰色页岩；洪二段主要为灰黑色、灰绿色页岩，顶部为灰绿色页岩夹泥质泥晶白云岩透镜体（藻礁）和薄层粉砂岩。

洪一段实际上是雾迷山组白云岩与洪二段页岩之间的过渡带。笔者认为，由于洪一段以碳酸盐岩而不是页岩为主，归雾迷山组比较合适。但为了尊重习惯，本书仍将其归洪水庄组。

5. 铁岭组（Jxt）

铁岭组与洪水庄组呈整合接触，对应于高振西等（1934）命名的"铁岭灰岩"，并在1959年召开的"第一次全国地层会议"中改称铁岭组，一直沿用至今。虽然鉴于铁岭组内的岩性具有明显的变化，且存在清楚的沉积间断和古地磁反转现象，陈晋镳等（1980）建议将其细分为下部的"代庄子亚组"和上部的"老虎顶亚组"，但是人们仍习惯用"铁一段"和"铁二段"替代，两者之间的平行不整合接触在区域上广泛存在，被称之为"铁岭上升"（杜汝霖和李培菊，1980）。

铁岭组在天津市蓟州区铁岭子村一带广泛出露，为一套以碳酸盐岩沉积为主，夹紫色铁质页岩和翠绿色海绿石页岩的沉积，以发育形态各异的叠层石为特征，总厚度约303m。

本书将该组自下而上分三段。铁一段与前人相同，将前人的铁二段细分为铁二段和铁三段。

铁一段（Jxt_1）厚155m，底部为灰白色中—薄层石英细砂岩，向上过渡为灰色含锰藻泥晶白云岩（藻礁）；主体为中厚层灰色叠层石白云岩夹灰绿色、翠绿色页岩；顶部为翠绿色海绿石页岩和含铁锰质暗棕色页岩。

铁二段（Jxt_2）厚60m，为云泥条带灰泥石灰岩夹竹叶石灰岩和少量叠层石石灰岩，局部见燧石结核。

铁三段（Jxt_3）厚88m，为叠层石石灰岩。叠层石类型多样，形态各异，包括层状、波状、柱状和墙状，是该区古元古宇叠层石类型最多、发育最好的层段。该段顶部为几米厚的云泥条带灰泥石灰岩。附近有采石坑，将叠层石石灰岩用作观赏石。地方政府要注意保护，防止乱挖乱采。

三、中元古界待建系（？）

中元古界待建系的建立主要是由于最近十几年在下马岭组中获取的一系列新的测年数据突破了原先的新元古界青白口系的年代格架，将下马岭组的年龄从1000—900Ma下拉至1400—1320Ma，应该归属于中元古代中部或国际地层表的延展纪早期（1400—1200Ma），从而导致青白口系的解体和重新定义。由于燕辽裂陷槽在蓟县系和青白口系之间仅存在下马岭组，且下马岭组代表的时限不足100Ma，不足以单独建系；而青白口系和龙山组缺乏新的年龄数据，它们的底界年龄仍分别暂定为1000Ma和900Ma，下马岭组与龙山组之间出现长达320—420Ma的沉积缺失。因此，地层工作者将下马岭组和其上缺失的地层暂统称为"待建系"，时限为1400—1000Ma，与国际地层表的中元古界延展系和狭带系相对应，且下马岭组仅代表延展系下部的部分地层。

本书认为，既然下马岭组因时限较短不足以建系，且属于中元古代，不妨将其归为蓟县系。

用"待建系"这个称呼很别扭。如果将来的工作完善了，问题弄清楚了，可以再改动。就像下马岭组原来归青白口系，现在又将其从青白口系划分出来一样。

下马岭组与下伏蓟县系铁岭组呈平行不整合接触关系，被前人称之为"芹峪上升"（陈晋镳等，1980）。下马岭组最早是由叶良辅（1920）命名于北京市门头沟区下马岭村，原指硅质灰岩之上，砂砾岩之下的一套粉砂岩、页岩夹白云岩的地层。后高振西等（1934）在天津市蓟州区城北将相当层位的地层称之为"下马岭页岩"。

下马岭组在天津蓟州区骆驼岭一带发育最好，厚165m，主体为一套灰绿色粉砂质页岩。该组的底部发育不稳定的细砾岩；下部为灰色、灰紫色砂岩与灰黑色粉砂岩和粉砂质页岩互层，砂岩发育交错层理。中部和上部主要为灰绿色粉砂质页岩。由于下马岭组岩性较软，易于风化，其分布区多形成中低型的山岭地貌。

四、新元古界青白口系（Qb）

新元古界青白口系主要分布在天津市蓟州区城北府君山一带，主要由下部的碎屑岩和上部的碳酸盐岩组成，总厚度385m。该系自下而上可划分为龙山组（Qbl）、景儿峪组（Qbj）和西井峪组（Qbx）。其中，西井峪组（Qbx）是金振奎等（2019）新建的组（详见后）。

青白口系的划分方案曾发生过多次变动。最早是由高振西等（1934）根据北京市门头沟区青白口村村名而命名的"青白口群"，并将其划分为下部的"下马岭页岩"和上部的"景儿峪灰岩"，但之后因不同的学者在"景儿峪灰岩"中采集到三叶虫（张文佑和李唐沁，1935；孙云铸，1957；王曰伦，1963），认为"景儿峪灰岩"中的角砾岩属于寒武系的底砾岩，并将原"景儿峪灰岩"的上部厚层灰岩划归寒武系，命名"福金山组"，后改称府君山组。北京市地质研究所于1975年编制华北地区区域地层表时，建议将原"景儿峪组"下部的砂页岩划分出来，并命名为"龙山组"。同时，如前面所述，由于在下马岭组中获取最新的定年数据，下马岭组也被下拉至待建系中，因此目前的青白口系仅包括龙山组和景儿峪组两个组，且与待建系和寒武系之间皆呈平行不整合接触或微角度不整合接触。

全国地层委员会于2014年发布的《中国地层表（2014）》中，将青白口系的底、顶界年龄分别定在1000Ma和780Ma。其底界年龄与国际地层表的新元古界拉伸系底界年龄一致，但顶界年龄比国际地层表的拉伸系顶界年龄（850Ma）新。由于龙山组和景儿峪组缺乏最新的年龄资料，只能按照前人根据两组中海绿石K—Ar定年资料所限定范围，即900—850Ma（钟富道，1977；于荣炳和张学淇，1984）。因此，龙山组与待建系下马岭组之间存在长达420Ma的地层缺失，景儿峪组与上覆下寒武统府君山组之间存在超过300Ma的地层缺失。

1. 龙山组（Qbl）

龙山组与下伏下马岭组间呈平行不整合接触，且被前人称为"蔚县上升"（陈晋镳等，1980）。龙山组最早是由郝诒纯（1954）在北京市昌平区龙山命名的"龙山砂岩"（乔秀夫，1976），并在1975年召开的华北区前寒武纪地层专题会议（太原）中采用"龙山组"一名。后因与南方"龙山群"重名，北京市地质研究院在1975年编制华北地区区域地层表时改称"长龙山组"，但1982年邢裕盛等考虑北京并无"长龙山"的地名，提出以该组发育较好的天津市蓟州区骆驼岭地名，将其替代为"骆驼岭组"（邢裕盛等，1989）。2014年出版的《中国岩石地层名称词典》中说明"长龙山组"一词停用，恢复原名称"龙山组"。因此，为了避免混淆，本书采用"龙山组"一词，与"骆驼岭组"同物异名。

龙山组在天津市蓟州区骆驼岭一带出露良好，为一套碎屑岩沉积，总厚度约118m。按照岩性组合变化，该组自下而上分两段：龙一段（Qbl_1）为含海绿石砂岩段，层内发育板状、楔状和槽状交错层理等，底部为灰绿色、灰黄色含海绿石细砾质长石质石英砂岩夹灰黄色泥质粉砂岩；龙二段（Qbl_2）为页岩段，以灰绿色和灰黑色页岩为主，水平层理发育，夹薄层海绿石细砂岩。顶部数米为紫红色泥岩。

2. 景儿峪组（Qbj）

景儿峪组一般认为与下伏龙山组呈整合接触，但因其底部出现含海绿石含细砾粗砂岩，也不排除可能存在过短暂的沉积间断。景儿峪组最早是由高振西等（1934）根据天津市蓟州区西井峪和东井峪的地名而命名的"景儿峪灰岩"，但目前的景儿峪组仅相当于原"景儿峪灰岩"的上部灰岩段，除去顶部"厚层灰岩（府君山组）"。

景儿峪组在天津市蓟州区府君山向斜的两翼近核部区域小范围出露，主要为一套碳酸盐岩夹薄层页岩沉积，总厚度约112m。该组底部含有十几厘米厚的含海绿石含细砾粗砂岩，下部主要为灰白色、灰紫色的中—薄层灰泥石灰岩；中部为灰色、蛋青色中—厚层灰泥石灰岩夹泥质灰泥灰岩；上部为灰色薄层泥质灰泥灰岩夹灰绿色页岩。

景儿峪组与上覆西井峪组块状角砾岩（原属下寒武统府君山组）间呈微角度不整合接触关系，代表著名的"蓟县运动"（孙云铸，1957；陈晋镳等，1980）。

五、新元古界南华系西井峪组（Nhx）

西井峪组是金振奎提出新建的组。该组的地层原来被划归下寒武统府君山组，为一套大小混杂的块状碳酸盐岩角砾岩，厚155m。金振奎带研究生在蓟州城区北侧的西井峪村附近研究后发现，该套地层为冰川沉积，为冰碛岩（详见第三章），其岩性特征与上覆下寒武统府君山组的中薄层白云岩完全不同，有天壤之别，因此认为有必要新建一个组。由于这套地层的剖面位于西井峪村附近，故名"西井峪组"。

西井峪组的角砾岩中，角砾的成分主要为白云岩，还有一些燧石及少量石灰岩；整体呈块状，无层理，无层面；角砾含量80%～90%，杂乱排列，大小混杂，分选差，磨圆差，普遍为棱角状和次棱角状，填隙物主要为白云质泥和粉砂。角砾岩的结构成熟度极低，为典型的冰川沉积。

西井峪组与下伏景儿峪组和上覆府君山组之间均呈显著的突变接触，与下伏景儿峪组之间被认为是平行不整合或微角度不整合，并被认为是"蓟县运动"所致（孙云铸，1957；陈晋镳等，1980）。

西井峪组与上覆府君山组之间呈突变接触，很可能是平行不整合接触。但因缺乏化石和其他定年矿物，两组之间是否缺失地层并不确定。

西井峪组的年代应是景儿峪组沉积之后、府君山组沉积之前，究竟是属于新元古代还是早寒武世，并没有确切证据，但很可能属于新元古代。金振奎等在这套角砾岩中并未找到任何三叶虫或其他化石。如果属于寒武系,应该有化石。而且角砾岩中常见燧石角砾,但在华北寒武系中并没有燧石结核或条带。这些证据表明其母岩不可能是寒武系。从角砾的成分看,其母岩主要是中元古界蓟县系雾迷山组,少量来自铁岭组。

本书暂将西井峪组划归南华系,因该期全球冰川发育,可能相当于上统的南沱组。

西井峪组冰碛岩的发现对恢复新元古代地球气候特征具有重要意义,其出现或许真与"雪球地球"有关。

第三节 大地构造演化

元古宙期间先后发生了哥伦比亚(Columbia)超大陆和罗迪尼亚(Rodinia)超大陆的聚合和裂解(Rogers and Santosh, 2003; Santosh et al., 2009; Santosh, 2010a),而华北克拉通在元古宙也经历了多期的伸展与裂谷事件,两者在发生时间和构造演化上存在明显的相关性和耦合性,且部分学者指出华北克拉通记录了两次超大陆裂解进程(Zhai and Liu, 2003),并提出燕山地区元古宙地层层序与超大陆旋回间可能的关系(孟庆任,2015),如图2-4所示。

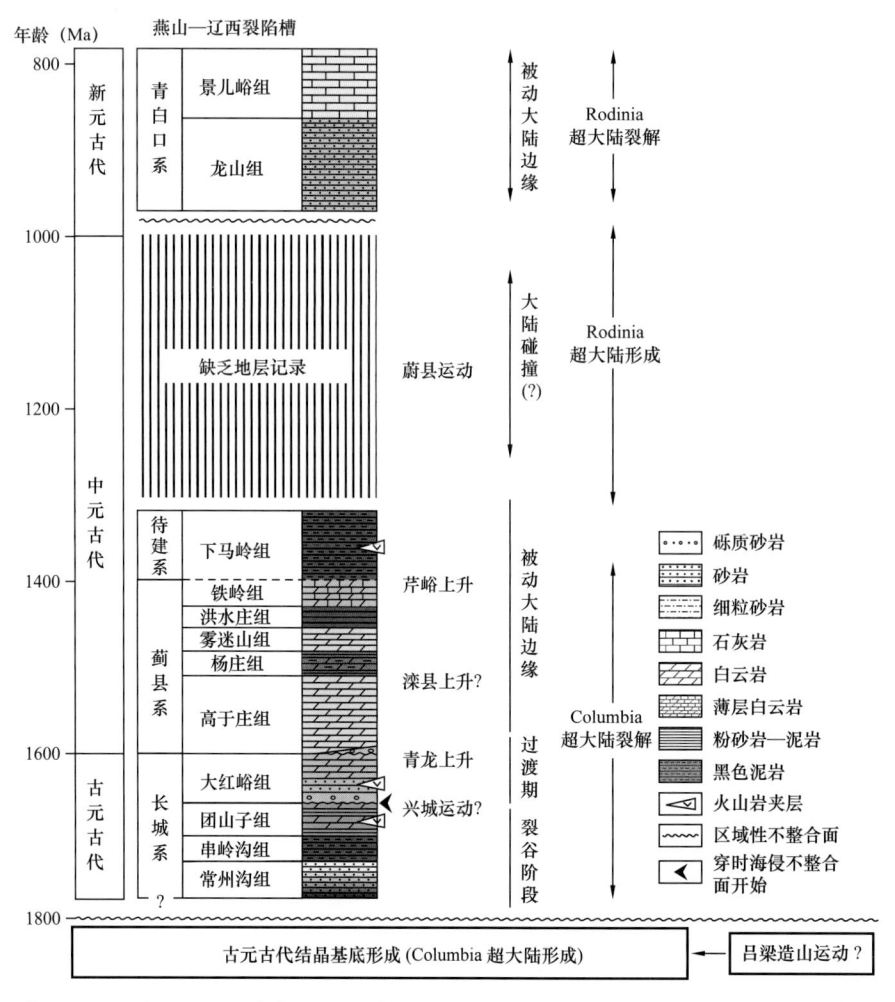

图2-4 燕山地区元古宙地层层序与超大陆旋回、区域构造活动之间的可能关系(据孟庆任等,2016)

Fig.2-4 Possible relationship among stratigraphic sequence of Proterozoic, supercontinent cycle and regional tectonics in Yanshan area(modified from Meng et al., 2016)

在经历 2.5—1.8Ga 的一系列聚合碰撞事件之后，华北东部地块与西部地块于 1.8Ga 发生碰撞，且通过贯穿华北的中部造山带而拼合，最终形成统一的克拉通（Wilde et al., 2002; Zhao et al., 2003; Kusky and Li, 2003; Kusky et al., 2007a, b; Kusky and Santosh, 2009; Santosh, 2010b），并作为统一块体并入哥伦比亚超大陆，自此华北板块进入了一个相对稳定发展的时期。这次构造事件被称为"吕梁运动"，是华北克拉通对哥伦比亚超大陆聚合事件的响应。

哥伦比亚超大陆从 1.6Ga 开始裂解，裂解过程可能一直持续至 1.2Ga（Zhao et al., 2003; Hou et al., 2008），而华北克拉通北缘出现的裂谷盆地完整地记录了整个裂解过程。自古元古代末开始，受伸展断裂控制，华北克拉通北缘开始出现燕辽裂陷槽，在初始阶段发生强烈断陷和快速充填，并在长城系内部形成两套由碎屑岩向碳酸盐岩过渡的地层单元，即下部长城系（自下而上包括常州沟组、串岭沟组和团山子组）和上部长城系大红峪组。

长城系常州沟组河流—滨岸相浅水碎屑岩记录了裂谷发育的初始阶段；串岭沟组的深水页岩代表裂谷盆地加速沉降的过程；团山子组的泥质白云岩、白云质泥岩以及叠层石、泥裂等的出现代表裂谷沉降速率降低，浅海环境发育。长城系整体记录了同裂谷阶段的强烈断陷和快速充填过程。

大红峪组发育石英砂岩和叠层石白云岩代表盆地不断扩张、广泛海侵，虽然夹杂着火山沉积，代表断陷和火山作用仍在继续，但强度明显减弱，热沉降开始，记录了从同裂谷期向后裂谷期过渡阶段的缓慢沉降和充填过程。

中元古代开始，盆地的热沉降作用明显增强，断陷作用基本停止，在被动大陆边缘背景下形成巨厚的碳酸盐岩夹泥岩沉积序列。直至下马岭组沉积阶段，发育厚层的深水暗色岩系，并夹有火山沉积，代表被动大陆边缘的进一步发展，华北克拉通北缘的沉降速率增大，并处于深水沉积环境。

新元古界青白口系与中元古界待建系间存在长达 420Ma 的沉积间断，其形成原因可能与华北克拉通和相邻地块在 1.3—1.0Ga 期间发生碰撞有关，地壳强烈挤压收缩，进而导致华北克拉通北缘大幅隆升，并遭受长期剥蚀（杜汝霖等，1979；曲永强等，2010）。该碰撞事件在时间上与罗迪尼亚超大陆的聚合相一致（Zhang, 2004）。新元古代的沉积可能与罗迪尼亚大陆裂解相关（Zhai et al., 2003；曲永强等，2010），代表伸展背景下大范围海侵而形成的龙山组河流相—滨浅海相碎屑沉积和景儿峪组台地相碳酸盐沉积。

此后，由于泛非洲运动导致冈瓦纳超大陆的聚合，地壳再次整体隆升、剥蚀，直至寒武纪早期才整体下降并接受沉积，形成景儿峪组与寒武系府君山组（本书将其下部单独划出，为新建组——西井峪组）间长达 220Ma 的沉积间断，并被称作"蓟县运动"，也标志着燕辽裂陷槽的最终闭合。

早寒武世中期开始华北地台整体沉降，发生广泛海侵，接受沉积，并一直延续至中奥陶世。此后，由于受加里东运动引起的区域性构造挤压作用的影响，华北地台再次隆升、剥蚀，并一直持续至早石炭世，形成广泛而持久的沉积间断，导致上奥陶统—下石炭统在华北地区的普遍缺失。晚古生代华北地台又开始沉降，重新接受沉积，尤其在晚石炭世—早二叠世，区域上沉积了一套含煤的海陆过渡相地层。晚古生代末期，地壳发生大幅度、快速沉降，并开始出现岩浆活动事件。

中生代三叠纪开始，燕山地区发生碱性岩体侵位事件，标志着燕山地区从海西造山阶段向陆内发展阶段过渡（阎国翰等，2000）。从晚三叠世开始，华北地台进入构造活跃的印支—燕山期。由于中亚蒙古大洋在海西期末闭合、消亡，华北板块与相邻的西伯利亚板块发生强烈碰撞，华北板块北缘发生异常激烈的构造—岩浆活动，形成大量的褶皱、断裂和火山及岩浆侵入，在区内主要形成 NNE、EW、NE 和 NW 向多组构造交切复合而成的复杂断块构造（潘家明，1992），其中马兰峪背斜就是在该背景下形成的结晶基底和盖层共同卷入褶皱作用而形成的厚皮式构造（杨付领等，2015）。

新生代主要受喜马拉雅运动的影响，华北克拉通主要受控于北西—南东方向的拉张应力场，使得燕山期形成的主断裂从挤压变成拉张，在区内以断块升降作用为主，从而奠定了现今地貌轮廓的基础。

第四节 资源概况

天津市蓟州区元古宇内蕴含的矿产资源较为丰富，主要包括锰矿、白云岩矿、石灰岩矿、含钾泥岩矿、铁矿、海泡石矿和黏土矿等。同时，区内发育大量的叠层石、臼齿灰岩等，现象奇特，形态优美，可以用作观赏石或建筑装饰材料。

一、锰矿

蓟州区的锰矿主要是以锰方硼石的形式赋存于蓟县系高于庄组二段的底部。由于锰方硼石矿物属于稀少矿物，而蓟州区的锰方硼石矿是目前世界上唯一且具有开采价值的锰方硼石矿床，矿体形态呈透镜状、扁豆状、饼状、筒状等，主要分布于东水厂、坝尺峪及前干涧等处（王秋舒等，2013）。

二、白云岩矿

白云岩在蓟州区的元古宇中广泛存在，是最为常见的岩石类型，可以用作碱性耐火材料或高炉炼铁的熔剂。

三、石灰岩矿

蓟州区内石灰岩主要出现在蓟县系的高于庄组、铁岭组和青白口系景儿峪组，但是具有工业用途的石灰岩主要来自铁岭组的三段，其氧化钙含量高，杂质含量低，是十分理想的水泥和石灰的原料。

四、含钾泥岩矿

蓟州区的含钾泥岩主要赋存于长城系团山子组顶部和大红峪组顶部的绿色、紫红色粉砂岩和粉砂质页岩之中，由于属于非水溶性钾矿，不仅可以用作钾肥的制作，还可以获得高附加值的碳酸钾、白炭黑、沸石分子筛及矿物聚合新型建材等。含钾层最好的地段位于下营镇团山子村南山一带，规模较大（闻秀明，2013）。

五、铁矿

蓟州区的铁矿主要赋存于待建系下马岭组底部的砂岩和粉砂质泥岩中，含矿层稳定，矿体呈透镜状，成分主要为赤铁矿，局部锰含量高，品位变化大，主要分布在铁岭一带，规模较小。

六、海泡石矿

蓟州区的海泡石矿主要赋存于蓟县系雾迷山组三段上部的灰白色含粉砂白云岩中，是目前世界上发现的最古老海泡石矿床。该区的海泡石矿包括两个矿化层，第一层矿层厚度较小，但质量较高；而第二层厚度较大，但质量较差；主要分布在夏庄子村北约 0.7 Km 处（陈一笠，1992）。

七、黏土矿

蓟州区的黏土质页岩主要分布在长城系串岭沟组、蓟县系洪水庄组、待建系下马岭组和青白口系龙山组中。这些黏土质页岩可用于工业砖瓦烧制的原料。其中，串岭沟组二段和洪水庄组的伊利石页岩还可以用作紫砂陶土，与江苏宜兴紫砂陶土成分相近，主要分布在道谷峪一带，规模属于大型黏土矿床。

第三章 考察路线描述及沉积相分析

第一节 概　　述

元古宇在天津市蓟州区北部山区从古元古界到新元古界连续出露。1984年国务院批准建立中—上元古界国家自然保护区。保护区北起长城脚下的常州村，南至蓟州城区北部的府君山，面积广、线路长。经长期建设，已配置有较为完善的考察便道、指示牌、地质现象解说牌等基础设施。

本指导书为了方便使用者灵活安排野外地质考察工作，依据蓟州区元古宇及上覆寒武系的划分情况，以组为单位设计了13条考察路线。

（1）下营镇常州村—青山岭村北长城系常州沟组（Chc）考察路线；

（2）下营镇青山岭村北—刘庄子村长城系串岭沟组（Chch）考察路线；

（3）下营镇团山子村西长城系团山子组（Cht）考察路线；

（4）下营镇大红峪沟长城系大红峪组（Chd）考察路线；

（5）下营镇大红峪沟—罗庄子镇翟庄村北蓟县系高于庄组（Jxg）考察路线；

（6）罗庄子镇翟庄村北—青山村—花果峪村蓟县系杨庄组（Jxy）考察路线；

（7）罗庄子镇青山村—磨盘峪村—二十里铺村蓟县系雾迷山组（Jxw）考察路线；

（8）渔阳镇小岭子村蓟县系洪水庄组（Jxh）考察路线；

（9）渔阳镇小岭子村—罗庄子镇铁岭子村南蓟县系铁岭组（Jxt）考察路线；

（10）罗庄子镇铁岭子村南—下庄子村待建系下马岭组（?x）考察路线；

（11）渔阳镇后寺沟青白口系龙山组（Qbl）考察路线；

（12）渔阳镇西井峪村北青白口系景儿峪组（Qbj）考察路线；

（13）渔阳镇府君山公园北侧南华系西井峪组（Nhx）—寒武系府君山组（$\epsilon_1 f$）考察路线。

以上线路的具体位置大致如图3-1所示。线路总体走向由北向南。该区元古宇总厚度9547m，考察路线总长度约34.4km，几乎全为平坦的柏油路。正常情况下，考察所有线路约需3天。其中常州沟组到串岭沟组约需0.5天，团山子组到高于庄组约需1天，杨庄组到雾迷山组约需0.5天，洪水庄组到铁岭组约需0.5天，下马岭组到西井峪组约需0.5天。快点的话，考察所有线路需2天。

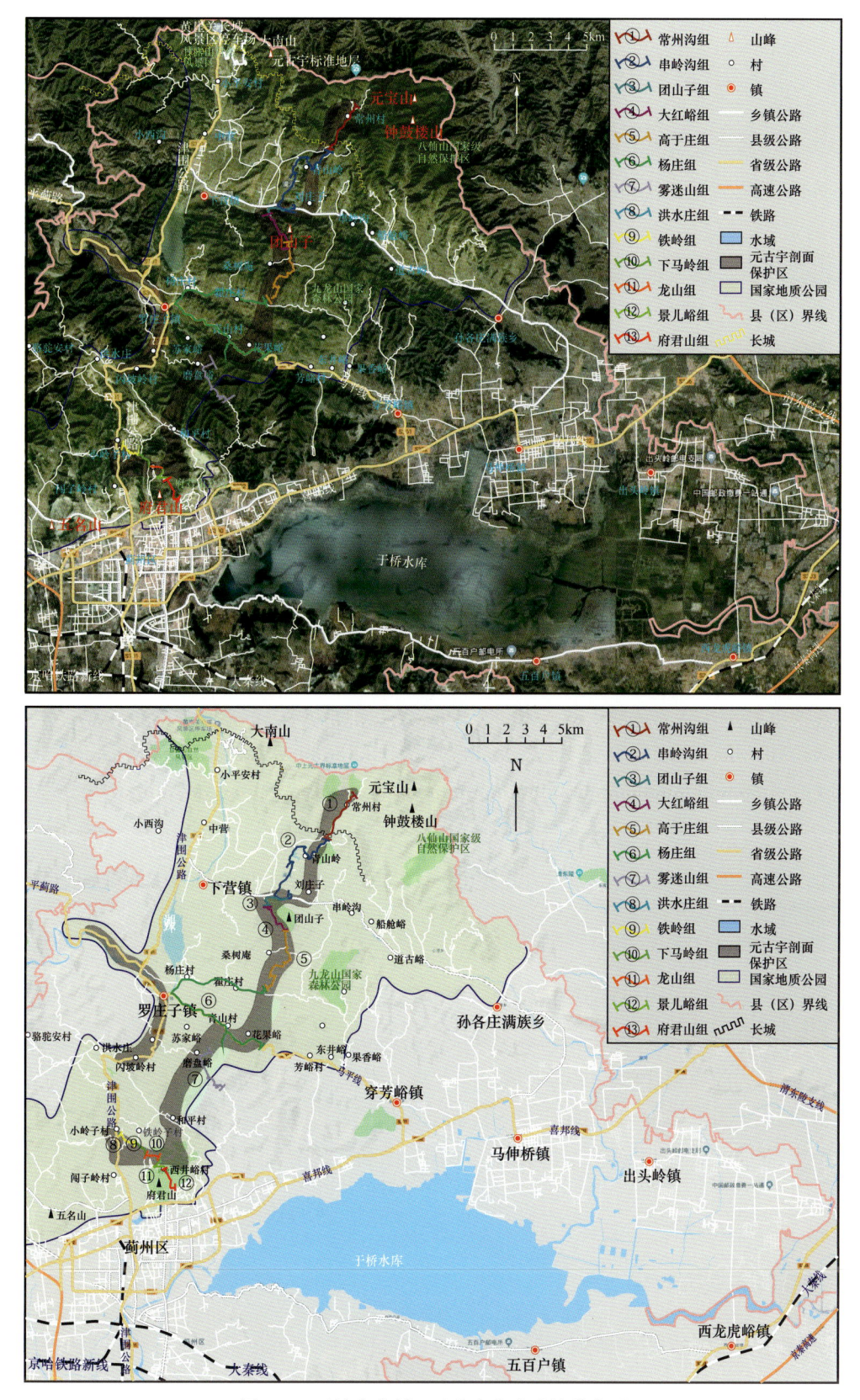

图 3-1 天津市蓟州区元古宇考察路线分布图

Fig.3-1 Geological survey routes of Proterzoic in Jizhou District, Tianjin

第二节　下营镇常州村—青山岭村北长城系常州沟组（Ch*c*）考察路线

古元古界长城系常州沟组为一套碎屑岩系，以砂岩为主，为天津市蓟州区最古老的沉积地层，总厚度859m。根据岩性变化和旋回，该组自下而上可以分为三段：

下部的常一段厚约260m，以紫红色厚层粗砂岩为主夹紫红色含砾粗砂岩、砾质粗砂岩和细砾岩。常见由含砾粗砂岩、砾质粗砂岩或细砾岩向上渐变为粗砂岩的沉积旋回，旋回的厚度多为数米。交错层理发育，层理的倾向自下而上都较稳定，为北北西。颜色红、岩性粗、磨圆较差、分选中等、层理倾向稳定是该段的特点。该段为裂谷型辫状河沉积（详见后）。

中部的常二段厚约220m，以乳白色中层至厚层石英细砂岩为主，下部夹一些含砾中—粗砂岩。砂岩多具块状构造，层理不发育或模糊，仅局部可见明显的交错层理。层面普遍波状起伏，不平整。颜色白、岩性细、磨圆好、分选好、层理不常见是该段的特点。常二段与常一段的接触处有断层，风化破碎严重，接触面上下被碎石覆盖。在地貌上，由常一段到常二段突变为陡崖，说明常二段岩层更坚硬，更抗风化。陡崖是常二段与常一段在地貌上的区别标志，陡崖底部为断层。该段为滨岸上部（即最浅正常浪基面之上的部位）沉积（详见后）。

上部常三段为灰白色中厚层石英细砂岩夹灰绿色薄层粉砂质页岩，其底界位于白常路与青常路交叉口北侧。该段下部的砂岩单层厚度较大，为数十厘米，属于中、厚层；上部的砂岩单层厚度较小，为几厘米到十几厘米，属于薄、中层。砂岩多具块状构造，层理不发育或模糊，仅局部可见明显的交错层理。中、厚层砂岩的层面普遍波状起伏，不平整，但薄层砂岩的层面较平整。泥质夹层厚度不大，多为数毫米到1cm左右，局部可达数厘米。自下而上，该段的变化规律是单层厚度变小，泥质夹层变多、变厚。层较薄、泥质夹层较多是该段的特点。该段为滨岸下部（即最浅正常浪基面与最深正常浪基面之间的部位）沉积，而且向上逐渐变深，向串岭沟组的滨外陆棚沉积过渡（详见第十五节）。

可见，整个常州沟组自下而上，颜色由氧化色变为弱还原色，粒度变细，层变薄，泥质夹层变多，由陆相变为海相，水体逐渐加深，为一个发育完整的海侵序列。唯一遗憾的是在这里未能见到河流相与滨岸相的接触界限和过渡关系，或许在其他地方能够找到。

常州沟组在天津市蓟州区最北部的黄崖关长城、九龙顶至八仙山一带出露较好，尤其在常州村至青山岭村北一带出露连续、完好，与下伏和上覆地层的界线明显，且该处交通较为便捷，是该组最佳的考察路线。沿线的常州沟组露头的信手剖面图和柱状图分别如图3-2和图3-3所示。

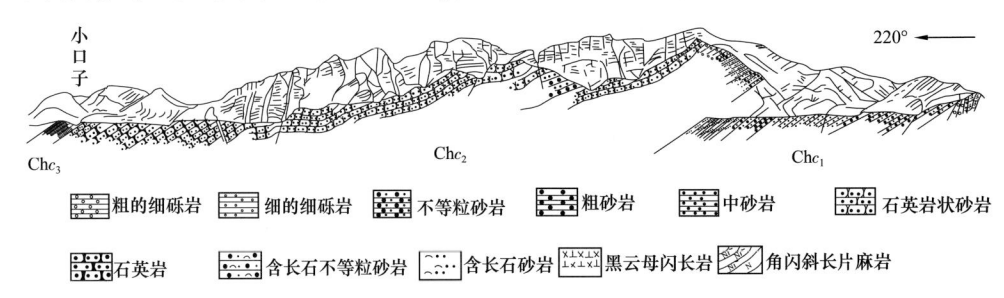

图3-2　天津市蓟州区古元古界长城系常州沟组（Ch*c*）信手剖面图（修改自天津地质矿产研究所，1964）

Fig.3-2　Sketch profile of Changzhougou Formation (Ch*c*), Changcheng System, Paleoproterozoic in the Jizhou District, Tianjin (modified after Tianjin Institute of Geology and Mineral Resources, 1964)

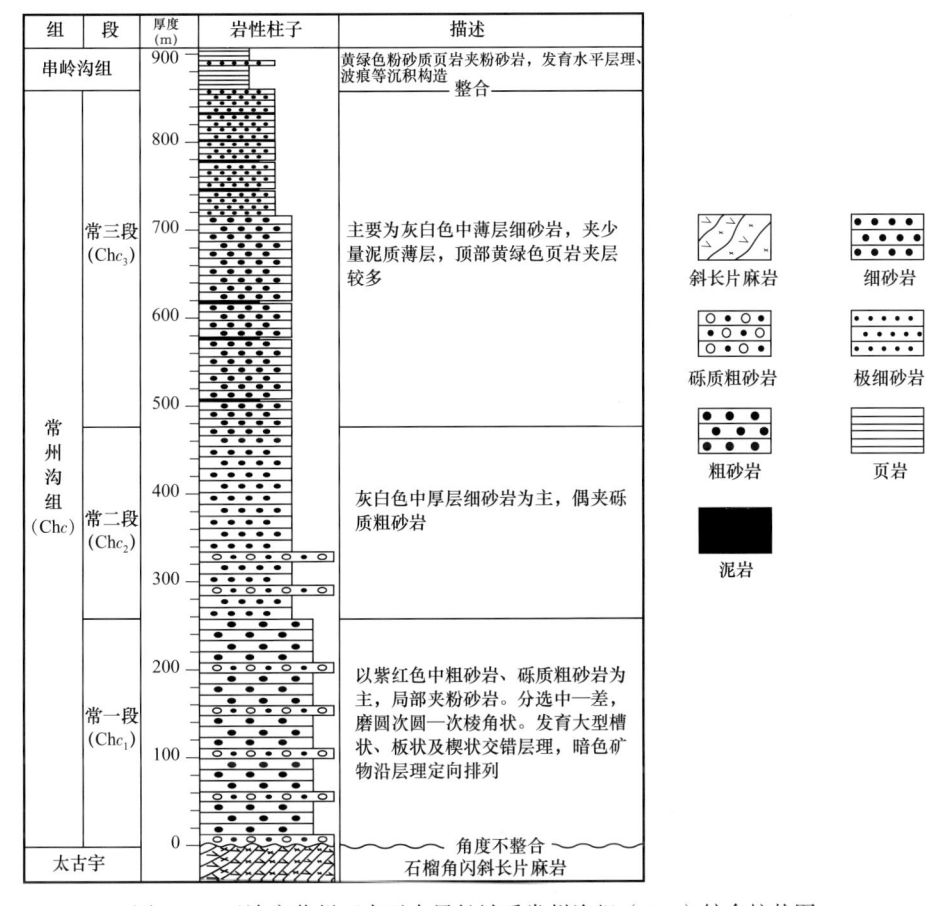

图 3-3　天津市蓟州区古元古界长城系常州沟组（Ch*c*）综合柱状图

Fig.3-3　Stratigraphic column of Changzhougou Formation（Ch*c*）, Changcheng System, Paleoproterozoic in the Jizhou District, Tianjin

一、线路位置

该路线位于天津市蓟州区下营镇常州村至青山岭村北一带，起点位于常州村西边山坡处（GPS坐标：E 117.476749°，N 40.164697°），终点位于青山岭村北东方向约 650m 处（GPS坐标：E 17.498056°，N 40.200278°），线路总长约 2470m，具体路线和考察点分布如图 3-4 所示。该线路的道路状况整体较好，适合中、小型车辆通行；考察路线的起始段位于常州村西侧山坡上，长度约 200m。山坡修有步行台阶，适宜步行考察。村内建有多处停车场，停车方便；其余考察线路皆位于道路两侧，但山区公路弯度高、坡度大，考察时应时刻注意往来车辆。

二、考察目的与任务

（1）观察古元古界长城系常州沟组与太古宇的接触关系及其岩性特征和沉积序列；

（2）观察常州沟组中的沉积旋回、典型的沉积构造（包括槽状、板状和楔状交错层理、冲刷面）、石英砾、泥砾、冲刷残余泥质层、假化石、断层、褶皱等；

（3）观察自下而上岩石的颜色、粒度、单层厚度、沉积构造等变化规律，分析沉积环境及其演化。

三、考察内容与考察点

该线路从蓟州区下营镇常州村至青山岭村北一带由底至顶依次考察长城系常州沟组的岩性变化和地层特征，在沿线选择有4个详细的考察点，分别包括太古宇与古元古界长城系分界点（图3-4中P1）、常州沟组一段下部的岩石类型、沉积构造和沉积序列（图3-4中P2）、常州沟组二段的岩石类型、沉积构造和沉积序列（图3-4中P3）、常州沟组三段沉积序列（图3-4中P4）。各考察点的详细描述如下。

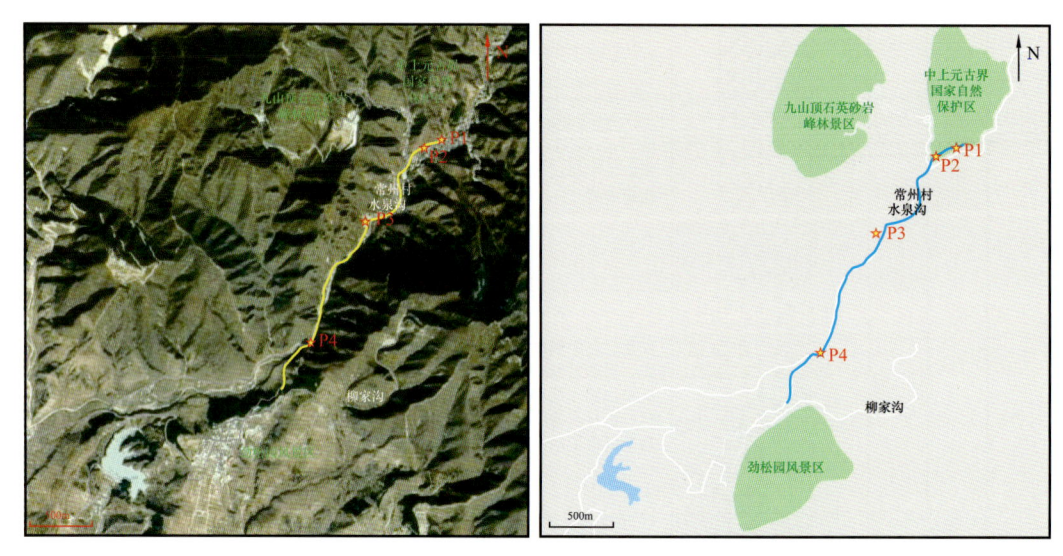

图3-4 蓟州区古元古界长城系常州沟组（Ch*c*）考察点位置及路线

Fig.3-4 Distribution of investigation stops and route of the Changzhougou Formation (Ch*c*) in the Jizhou, Tianjin

1. 考察点P1：太古宇（Ar）与古元古界长城系（Ch）分界

位置：该考察点位于常州沟村西侧半山坡步行台阶旁太古宇与元古宇分界碑处，GPS点位：E 117.520037°，N 40.219031°。

考察内容：（1）太古宇片麻岩与常州沟组之间的角度不整合界面及风化壳发育情况；（2）含砾砂岩中的层理构造和砾石的成分及磨圆；（3）讨论长期风化剥蚀的不整合面上为什么没有较厚风化壳，为什么没有大量砾石富集；（4）讨论常州沟组底部含砾砂岩的沉积环境。

现象描述：太古宇片麻岩（Ar）与常州沟组（Ch*c*）底部紫红色含砾砂岩的角度不整合界面（图3-5A）。常州沟组底部为紫红色含砾粗砂岩，分选中等，磨圆以次棱角为主。交错层理发育，层理面上常富集暗色矿物，镜下观察为磁铁矿。该套含砾砂岩为辫状河河道沉积（详见第十五节）。

2. 考察点P2：常州沟组一段下部的岩石类型、沉积构造和沉积序列

位置：该考察点距上一考察点向西南方向约20m，GPS点位：E 117.517468°，N 40.217979°。

考察内容：（1）岩石类型；（2）交错层理的类型及古水流方向；（3）冲刷面和撕裂泥砾；（4）河道沉积序列和沉积旋回；（5）讨论沉积环境。

现象描述：该考察点是一段几十米厚的地层，其底部位于考察点1西南方向约20m处，顶部为该段考察路线的结束处，这里在地貌上为一山沟，砂岩层面发育波痕，旁边为一水池（无水）和农房。这段地层是常州沟组一段岩石类型、沉积构造和沉积序列的典型代表。这里沉积构造类型丰富、清晰，且能见到发育较好的总体向上变细的米级沉积序列（图3-6和图3-7）。地层产状240°∠33.5°。

图 3-5 天津市蓟州区太古宇（Ar）与古元古界长城系常州沟组（Chc）分界处及常一段地质现象

Fig.3-5 Geological phenomena near the boundary between the Archean (Ar) and the Changzhougou Formation (Chc), Changcheng System, Paleoproterozoic, and the Chc$_1$ in the Jizhou District, Tianjin

（A）太古界与常州沟组之间的不整合界面，属于考察点 P1；（B）常州沟组一段下部紫红色砂岩中的大型板状交错层理，层理倾向北北西，该点距元古宇 / 太古宇分界几十米，小树干为比例尺，属于考察点 P2；（C）常州沟组一段下部紫红色砂岩中的"假羽状交错层理"，纹层倾向相反是由于下部槽状交错层理的右侧被上部槽状交错层理的左侧切割、叠置导致的，不是真正的羽状交错层理，该点距元古宇和太古宇分界几十米，属于考察点 P2；（D）常州沟组一段下部紫红色砂岩中的板状交错层理，照片下部层理的纹层与层系底面高角度相交，上部层理的纹层则与层系底面相切，反映流速比下部的高，该点距离元古宇 / 太古宇分界几十米，属于考察点 P2；（E）常州沟组一段下部紫红色砂岩中的紫红色泥砾（照片左上部），该点距离元古宇 / 太古宇分界几十米，属于考察点 P2；（F）常州沟组一段下部的河道紫红色砂岩夹的河漫滩薄层泥质粉砂岩（阴影处），这种夹层在以发育大套砂岩为特征的常州沟组中少见，该点距离元古宇 / 太古宇分界几十米，属于考察点 P2；（G）常州沟组一段下部紫红色砂岩露头。该段地层距离元古宇 / 太古宇分界几十米，属于考察点 P2；（H）常州沟组一段下部考察点 P2 的尽头，对面远处的陡崖为常二段砂岩，其底面为常一段与常二段的分界，有断层

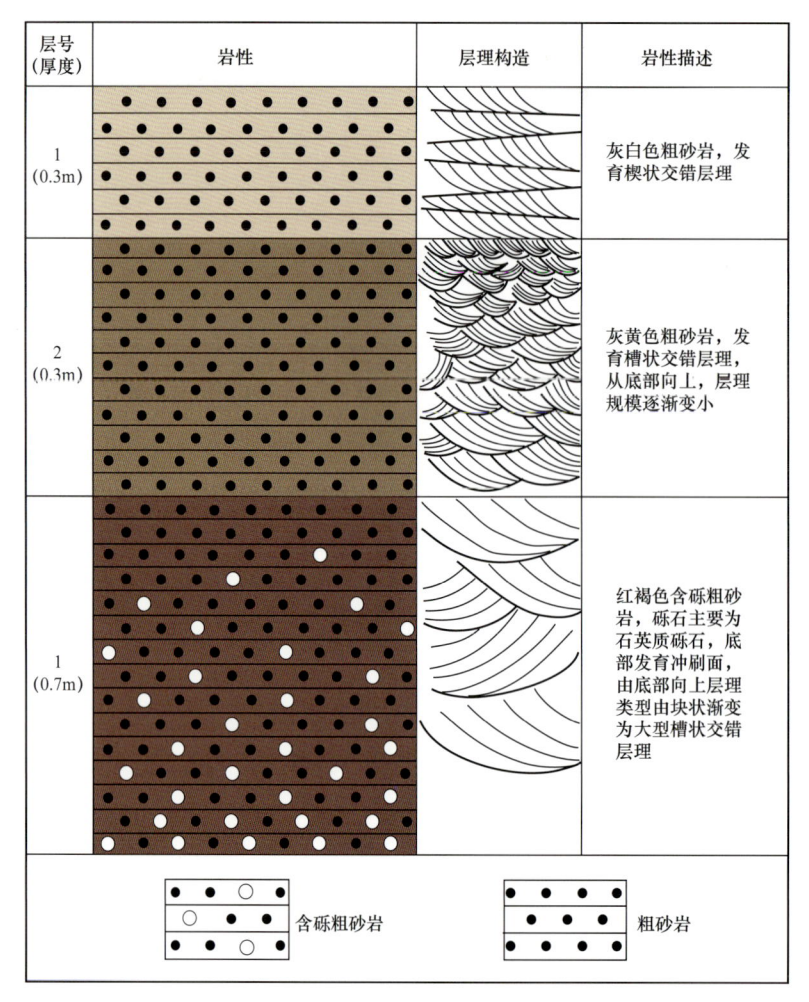

层号 （厚度）	岩性	层理构造	岩性描述
1 (0.3m)			灰白色粗砂岩，发育楔状交错层理
2 (0.3m)			灰黄色粗砂岩，发育槽状交错层理，从底部向上，层理规模逐渐变小
1 (0.7m)			红褐色含砾粗砂岩，砾石主要为石英质砾石，底部发育冲刷面，由底部向上层理类型由块状渐变为大型槽状交错层理

含砾粗砂岩　　粗砂岩

图 3-6　常州沟组一段河道沉积序列示意图

Fig.3-6　Schematic diagram of the lithofacies association in the Chc_1

底部砾质砂岩，向上粒度变细，交错层理规模变小

常一段可以见到的岩石类型主要包括：紫红色细砾岩、浅红色含砾粗砂岩和石英砂岩、灰白含砾石英砂岩等。砾石直径多为几毫米至1cm左右，成分主要为石英岩。常见的沉积构造包括槽状、楔状和板状交错层理，且层理的纹层倾向均为北北西，指示古水流流向北北西（图3-5B至D）。常一段为多期砂砾质辫状河河道沉积叠置（详见第十五节）。

在该段地层中下部，露头最清晰处，偶见砂岩中夹的灰紫色泥质条带，为侵蚀残留的河漫滩沉积。砂岩中偶见撕裂状泥砾（图3-5E），为河流相沉积中的常见特征。

3. 考察点 P3：常州沟组二段的岩石类型、沉积构造和沉积序列

位置：该点位于常州村南白常路西北侧的一个很短的小道上，GPS点位：E 117.512945°，N 0.213335°。

考察内容：（1）岩石类型；（2）交错层理的类型及古水流方向；（3）与常一段的岩性差异；（4）讨论常二段与常一段沉积环境的差异。

现象描述：该考察点为常州沟组二段，主要发育乳白色石英细砂岩，下部夹有一些紫红色和乳白色含砾中—细砂岩和细砾岩（图3-8和图3-9）。该点可见乳白色中—细砂岩中发育的交错层理。常二段颜色变为弱还原色为主，分选好，磨圆好，为海洋滨岸沉积（详见第十五节）。

图 3-7　常州沟组一段岩相序列主要岩石类型及镜下特征

Fig.3-7　Major lithofacies sequence and microscopic features of the Chc_1

（A）常州沟组一段下部向上变细的沉积序列，由含砾砂岩向上变为砂岩，该点距离元古宇和太古宇分界几十米，属于考察点 P2；（B）常州沟组一段紫红色砾质粗砂岩显微照片，单偏光；（C）常州沟组一段下部的紫红色含砾砂岩中的槽状交错层理，暗色纹层富磁铁矿，该点属于考察点 P2；（D）常州沟组一段下部浅红色中—薄层含砾粗砂岩显微照片，单偏光；（E）常州沟组一段灰白色粗砂岩，层面发育波痕构造，具楔状、板状交错层理，属于考察点 P2；（F）常州沟组一段灰白色粗砂岩显微照片，单偏光；（G）常州沟组一段下部两期河道沉积叠置，两者之间为突变冲刷面，上部河道沉积为紫红色含砾和砾质砂岩，下部砾少，上部砾多，反映水流变强；下部河道沉积为紫红色砂岩，该点距离元古宇和太古宇分界几十米，黑色榔头把为比例尺，属于考察点 P2；（H）常州沟组一段下部的紫红色砾质砂岩（图 3-7F 的放大），砾石为石英岩，直径大都在 1cm 左右，磨圆为次棱角到次圆，该点属于考察点 P2

图 3-8　常州沟组二段下部灰白色砂岩中的交错层理，考察点 P3
Fig.3-8　Crossbedding in gray sandstone of Chc_2

图 3-9　常州沟组二段中部石英细砂岩镜下照片
Fig.3-9　Microscopic photo of quartz sandstone in middle Chc_2.（A）Plane polarized；（B）Cross polarized, same view to A. Well sorted, well rounded, quartz cemented. Stop P3
（A）单偏光；（B）视域同 A，正交光。颗粒分选好，磨圆好，硅质胶结。考察点 P3

4. 考察点 P4：常州沟组三段沉积序列

位置：该点位于白常路与青常路交叉口附近（图 3-10A），GPS 点位：E 117.506340°，N 40.204052°。

考察内容：（1）岩石类型；（2）砂岩单层厚度变化及页岩夹层的数量变化；（3）交错层理的类型及古水流方向；（4）层面波状起伏的原因；（5）向上变粗的沉积旋回；（6）假化石和"李泽刚格环"；（7）小型褶皱；（8）讨论常三段与常二段沉积环境的差异及演化；（9）讨论常州沟组形成巨厚砂岩沉积的条件。

现象描述：常州沟组三段下部为灰白色中—厚层石英细砂岩夹少量页岩薄层，砂岩层面凹凸不平，为风暴形成的波痕（图 3-10B 和 C）；上部为薄层石英细砂岩夹粉砂质页岩（图 3-11B-D）。在这个考察点处，可以看到向上变粗的沉积序列，表现为自下而上层变厚，粒度变粗（图 3-10B 和 C）。有些砂岩呈透镜状。砂岩内部可见大型交错层理，倾向南。在该段中部的砂岩节理面上可见到"假化石"，像叠层石（图 3-11A），是地下水沿节理渗流时，铁锰质在节理面上沉淀形成的。该段上部薄层细砂岩中有小型褶皱，局部岩层近乎直立（图 3-11B）。

图 3-10 常州沟组三段沉积序列

Fig.3-10 Sedimentary sequence of the Chc_3

（A）常州沟组二段与三段分界，位于指路牌附近，属于考察点 P4；（B）常州沟组三段中向上变粗的沉积序列，表现为自下而上层变厚，粒度略有变粗。层面起伏不平，乃风暴形成的波痕。榔头为比例尺，属于考察点 P4；（C）常州沟组三段中向上变粗的沉积序列，表现为自下而上层变厚，粒度略有变粗。榔头为比例尺，属于考察点 P4

图 3-11　常州沟组三段典型岩性类型及镜下特征

Fig.3-11　Typical rock types and microscopic characteristics in the the Chc_3

（A）砂岩节理面上的"假化石"，形态像半球状叠层石，常三段中部，属于考察点 P3，榔头为比例尺；（B）薄层细砂岩，褶皱变形，常三段上部，属于考察点 P3；（C）薄层细砂岩和粗粉砂岩夹页岩，常三段顶部，属于考察点 P3；（D）泥质石英细砂岩，单偏光，常三段顶部

常三段中灰绿色页岩夹层变多，仍为海洋滨岸沉积，但水体比常二段深，为滨岸带下部。从常二段到常三段，水体逐渐变深，最终过渡为上覆串岭沟组的页岩陆棚沉积（详见第十五节）。

第三节　下营镇青山岭村北—刘庄子村长城系串岭沟组（Chch）考察路线

古元古界长城系串岭沟组为一套以粉砂质伊利石页岩为主的地层，夹少量粉砂岩及白云岩，局部出现火山岩岩墙与岩床。该组总厚度约 889m，根据岩性组合特征，可细分为三段：下部串一段以灰绿色、黄绿色粉砂岩和粉砂质伊利石页岩互层为主，且向上粉砂岩含量逐渐降低；中部串二段以灰黑色、灰绿色页岩为主，夹杂少量粉砂岩；上部串三段以灰黑色伊利石页岩夹薄层粉砂岩和含铁白云岩为主。

串岭沟组在天津市蓟州区下营镇青山岭村、郭家沟及刘庄子一带广泛出露，但是由于页岩岩性较软，多形成沟谷地貌，且植被繁茂。沿青常路和郭马路旁地层出露良好，是该组考察的最佳路线。此处串岭沟组的信手剖面及地层柱状图分别如图 3-12 和图 3-13 所示。

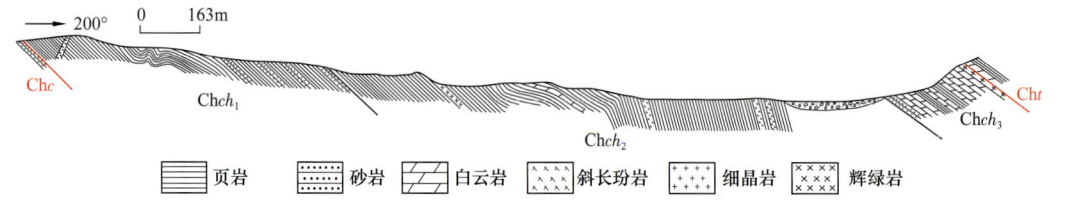

图 3-12　天津市蓟州区下营镇青山岭村北—刘庄子长城系串岭沟组信手剖面图（修改自天津地质矿产研究所，1964）

Fig.3-12　Sketch profile of the Chuanlinggou Formation, Changcheng System, Paleoproterozoic from north of Qingshanling to Liuzhuangzi, Xiaying Town, Jizhou District, Tianjin (modified after Tianjin Institute of Geology and Mineral Resources, 1964)

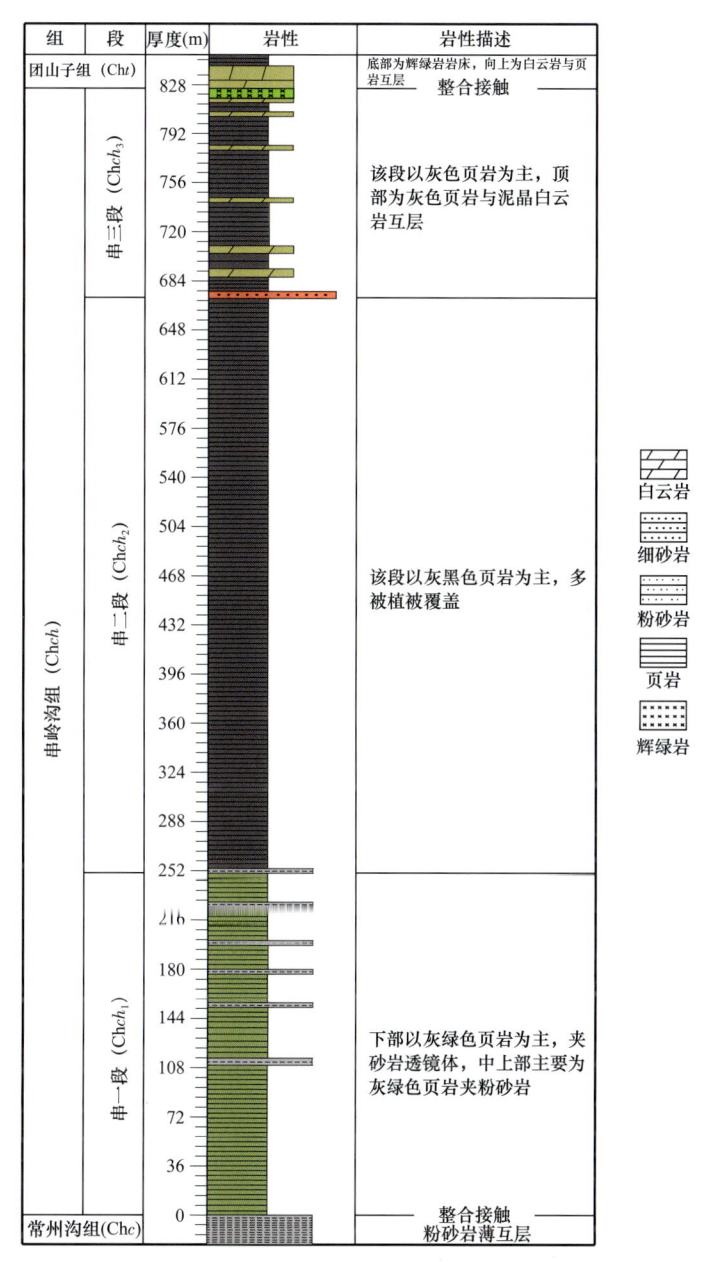

图 3-13　天津市蓟州区郭家沟长城系串岭沟组（Chch）地层柱状图（修改自 Yang et al., 2016）

Fig.3-13　Stratigraphic column of the Changcheng System Chuanlinggou Formation (Chch) in Goujiagou, Jizhou District, Tianjin (modified after Yang et al., 2016)

一、线路位置

该路线位于天津市蓟州区下营镇青山岭村北至团山子村一带，起点位于青山岭村东北角（GPS 坐标：E 117.498056°，N 40.200278°），终点位于马营公路旁团山子村西侧（GPS 坐标：E 117.474444°，N 40.176389°），路线总长度约 5600m，具体路线和考察点分布如图 3-14 所示。该路线道路总体较为宽阔，大、小型车辆皆可通行，且日常车辆较少，停车方便，但道路坡度大、弯度高，应低速、谨慎驾驶，考察时注意往来车辆。

二、考察目的与任务

（1）观察串岭沟组岩性特征与地层序列；
（2）分析串岭沟组沉积环境及其演化，尤其是串一段、串二段和串三段沉积环境之间的差异。

三、考察内容与考察点

该路线沿青山岭村至团山子村由底至顶依次考察串岭沟组的岩性特征与地层序列，并在路线上选择了5个详细的考察点，分别包括常州沟组与串岭沟组分界考察点（图3-14中P1）、串一段下部沉积特征考察点（图3-14中P2）、上部的"泥裂"（?）和波痕考察点（图3-14中P3）、串二段黑色页岩考察点（图3-14中P4）及串三段黑色页岩夹薄层泥晶白云岩考察点（图3-14中P5）。

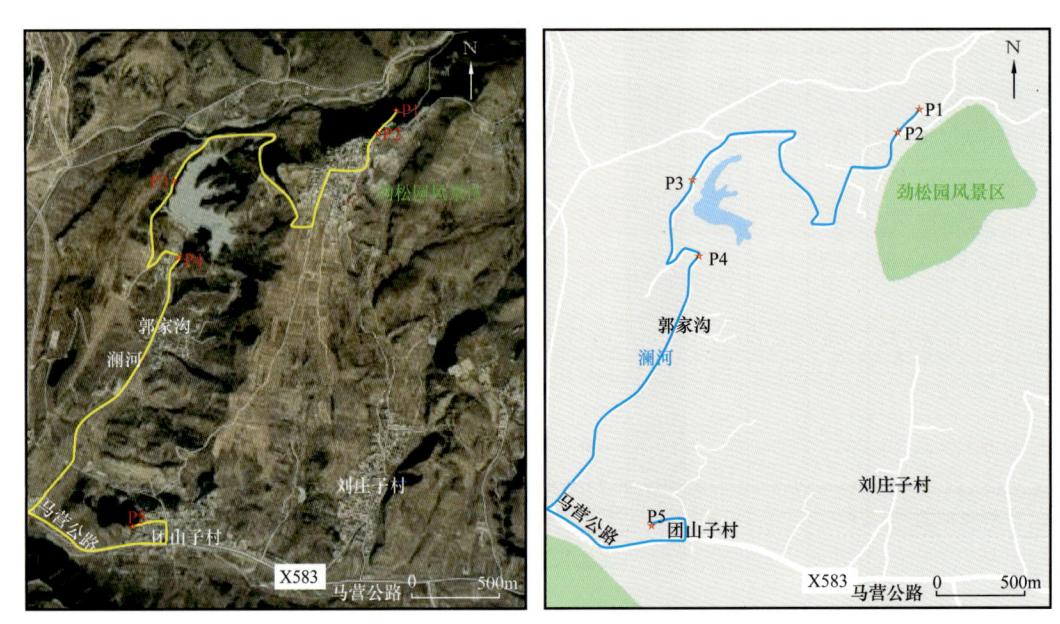

图3-14 天津市蓟州区长城系串岭沟组（Ch*ch*）考察点位置及路线

Fig.3-14 Location of investigation stops and route of the Changcheng System Chuanlinggou Formation (Ch*ch*) in Jizhou District, Tianjin

1. 考察点P1：常州沟组（Ch*c*）与串岭沟组（Ch*ch*）分界

位置：青山岭村东北角常州沟组（Ch*c*）与串岭沟组（Ch*ch*）分界碑处。GPS点位：E 117.498056°，N 40.200278°。

考察内容：（1）常州沟组与串岭沟组分界；（2）观察常州沟组顶部的砂岩如何过渡到串岭沟组下部的页岩；（3）串一段底部薄层细砂岩的波痕类型、成因及指示的古水流方向。

现象描述：该点为古元古界长城系常州沟组（Ch*c*）与串岭沟组（Ch*ch*）分界，两组之间为整合接触（图3-15A），但由于串岭沟组底部地层软，因风化而覆盖，覆盖段厚数米（图3-15B）。串岭沟组一段以灰绿色页岩为主（图3-15C），底部有几米厚的细砂岩与页岩薄互层，细砂岩发育波痕，顶面波状起伏（图3-15D）；向上变为灰绿色页岩（图3-15E）。串岭沟组一段总体为滨外陆棚沉积，水体安静，水底贫氧但不缺氧，因为颜色为灰绿色，代表弱还原环境（详见第十五节）。

图 3-15 长城系串岭沟组下部地质现象

Fig.3-15 Geological phenomena in lower Chuanlinggou Formation, Changcheng System

（A）常州沟组与串岭沟组分界，串岭沟组考察点 P1；（B）常州沟组与串岭沟组分界附近的界碑。串岭沟组底部风化严重，被覆盖，串岭沟组考察点 P1；（C）串岭沟组下部的灰绿色页岩，倾向南，串岭沟组考察点 P1；（D）串岭沟组下部的细砂岩与页岩薄互层，位于串岭沟底界附近，向上变为大套灰绿色页岩（见 E），串岭沟组考察点 P1；（E）串岭沟组下部的粗粉砂岩与页岩薄互层（图 3-15D）向上变为大套灰绿色页岩，串岭沟组考察点 P1；（F）串岭沟组下部页岩中的"假泥裂"（红色箭头所示），串岭沟组考察点 P2

2. 考察点 P2：串岭沟组一段（Chch_1）页岩沉积特征

位置：青山岭村东北角，GPS 点位：E 117.496111°，N 40.198611°。

考察内容：（1）串一段页岩的特征；（2）分析灰绿色页岩的沉积环境；（3）观察页岩中的节理，并讨论其在油气运移中的作用。

现象描述：该点是串岭沟组一段（Chch_1）典型页岩考察点，页岩颜色以灰绿色为主。页岩中夹有少量薄层粉砂岩透镜体，应为风暴流搬运沉积形成。更为特殊的是，在下部页岩中常见"假泥裂"，其内好像"充填"粉砂，实为页岩之下粉砂纹层中波痕的波脊上披覆的页岩纹层被剥蚀导致的，使下伏的粉砂层裸露（图 3-15F）。该处的页岩为页岩陆棚沉积（详见第十五节）。

3. 考察点 P3：串岭沟组页岩假泥裂和波痕特征

位置：郭家沟水库左侧公路旁，GPS 点位：E 117.480000°，N 40.195278°。

考察内容：（1）串岭沟组页岩中的"假泥裂"、波痕和藻类化石；（2）讨论"假裂"的成因；（3）分辨页岩的原生色和次生色。

现象描述：该点岩性为灰色页岩，在页岩的层面上发育假泥裂和波痕等沉积构造（图3-16A和B）。页岩风化后表面呈褐色。该处的页岩为页岩陆棚沉积。

图3-16 长城系串岭沟组中上部地质现象

Fig.3-16 Geological phenomena in middle-upper Changcheng System Chuanlinggou Formation

（A）页岩（风化面呈黄褐色）表面发育的泥裂，串岭沟组二段，考察点P3；（B）灰色页岩表面的波痕，串岭沟组二段，考察点P3；（C）黑色页岩，串岭沟组二段，考察点P4；（D）黑色页岩与泥晶白云岩（风化面呈褐色）薄互层，串岭沟组三段上部，考察点P5

4. 考察点P4：串岭沟组二段黑色页岩特征

位置：郭家沟翠玉坊右侧公路旁，GPS点位：E 117.480556°，N 40.191111°。

考察内容：（1）页岩的颜色；（2）页岩产状的变化，是否存在褶皱；（3）分析串二段页岩与串一段页岩的差别及沉积环境的差异。

现象描述：该点主要观察串岭沟组二段（$Chch_2$）典型的黑色页岩，风化比较严重，呈碎片状（图3-16C）。页岩产状发生变化，存在褶皱。串二段颜色比串一段暗，还原性更强，页岩更纯，因此水体更深，为缓斜坡—盆地沉积（详见第十五节）。见到的褶皱可能是滑塌褶皱。

5. 考察点P5：串岭沟组三段特征

GPS点位：马营公路旁北侧团山子村西侧小山包。GPS点位：E 117.474444°，N 40.176389°。

考察内容：（1）串岭沟组向上覆团山子组过渡时岩性发生的变化；（2）页岩中所夹的薄层泥晶白云岩的特征；（3）分析白云岩薄层的成因。

现象描述：该点为古元古界长城系串岭沟组三段（$Chch_3$）典型地层，主要为黑色粉砂质页岩夹薄层含铁泥晶白云岩（图3-16D）。白云岩层风化后呈褐色。串三段仍为缓斜坡—盆地沉积，但向上水体逐渐变浅，因为白云岩夹层变多，灰绿色页岩增多，并最终过渡为团山子组的碳酸盐岩（详见第十五节）。

第四节　下营镇团山子村西长城系团山子组（Cht）考察路线

古元古界长城系团山子组为一套以碳酸盐岩为主夹碎屑岩的混积地层。该组总厚度约 518m，根据岩性组合的差异，自下而上可以细分为四段：团一段以深灰色（风化后黄褐色）中薄层泥晶白云岩、泥质泥晶白云岩、白云质泥岩互层为主，水平层理发育。团二段以深灰色（风化后黄褐色）中—厚层含铁泥晶白云岩夹泥质泥晶白云岩为主，发育水平层理。团三段以薄—中层白云质砂岩、砂质白云岩和泥质泥晶白云岩互层为主，夹多套叠层石泥晶白云岩，白云质砂岩中小型交错层理和波痕发育，泥晶白云岩中泥裂十分发育。团四段以紫红色含粉砂泥质泥晶白云岩和灰白色白云质粉、细砂岩薄互层为主。

团山子组在天津市蓟州区下营镇团山子村、大红峪沟、船舱峪及道谷峪一带广泛出露，尤其以团山子村西侧跨马营公路至大红峪沟一带出露良好，分布较为连续，是该组的最佳考察路线。此处团山子组的信手剖面图及地层柱状图如图 3-17 和图 3-18 所示。

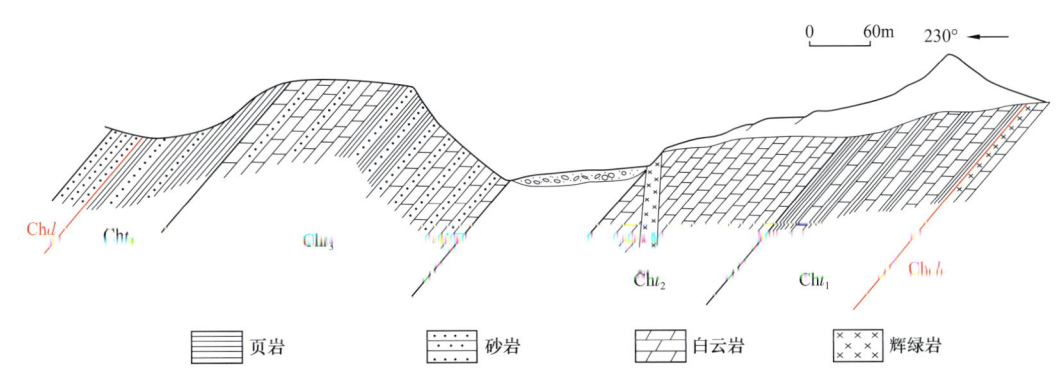

图 3-17　天津市蓟州区下营镇团山子村西古元古界长城系团山子组（Cht）信手剖面图（修改自天津地质矿产研究所，1964）

Fig.3-17　Sketch profile of the Tuanshanzi Formation（Cht）, Changcheng System, Paleoproterozoic to the west of Tuanshanzi Village, Xiaying Town, Jizhou District, Tianjin（modified after Tianjin Institute of Geology and Mineral Resources, 1964）

一、线路位置

该路线位于天津市蓟州区下营镇团山子村附近，路线的起点位于马营公路旁团山子村西侧（GPS 坐标：E 117.474444°，N 40.176389°），终点在大红峪沟营树路西侧（E 117.466389°，N 40.174167°），路线总长度约 1670 m，具体路线和考察点分布如图 3-19 所示。该路线道路较为宽阔，可乘车到达考察点附近，且停车方便。考察线路部分位于马营公路和营树路旁，观察时应注意来往车辆；其余路线都修有步行台阶，适合步行观测。

二、考察目的与任务

（1）观察古元古界长城系团山子组岩性特征与地层序列；
（2）观察波痕、交错层理、叠层石、泥裂等典型沉积构造；
（3）分析团山子组各段的沉积环境及其演化。

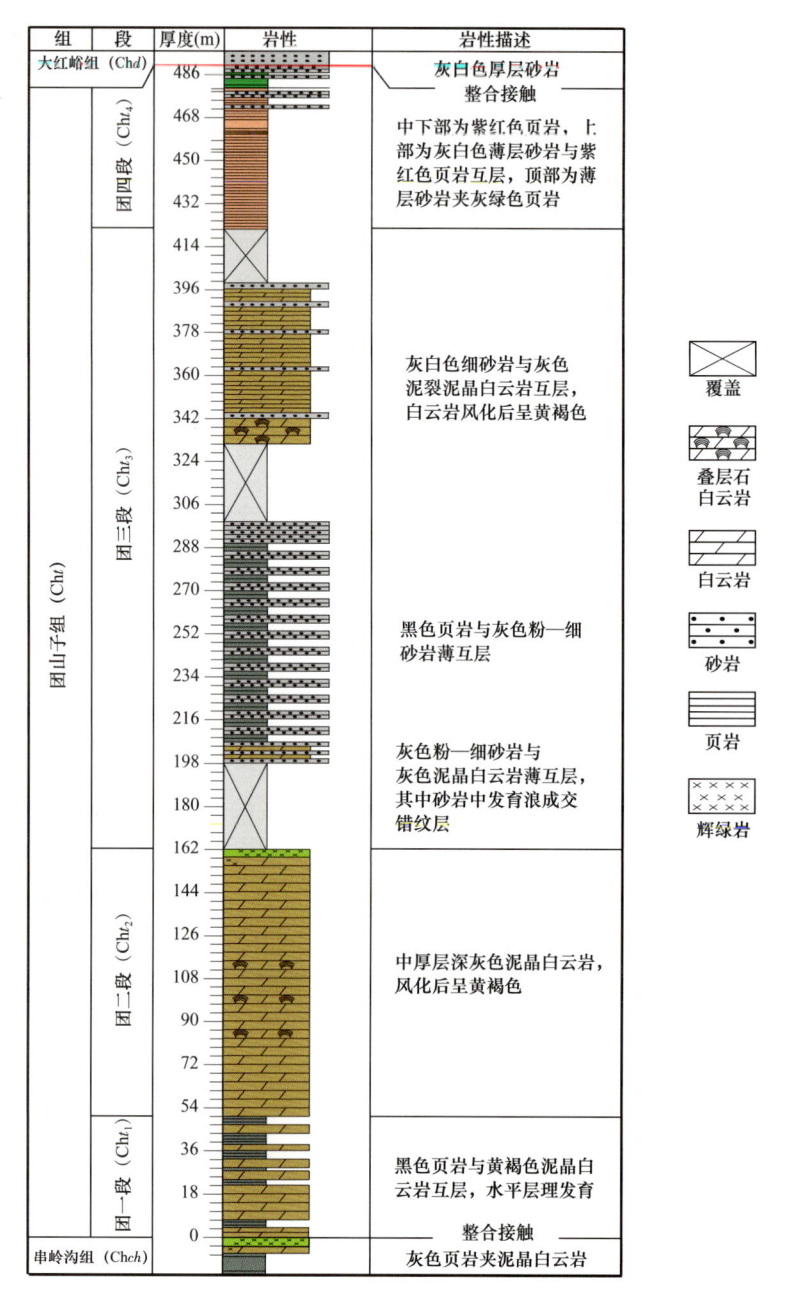

图 3-18 天津市蓟州区下营镇团山子村西古元古界长城系团山子组地层柱状图

Fig.3-18 Stratigraphic columns of the Tuanshanzi Formation, Changcheng System, Paleoproterozoic in the west of Tuanshanzi Village, Xiaying Town, Jizhou District, Tianjin

三、考察内容与考察点

该路线沿下营镇团山子村西马营公路两侧分布，其中团山子四段地层出露在营树路两侧。沿道路两侧可以观察到较为完整的团山子组岩性与地层特征，并选择了 7 个考察点，包括：串岭沟组与团山子组分界考察点（图 3-19 中 P1）、团二段含铁白云岩考察点（图 3-19 中 P2）、团三段下部白云岩与砂岩互层考察点（图 3-19 中 P3）、团三段中部黑色页岩与砂岩考察点（图 3-19 中 P4）、团三段中上部厚层砂岩考察点（图 3-19 中 P5）、团三段上部白云岩夹砂岩考察点（图 3-19 中 P6）及团四段紫红色页岩与砂岩考察点（图 3-19 中 P7）。各考察点的具体描述如下。

图 3-19 天津市蓟州区下营镇团山子村西长城系团山子组考察点位置及路线

Fig.3-19 Location of investigation stops and route of the Tuanshanzi Formation, Changcheng System in the west of Tuanshanzi Village, Xiaying Town, Jizhou District, Tianjin

1. **考察点 P1：长城系串岭沟组（Ch*ch*）与团山子组（Ch*t*）分界**

位置：马营公路旁北侧团山子村西侧小山包长城系串岭沟组（Ch*ch*）与团山子组（Ch*t*）分界碑处，GPS 点位：E 117.474444°，N 40.176389°。

考察内容：（1）团山子组与串岭沟组的分界；（2）分界处的辉绿岩侵入体；（3）团一段的岩石类型、沉积构造及沉积序列；（4）沉积环境分析。

现象描述：该点为古元古界长城系串岭沟组（Ch*ch*）与团山子组（Ch*t*）的分界，两组之间为整合接触关系，在分界处发育一套辉绿岩岩脉（图 3-20A 和 B）。分界面之上为团山子组一段下部地层，主要为深灰色（风化后黄褐色）泥晶白云岩与粉砂质页岩互层，其中白云岩多为厚层，水平层理发育（图 3-20C 和 D）。粉砂质页岩（图 3-20E 和 F）由该段底部的灰黑色向上逐渐变为灰绿色。团一段为水体较深的缓斜坡沉积（详见第十五节）。

2. **考察点 P2：团山子组二段中厚层含铁白云岩**

位置：马营公路北侧含铁白云岩石碑旁，GPS 点位：E 117.475556°，N 40.176667°。

考察内容：（1）白云岩新鲜面颜色与风化面颜色的差别及原因；（2）为什么有的白云岩具块状构造，而有的水平层理发育？（3）讨论串二段与串一段岩性差异和沉积环境差异；（4）讨论串二段上部的辉绿岩侵入体对白云岩的影响。

现象描述：该点为团山子组二段中—厚层含铁含粉砂白云岩考察点（图 3-21A 和 B）。由于白云岩中普遍含铁，所以风化面呈黄褐色。在显微镜下观察发现，该套白云岩含较多零散分布的石英粉砂（图 3-21C），局部粉砂纹层与泥晶白云石纹层交互（图 3-21D）。上部有辉绿岩侵入体，烘烤边不明显。团二段为开阔台地与缓斜坡交互沉积（详见第十五节）。

3. **考察点 P3：团山子三段下部岩石特征**

位置：马营公路西侧河沟谷对面，GPS 点位：E 117.474444°，N 40.176389°。

考察内容：（1）描述岩石类型；（2）观察薄层细砂岩中的波痕、交错层理，分析古水流方向；（3）分析沉积环境。

现象描述：该点为团山子组三段下部的典型岩石类型考察点，以灰白色细砂岩与黄褐色泥晶白云岩薄互层沉积为主。细砂岩呈透镜状、条带状或薄层状，内部发育小型不对称或对称交错层理（图 3-22A—C）。团三段下部主要为碳酸盐岩—碎屑岩混合潮坪沉积（详见第十五节）。

图 3-20　长城系团山子组一段地质现象及其岩石显微照片

Fig.3-20　Geological phenomena in Cht₁ of the Tuanshanzi Formation and photomicrographs of the rocks

（A）长城系团山子组与串岭沟组分界处的辉绿岩岩床，考察点 P1；（B）辉绿岩，正交光，考察点 P1；（C）泥晶白云岩、泥质白云岩和白云质泥岩薄互层，团山子组一段底部，考察点 P1；（D）薄层泥晶白云岩，单偏光，团山子组一段底部，考察点 P1；（E）黑色粉砂质页岩，团一段中部，考察点 P1；（F）黑色粉砂质页岩显微照片，含有较多石英粉砂，单偏光，团一段中部，考察点 P1

4. 考察点 P4：团山子组三段中部岩性特征

位置：团山子村马营公路向西拐弯处南侧河谷对面，GPS 点位：E 117.469167°，N 40.177222°。

考察内容：（1）岩石类型及其特征；（2）与黑色页岩互层的灰白色细砂岩薄层中的沉积构造；（3）古水流方向分析；（4）沉积环境分析；（5）观察玄武岩层中气孔构造，观察有无烘烤边。

现象描述：该点为团山子三段中部典型岩性考察点，主要岩性为黑色页岩与灰白色细砂岩薄互层（图 3-22D），局部为泥晶白云岩与砂岩互层（图 3-22E）。见数层玄武岩，具有气孔构造。团三段中部主要是砂—泥混合的局限潮下带沉积（详见第十五节）。

5. 考察点 P5：团山子组三段中上部岩性特征

位置：营树路东侧中上元古界景区入口处石碑的东北方向约 50m 处，GPS 点位：E 117.465833°，N 40.176667°。

考察内容：（1）观察叠层石类型及特征；（2）观察砂岩夹层的特征；（3）分析沉积环境。

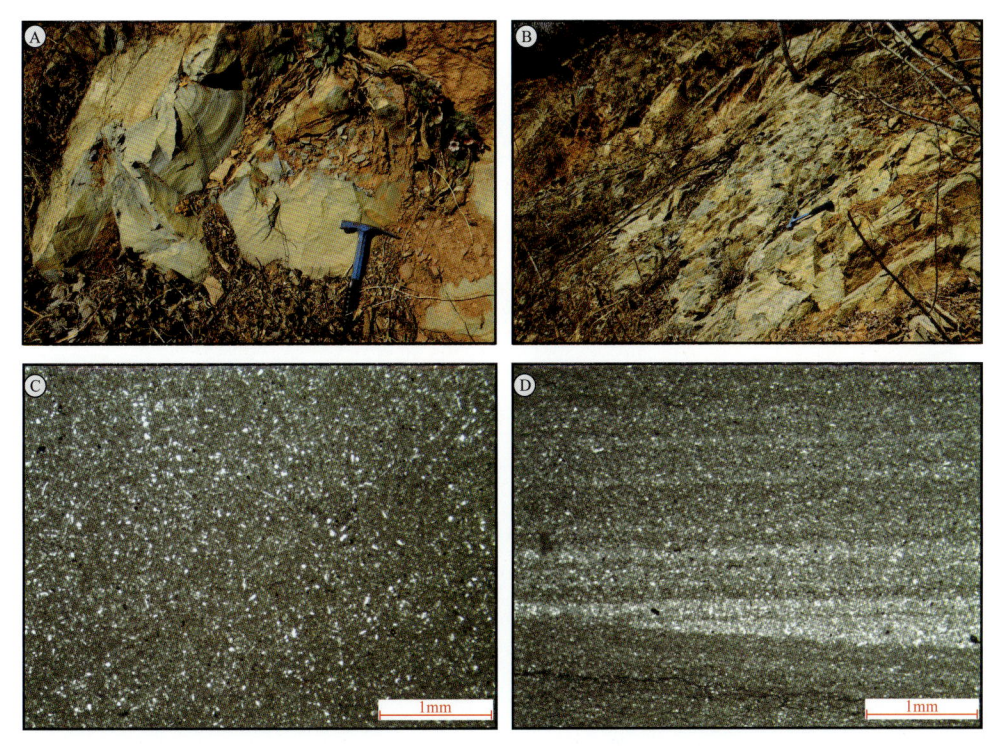

图 3-21 长城系团山子组二段（Cht_2）顶部地质现象及显微照片

Fig.3-21 Geological phenomena of upper part of the Cht_2 and photomicrographs of the rocks

（A 和 B）厚层含铁含粉砂泥晶白云岩，团二段，考察点 P2；（C 和 D）含铁含粉砂泥晶白云岩显微照片，单偏光，团二段，考察点 P2

现象描述：该点为团山子组二段中上部地层，以中—厚层泥晶白云岩与黄灰色细砂岩互层为主。叠层石多呈层状、波状或半球状。团三段中部为碳酸盐岩与碎屑岩混合台地沉积（详见第十五节），主要是潮坪（叠层石和纹理发育为特征）、开阔（或局限）台地（块状构造为特征）、沙滩。

6. 考察点 P6：团山子组三段顶部岩性特征

位置：营树路东侧中上元古界景区石碑东南侧格鲁那叠层石碑旁，GPS 点位：E 117.466111°，N 40.175833°。

考察内容：（1）观察叠层石类型及特征；（2）观察白云岩泥裂的平面形态和剖面形态；（3）分析白云岩的成因；（4）观察砂岩夹层的特征；（5）分析沉积环境以及碳酸盐岩与砂岩交替出现的机理，建立碳酸盐岩与碎屑岩混积模式。

现象描述：该点为团山子三段顶部地层，岩性主要为中厚层黄褐色泥晶白云岩与中厚层灰白色细砂岩互层。泥晶白云岩中发育泥裂，是典型潮上带沉积（图 3-23A）（详见第十五节）。砂岩为沙滩沉积。

7. 考察点 P7：团三子组四段岩石特征

位置：营树路西侧团山子组与大红峪组分界碑处，GPS 点位：E 117.466389°，N 40.174167°。

考察内容：（1）观察岩石类型及特征；（2）观察薄层砂岩中的层理和层面构造，分析古水流方向；（3）观察自下而上砂/泥比的变化，以及与上覆大红峪组的过渡关系；（4）分析沉积环境及其纵向演化。

现象描述：该段为团山子组四段典型地层，其下部发育大套紫红色页岩夹薄层砂岩。自下而上，紫红色页岩变少、变薄，灰白色细砂岩变多、变厚（图 3-23B）。该段是典型的潮上坪（紫红色页岩为主）和潮间砂泥混合坪（页岩与砂岩薄互层）。自下而上由潮上坪变为潮间坪（详见第十五节）。

图 3-22 团山子组三段（Cht_3）地质现象及其岩石显微照片

Fig.3-22 Geological phenomena of the Cht_3 and photomicrographs of their rocks

（A）白云质砂岩与白云岩互层现象，砂岩内发育交错层理，团三段下部，考察点 P3；（B）白云质砂岩，发育明显的交错层理，单偏光，团三段下部，考察点 P3；（C）白云质砂岩显微照片，正交光，团三段下部，考察点 P3；（D）砂岩与页岩互层，团三段，考察点 P4；（E）砂岩与白云岩互层显微照片，单偏光，团三段，考察点 P4

图 3-23 团山子组三段（Cht_3）顶部泥裂及团四段（Cht_4）岩石特征

Fig.2-23 Mud cracks in the upper Cht_3 and characteristics of the Cht_4

（A）泥晶白云岩的泥裂，团三段顶部，考察点 P6；（B）灰白色白云质细砂岩与紫红色白云岩互层，团四段，考察点 P7

第五节　下营镇大红峪沟长城系大红峪组（Chd）考察路线

　　大红峪组为一套火山—沉积岩系，以沉积岩为主，火山岩仅占其中一部分。根据岩性组合的差异，该组可以细分为三段：下部大一段以乳白色石英砂岩为主，夹紫红色粉砂岩、砂质白云岩和白云质砂岩及蓝绿色页岩；中部大二段为火山岩和火山角砾岩集中发育段，夹少量石英砂岩；上部大三段主要发育灰白色硅质锥状叠层石白云岩。

　　虽然大红峪组的火山岩和火山角砾岩仅在大二段发育，但火山活动范围广、持续时间长，在空间上沿东西方向延展，且在下营镇由西向东的高各庄东、苦梨峪、大红峪沟、南山、岛子峪一带出露。各剖面上火山岩的喷发方式、厚度和岩性存在明显的差异，其中西部火山活动强烈，持续时间长，厚度大，以火山熔岩为主，而中部和东部厚度相对薄，以火山碎屑岩为主。总体而言，大红峪组在大红峪沟的出露最为齐全、连续，是该组考察的最佳路线。此处大红峪组的信手剖面图及地层柱状图如图 3-24 和图 3-25 所示。

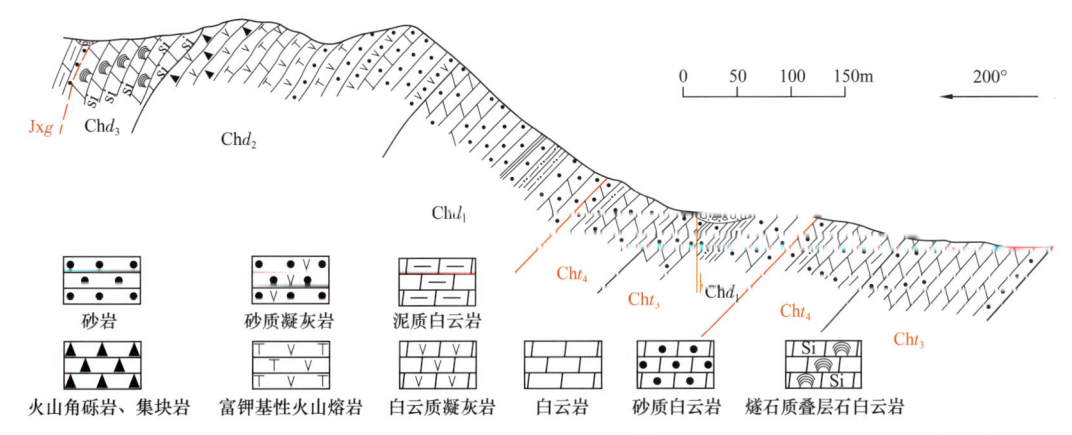

图 3-24　天津市蓟州区下营镇大红峪沟古元古界长城系大红峪组信手剖面图（引自天津地质矿产研究所，1964）

Fig.3-24　Sketch profile of Dahongyu Formation, Changcheng system, Paleoproterozoic in Dahongyugou Valley, Xiaying Town, Jizhou District, Tianjin (Tianjin Institute of Geology and Mineral Resources, 1964)

一、线路位置

　　该路线位于天津市蓟州区下营镇大红峪沟，起点位于马营公路与营树路交叉口向南约 420m 处（GPS 坐标：E 117.466482°，N 40.174527°），终点位于大红峪沟营树路岔口处（GPS 坐标：E 117.476749°，N 40.164697°），路线总长度约 1613m。具体路线及考察点的分布如图 3-26 所示。该路线道路较为宽阔，大小型车辆皆可通行，且日常车辆较少，停车方便。

二、考察目的与任务

（1）观察古元古界长城系大红峪组岩性特征与地层序列；

（2）观察大红峪组与团山子组的整合过渡接触关系；

（3）观察大红峪组二段富钾火山岩和火山碎屑岩的岩石特征，如气孔、杏仁等构造；

（4）观察大红峪组二段（火山岩和火山碎屑岩）与三段（含燧石叠层石白云岩）间的接触关

系及岩性变化；

（5）分析沉积环境，建立沉积模式。

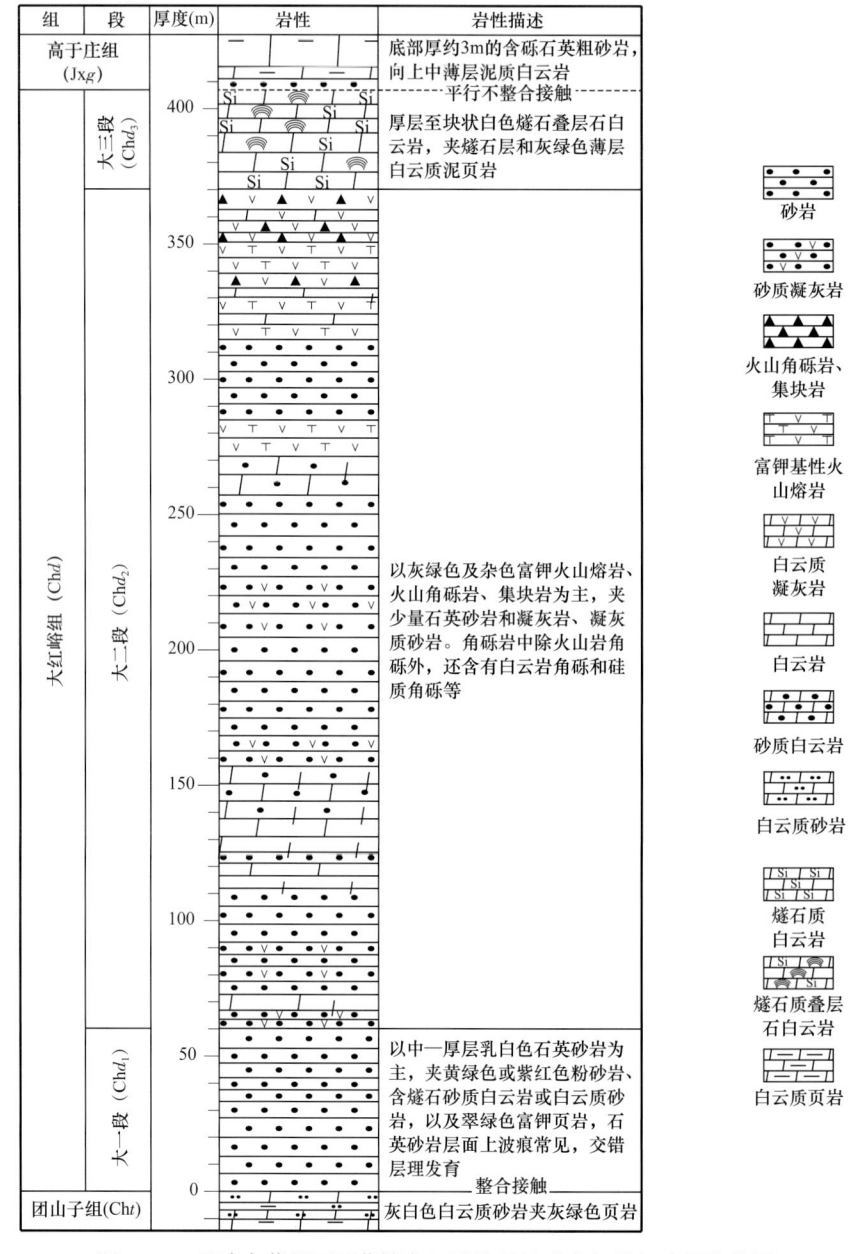

图 3-25 天津市蓟州区下营镇大红峪沟长城系大红峪组地层柱状图

Fig.3-25 Stratigraphic column of Changcheng System Dahongyu Formation in Dahongyugou Valley, Xiaying Town, Jizhou District, Tianjin

三、考察内容与考察点

该路线沿大红峪沟由底至顶依次考察大红峪组的岩性与地层特征，并在路线中选择了 3 个详细考察点，分别为团山子组与大红峪组分界考察点（图 3-26 中 P1）、火山碎屑岩岩石特征考察点（图 3-26 中 P2）及富钾火山熔岩和火山角砾岩考察点（图 3-26 中 P3）。各考察点的具体描述如下。

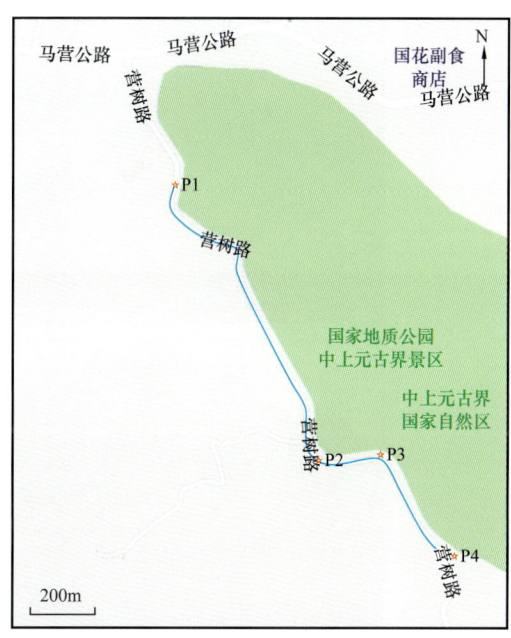

图 3-26 天津市蓟州区长城系大红峪组考察点位置及路线

Fig.3-26 Distribution of investigation stops and route of Changcheng System Dahongyu Formation in Jixian District, Tianjin

1. 考察点 P1：长城系团山子组（Ch*t*）与大红峪组（Ch*d*）分界碑处

位置：该考察点位于马营公路与营树路交叉口向南约 420m 团山子组与大红峪组分界碑处，GPS 点位：E 117.466482°，N 40.174527°。

考察内容：（1）大红峪组与团山子组的界限及其上下岩性变化特征；（2）砂岩中的沉积构造；（3）砂岩的沉积环境。

现象描述：该点为古元古界长城系团山子组（Ch*t*）和大红峪组（Ch*d*）之间的分界，分界面之下是团山子组顶部的页岩与灰白色石英砂岩薄互层，分界面之上为大红峪组一段底部的厚层乳白色石英砂岩，两者之间为整合过渡关系（图 3-27A 和 B）。石英砂岩层面上发育多组不同方向的浪成波痕（图 3-27C），其中厚层石英砂岩中平行层理发育（图 3-27D）。砂岩为潮下浅滩沉积（详见第十五节）。

2. 考察点 P2：大红峪组二段（Ch*d*$_2$）典型火山角砾岩

位置：该考察点为距上一考察点向西南约 1km 处火山角砾岩考察点，GPS 点位：E 117.471845°，N 40.167216°。

现象描述：该点为大红峪组大二段（Ch*d*$_2$）的典型火山角砾岩（图 3-28A）。角砾大小不一，成分复杂，主要包括燧石白云岩、火山岩、凝灰岩等。

3. 考察点 P3：大红峪组二段（Ch*d*$_2$）富钾火山岩和火山角砾岩

位置：该考察点距上一考察点向东约 200m，GPS 点位：E 117.474147°，N 40.167324°。

现象描述：该点为大红峪组大二段（Ch*d*$_2$）顶部的富钾火山熔岩和火山角砾岩，火山熔岩内气孔、杏仁构造发育（图 3-28B 和 C）；火山角砾岩中角砾大小不一，角砾成分主要为杏仁状火山岩（图 3-28D）。

图 3-27　长城系团山子组（Ch*t*）与大红峪组（Ch*d*）分界处地质现象

Fig.3-27　Geological phenomena near the boundary between Tuanshanzi Formation（Ch*t*）and Dahongyu Formation（Ch*d*） of Changcheng System

（A）长城系团山子组（Ch*t*）与大红峪组（Ch*d*）之间的过渡整合接触面，界面之下为团山子组顶部的白云质页岩与灰白色石英砂岩互层，分界之上为大红峪组一段底部厚层乳白色石英砂岩，考察点 P1；（B）团山子组顶部紫红色白云质砂岩向灰白色石英砂岩过渡，考察点 P1；（C）大一段（Ch*d*$_1$）底部乳白色石英砂岩岩层中发育几组不同方向的浪成波痕，考察点 P1；（D）大一段（Ch*d*$_1$）底部厚层乳白色石英砂岩中发育平行层理，考察点 P1

图 3-28　长城系大红峪组大二段（Ch*d*$_2$）火山角砾岩和火山熔岩特征

Fig.3-28　Characteristics of volcanic breccias and lavas in the Ch*d*$_2$

（A）火山角砾岩，角砾成分复杂，主要包括含燧石白云岩、火山岩和凝灰岩等，大红峪组二段，考察点 P2；（B）富钾火山岩的气孔杏仁构造，红色箭头指示杏仁构造，大红峪组二段，考察点 P3；（C）富钾火山岩的气孔构造，大红峪组二段，考察点 P3；（D）火山角砾岩，角砾主要为杏仁状的火山岩，大红峪组二段，考察点 P3

第六节　下营镇大红峪沟—罗庄子镇翟庄村北蓟县系高于庄组（Jxg）考察路线

中元古界蓟县系高于庄组为一套以碳酸盐岩占绝对优势的地层，与下伏古元古界长城系大红峪组呈平行不整合接触，总厚度约1770m。陈晋镳等（1980）将其划分为四个亚组。本书按照岩性组合的差异，自下而上将其细分为六个岩性段：高一段以泥晶白云岩和含叠层石白云岩为主，夹灰绿色页岩，底部有厚约3m的石英砂岩；高二段以含锰粉砂质页岩和含锰粉砂质白云岩、含锰粉砂岩为主；高三段为中—厚层泥晶白云岩；高四段以中—薄层灰质白云岩、白云质灰岩及其薄互层为主，底部为瘤状灰岩，顶部有臼齿灰岩；高五段下部以沥青质纹层状或团块状结晶灰岩为主，上部以包粒（为核形石）粗晶白云岩、波状纹层状中细晶白云岩、水平纹层状细晶白云岩组成的旋回组成；高六段为中厚层含燧石结核和条带的泥粉晶白云岩。

高于庄组沿大红峪沟—翟庄村北一带出露较好，且交通较为便利，是该组最佳的考察路线。沿线露头的信手剖面图和地层柱状图分别如图3-29和图3-30所示。

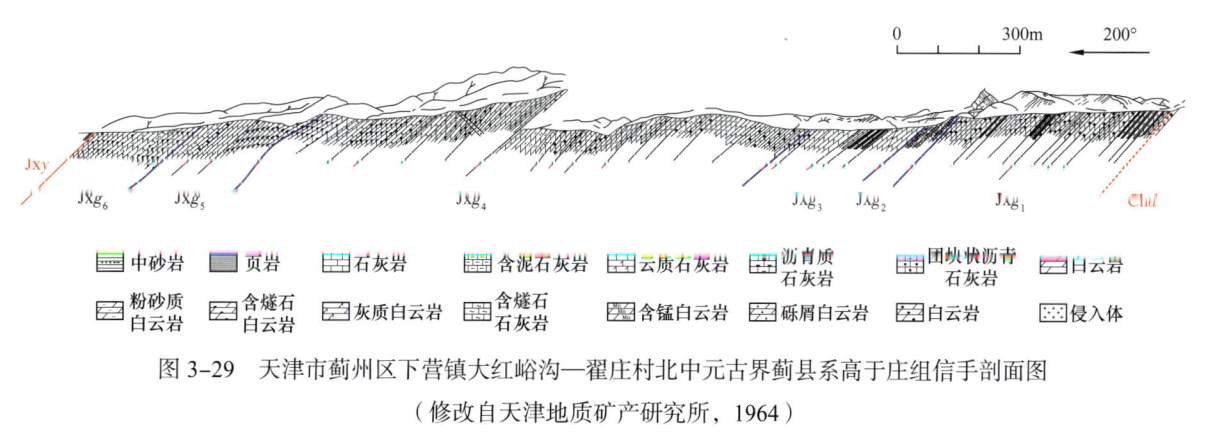

图3-29　天津市蓟州区下营镇大红峪沟—翟庄村北中元古界蓟县系高于庄组信手剖面图
（修改自天津地质矿产研究所，1964）

Fig.3-29　Sketch profile of the Gaoyuzhuang Formation, Jixian system, Mesoproterozoic from Dahongyugou Valley to north of Zhaizhuang Village in Xiaying Town, Jizhou District, Tianjin (modified from Tianjin Institute of Geology and Mineral Resources, 1964)

一、线路位置

该路线位于天津市蓟州区下营镇和罗庄子镇，起点位于下营镇大红峪沟营树路岔口处（GPS坐标：E 117.476749°，N 40.164697°），终点在罗庄子镇翟庄村北（GPS坐标：E 117.467395°，N 40.141950°），路线总长度约4107.5m。具体路线和考察点分布如图3-31所示。该路线前半段（营树路段）较为宽阔，大小车辆皆可通行，考察点沿路分布；后半段（马翟路段）地貌起伏较大，道路情况不允许车辆通行，但地质公园内设有步行台阶、地质现象解释牌等，适于步行考察，车辆可绕至终点附近的成龙路与马翟路交叉口等候。

二、考察目的与任务

（1）观察中元古界蓟县系高于庄组岩性特征与地层序列；

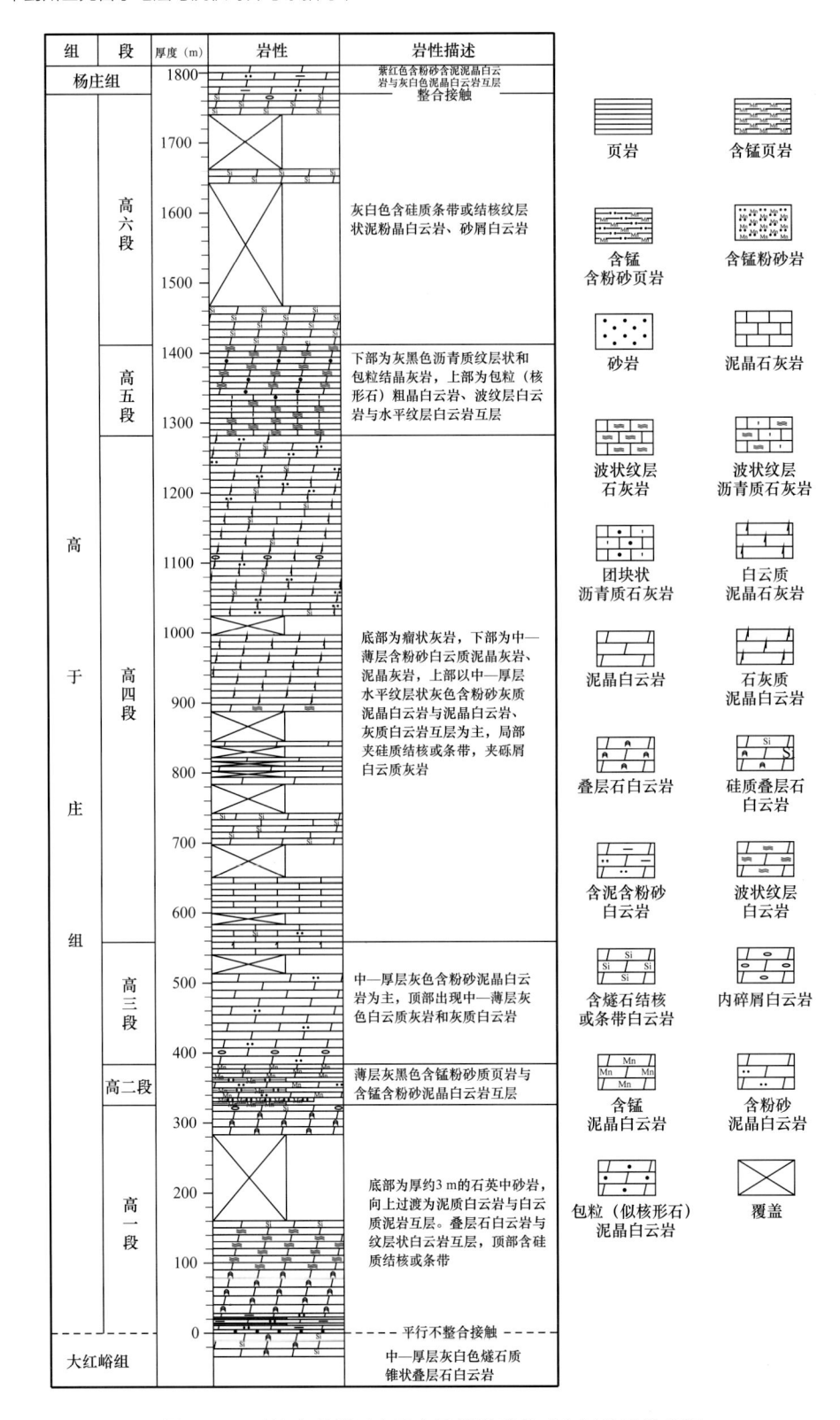

图 3-30 天津市蓟州区中元古界蓟县系高于庄组地层柱状图

Fig.3-30 Stratigraphic column of the Gaoyuzhuang Formation, Jixian System, Mesoproterozoic in Jizhou District, Tianjin

（2）观察蓟县系高于庄组与下伏长城系大红峪组间的平行不整合接触关系；

（3）观察叠层石、圆脊波痕、瘤状、白齿等典型构造；

（4）分析沉积环境及其演化。

三、考察内容与考察点

该路线沿大红峪沟至翟庄村北由底至顶依次考察高于庄组的岩性与地层特征，并在路线中选择了9个详细考察点，分别为大红峪组与高于庄组分界点（图3-31中P1）、高一段与高二段分界点（图3-31中P2）、锰硼矿床（图3-31中P3）、圆脊波痕（图3-31中P4）、瘤状灰岩（图3-31中P5）、高四段与高五段分界点（图3-31中P6）、高五段典型岩性（图3-31中P7和P8）及高六段典型岩性（图3-31中P9）等。各考察点的具体描述如下。

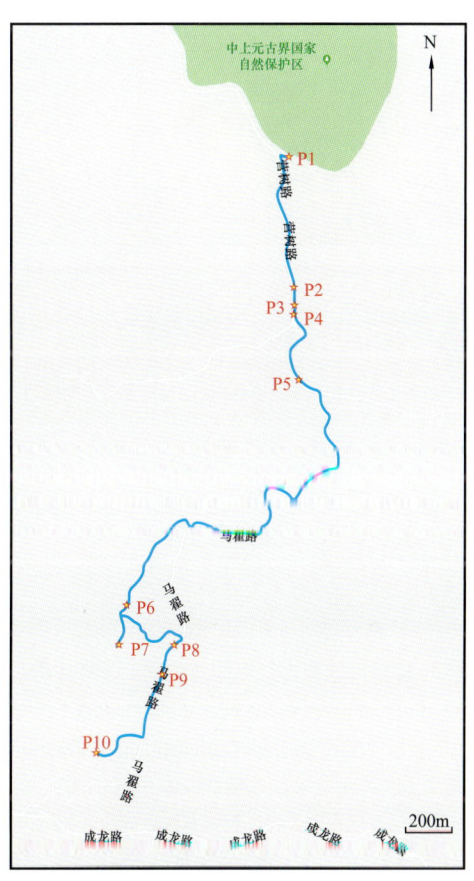

图 3-31　天津市蓟州区蓟县系高于庄组考察点位置及路线

Fig.3-31　Distribution of investigation stops and route of the Jixian System Gaoyuzhuang Formation in Jixian District, Tianjin

1. 考察点P1：长城系大红峪组（Chd）与蓟县系高于庄组（Jxg）分界

位置：该考察点位于营树路岔口处长城系大红峪组与蓟县系高于庄组分界牌，处GPS点位：E 117.476749°，N 40.164697°。

考察内容：（1）高于庄组与下伏大红峪组的平行不整合；（2）大红峪组三段具有锥状叠层石的硅质白云岩；（3）底砂岩中的波痕及成因，据其走向推测海岸线走向；（4）不整合面上、下地层的沉积环境分析。

现象描述：该点为大红峪组（Chd）与高于庄组（Jxg）的分界，即古元古界长城系与中元古界蓟县系的分界，两组之间为平行不整合接触（图3-32A和B）。分界面之上为高于庄组底部的厚约3m的含细砾粗砂岩及长石质石英细砂岩，砂岩表面发育对称浪成波痕。波痕较为连续，波脊延伸方向大致为290°（图3-32C和D），为潮下砂坪沉积（详见第十五节）。分界面之下为大红峪组含锥状叠层石燧石白云岩，叠层石多被硅化，可见大型锥状叠层石，代表浅潮下及潮间坪沉积；叠层石顶部被高于庄组底部砂岩截切，表明曾受到侵蚀（图3-32B）。界面上局部可见薄层的风化壳，代表一次沉积间断，前人称之为"青龙上升"（陈晋镳等，1980）。

图3-32 长城系大红峪组（Ch*d*）与蓟县系高于庄组（Jx*g*）分界处地质现象及其岩石显微照片

Fig.3-32 Geological phenomena near the boundary between the Changcheng System Dahongyu Formation（Ch*d*）and the Jixian System Gaoyuzhuang Formation（Jx*g*），and photomicrographs of the rocks

（A和B）长城系大红峪组（Ch*d*）与蓟县系高于庄组（Jx*g*）之间平行不整合接触面，界面之上为高于庄组石英砂岩，界面之下为大红峪组含锥状叠层石燧石白云岩，叠层石锥顶被砂岩截切，考察点P1；（C）高于庄组底部含长石石英砂岩，表面发育对称的浪成波痕，波脊延伸方向约290°，高一段底部，考察点P1；（D）高于庄组底部含长石石英中砂岩的显微照片，正交光，高一段底部，考察点P1

2. 考察点P2：高于庄组一段（Jxg_1）与二段（Jxg_2）分界及高二段岩性特征

位置：该考察点位于桑树庵村村委会旁的高于庄组一段与二段分界牌，处GPS点位：E 117.477235°，N 40.159696°。

考察内容：（1）岩石类型及特征；（2）页岩中的沉积构造；（3）高一段顶部岩性及其波状层面的成因；（4）高二段与下伏高一段和上覆高三段之间的突变接触；（5）高一段顶部燧石结核和条带的特征和成因；（6）高二段的岩石类型；（7）分析高一段和高二段的沉积环境及环境巨变的原因。

现象描述：该点为高于庄组一段（Jxg_1）与二段（Jxg_2）分界，两段之间呈整合接触关系（图3-33A）。界面之下的高一段的顶部为深灰色含锰燧石结核和条带的白云岩（图3-33B），为台地上的局限台地和潮坪沉积；界面之上高二段底部为深灰色含锰粉砂质页岩、含锰泥质粉砂岩（图3-33C）夹薄层泥质泥晶白云岩（图3-33D），为深水陆棚沉积（详见第十五节）。

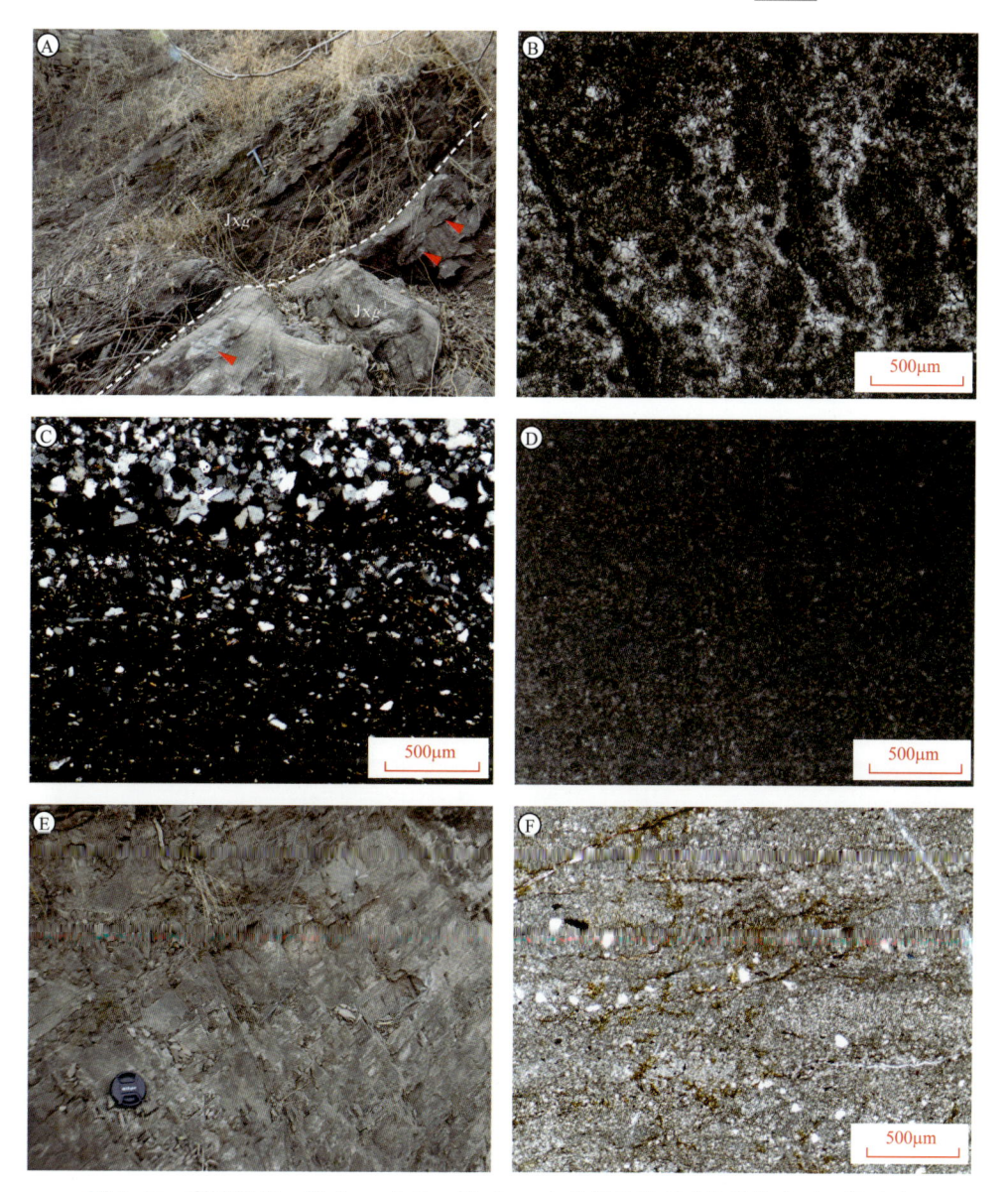

图 3-33 高于庄组一段（Jxg_1）和二段（Jxg_2）分界处地质现象及其岩石显微照片

Fig.3-33 Geological phenomena near the boundary between the Jxg_1 and the Jxg_2, and photomicrographs of the rocks

（A）高于庄组一段和二段分界，界面之下为高一段底部的含锰含燧石结核和条带（红色箭头标记）藻白云岩，界面之上为高二段底部含锰粉砂质页岩、含锰泥质粉砂岩与含锰含粉砂泥晶白云岩、泥晶白云岩薄互层，考察点 P2；（B）高一段顶部藻白云岩，单偏光，考察点 P2；（C）高二段底部含锰粉砂质页岩与含锰泥质粉砂岩薄互层，正交光，考察点 P2；（D）高二段中部泥晶白云岩，单偏光，考察点 P2；（E）高二段顶部页岩夹薄层泥质泥晶白云岩，伴生锰方硼矿，考察点 P3；（F）高二段顶部薄层含粉砂泥晶白云石，单偏光，考察点 P3

3. 考察点 P3：锰（硼）矿床

位置：该考察点位于桑树庵村村委会锰（硼）矿床牌处，GPS 点位：E 117.477340°，N 40.158943°。

考察内容：锰方硼石矿的特征。

现象描述：此处距上点向南约 50m，处于高于庄二段（Jxg_2）上部，岩性为灰褐色含锰粉砂质泥晶白云岩（图 3-33E 和 F），伴有锰方硼石，构成著名的"蓟县式锰（硼）矿"。关于该处锰方硼石矿床的成因，前人主要存在以下观点：（1）一些学者认为锰硼岩系形成于最大海退期的潟湖相，微晶白云岩中的有机质和黄铁矿在空间上和时间上都与锰方硼石有关（天津市区域地质志，1992）；

（2）王培君（1996）认为蓟州区锰方硼矿符合沉积型硼矿床的二元结构模型，硼的来源与火山喷发有关；（3）Fan et al. (1999) 认为其成矿物质来源于海底火山喷发及少量早元古界含硼沉积物的风化；（4）肖荣阁等（2002）通过对锰方硼石的硼同位素进行研究，认为属于海底热水沉积矿物；（5）王秋舒（2013）通过对锰方硼石的矿物学特征的研究，判断硼和锰来源于海底火山喷发，成矿温度低于400℃，弱酸性、高盐度、厌氧环境，推测其是海底热液沉积而形成。

4. 考察点 P4：高三段岩性变化及圆脊波痕

位置：该考察点位于桑树庵村封山育林区宣传牌岔路口处，GPS点位：E 117.47725°，N 40.158758°。

考察内容：（1）白云岩顶面的圆脊波痕；（2）高三段岩石类型及沉积旋回；（3）高三段沉积环境。

现象描述：此处距离上点向南约28m处，位于高于庄组三段之中，岩性为中—厚层白云岩，表面发育圆脊波痕（图3-34A）。圆脊波痕的特点是波痕断面的形状为对称的正弦曲线状，不能指示水流方向，为风暴浪形成的。整个高三段为中厚层含铁泥晶白云岩与纹理发育的泥晶白云岩互层，夹灰绿色页岩。水平纹理为层状叠层石。见少量燧石结核和条带，多集中在纹理发育的泥晶白云岩中。整个高三段为台地上的潮坪和局限（或开阔）台地交互。页岩代表海侵事件，为较深水页岩陆棚沉积（详见第十五节）。

图 3-34 圆脊波痕和瘤状石灰岩

Fig.3-34 Round-ridge ripple marks and nodular limestone

（A）圆脊波痕，高于庄组三段（Jxg_3），考察点P4；（B）瘤状灰岩，高于庄组四段（Jxg_4）底部，以其出现作为高四段（Jxg^4）与高三段（Jxg^3）的分界标志，考察点P5

5. 考察点 P5：瘤状灰岩

位置：该考察点位于桑树庵村村委会向南约340m处，GPS点位：E 117.477429°，N 40.156295°。

考察内容：（1）瘤状石灰岩的特征；（2）瘤状石灰岩的成因。

现象描述：此处发育瘤状石灰岩，以其出现作为高四段与高三段的分界标志，是石灰岩段（高四段）与白云岩段（高三段）的分界。瘤状石灰岩之下主要为高三段顶部的中—薄层泥晶白云岩和含灰（或灰质）白云岩互层，向上逐渐过渡为中薄层石灰岩。瘤状构造是指岩层中出现的如瘤的团块、半球形的透镜状岩石，常顺层分布，横向上可连续成层（图3-34B）。瘤状石灰岩的瘤状体成分与其周边岩石（基质）略有区别，其中瘤状体主要由方解石（约90%）、白云石（5%～10%）和少量有机质（<5%）、陆源粉、细砂等组成；而基质主要呈浅灰色、黄色，围绕瘤状体的周围分布，成分主要为泥—粉晶方解石，其晶粒明显较瘤状体细，黏土含量较瘤状体高很

多；部分瘤状体和基质界线处发育缝合线，缝合线起伏程度不大（几毫米或几厘米）。前人认为瘤状构造的成因主要有以下几种：（1）在较深水盆地和台缘斜坡下部，由于周期性重力作用而形成海底底流作用，海底底流带来的$CaCO_3$不饱和的流体部分溶解沉积于海底的碳酸盐岩，并将剩余富含泥质物残留下来而形成的（高计元，1988）；（2）沉积成岩作用成因（张继庆，1986）；（3）压溶成岩作用成因，即由于在上覆压力或构造应力的作用下，压溶作用将弱固结的、结构成分不均一的含泥灰岩分割成大小不一、形状多变的瘤状体而形成，沿瘤状体和基质分界面（即溶解面）上产生凹凸不平的不溶残留物膜，在断面上表现为波状起伏或齿状的缝合线（Wanless，1979；郭福生，1989）；（4）周期性海底底流溶解作用和压溶作用共同作用而形成（夏丹等，2009）。

本书认为，石灰岩瘤是微生物成因的，是微生物局部繁殖、生长形成的透镜体，只不过这些透镜体较小。

6. 考察点P6：高于庄组四段（Jxg_4）及其与五段（Jxg_5）的分界

位置：界限位于营树路与马翟路岔口沿马翟路前行约2.4km处（原高于庄组三段与四段分界牌处），GPS点位：E 117.468656°，N 40.147629°。高四段的观察，前半段可沿小柏油路观察，后半段则需沿新修建的台阶路观察，行走2km左右，车不能通行。越野车可以沿山脊的土路继续前行，但路两边露头不好。

考察内容：（1）高四段的岩石类型；（2）沉积旋回；（3）"臼齿构造"及其成因；（4）高四段沉积环境。

现象描述：高四段以石灰岩发育、燧石结核稀少、缺乏叠层石为特征，有些含有白云质。该段岩层面大都平整。主要由三种岩石类型：无层理、具块状构造的中厚层灰泥石灰岩（可含白云质）、水平层理发育的灰泥石灰岩（可含白云质）、页状泥质灰泥石灰岩（可含白云质）。这三种岩石组成多个米级沉积旋回。这些旋回主要有两类，自下而上为：页状泥质灰泥石灰岩—水平层理发育的灰泥石灰岩—具块状构造的中厚层灰泥石灰岩，或水平层理发育的灰泥石灰岩—具块状构造的中厚层灰泥石灰岩。燧石结核主要出现在具块状构造的中厚层灰泥石灰岩中，而且多呈管壳状，内部未硅化。这种形状的燧石结核可能与虫孔或一种大型蠕虫类生物有关，生物的皮肉部分腐烂产生酸性微环境，导致选择性硅化。含燧石结核的石灰岩普遍缺乏层理，也说明有生物扰动。由此推测该时期已有软体动物。

高四段普遍缺乏叠层石，可能是水体较深所致，层间平整成水平层理为高的石灰岩，沉积于较深的环境，波浪不能影响到海底。有些页状泥质灰泥石灰岩具有浪成波痕，有些具块状构造的中厚层灰泥石灰岩顶面波状起伏（实际上就是波痕），说明波浪有时能影响到海底。因此，该段的沉积环境总体较深，在风暴浪基面附近，为较深水开阔台地到缓斜坡沉积。接近高四段顶部时，水体变浅。

高于庄组四段（Jxg_4）与五段（Jxg_5）的分界以臼齿灰岩的出现作为标志，臼齿灰岩被划归高四段。分界面以下为高四段上部的灰色中厚层含燧石结核白云质灰岩，分界面以上迅速过渡为高五段底部的纹层状和包粒状沥青质结晶灰岩（图3-35D-F）。高四段顶部中薄层泥晶灰岩中含有大量"臼齿构造"。臼齿构造表现为泥晶灰岩中形态各异的脉体，脉体长约数厘米至十余厘米，中部较粗，向两端逐渐尖灭，脉体侧部可见更为细小的脉体分支和孤立斑点（图3-35A和B）。通常脉体相对围岩的泥晶灰岩而言晶粒较粗，多由粉晶方解石（或白云石）组成（图3-35C）。关于"臼齿构造"的成因，主要有以下观点：（1）认为其形成与藻类生物的生命活动有关（James et al.，1998；

Frank and Lyons, 1998; 刘燕学等, 2003);(2) 认为是地震诱发的液化作用导致尚未固结的灰泥沉积物发生水下收缩作用有关(Pratt, 1988; Fairchild *et al.*, 1997)。

图 3-35　高于庄四段(Jxg$_4$)与五段(Jxg$_5$)分界处地质现象及岩石显微照片

Fig.3-35　Geological phenomena near the boundary between Jxg$_4$ and the Jxg$_5$, and photomicrographs of their rocks

(A 和 B) 臼齿灰岩,高于庄组四段顶部,泥晶方解石脉(红色箭头标记)在平面上中间宽,向两端尖灭,在剖面上分布不规则,且呈较多的细小分支和孤立斑点状产出,考察点 P6;(C)臼齿灰岩,单偏光,臼齿晶粒较围岩泥晶灰岩的晶粒粗,臼齿主要由细粉晶方解石组成,高于庄组四段顶部,考察点 P6;(D)纹层状沥青质结晶灰岩,高五段底部,考察点 P6;(E)纹层状沥青质结晶灰岩,单偏光,方解石晶粒以细晶为主,纹层之间为孔隙或部分被白云石充填,考察点 P6;(F)核形石沥青质结晶灰岩,高五段底部,考察点 P6

本书认为,臼齿石灰岩是一种特殊的、未充分发育的泥裂,是很浅的局限台地沉积的灰泥沉积物,经过水下沉积、水上短暂暴露形成泥裂,又淹没于水下使泥裂被充填而形成的。例如,极浅水潟湖沉积的灰泥,在异常的特低潮期间暴露,形成泥裂;在泥裂尚未发育成型时又涨潮,淹没于水下,泥裂被灰泥充填。由于泥裂缝内滞流、还原性的微环境,使泥裂缝中的沉积物颜色更暗。

高五段底部重结晶灰岩中的纹层状构造就是层状叠层石(图 3-35D)。纹层主要由细晶方解石组成,是经过重结晶形成的;纹层间为孔隙或部分被白云石充填(图 3-35E);而包粒状结晶灰岩中的包粒应为核形石(图 3-35F)。

7. 考察点 P7：高于庄组五段（Jxg_5）典型岩性组合

位置：该考察点距上一考察点向南 50m 半山坡处，GPS 点位：E 117.468379°，N 40.146352°。

考察内容：（1）高五段岩石类型及沉积旋回的特征；（2）"虾米石"的特征及成因；（3）波状纹层的特征及成因；（5）油浸砂糖状白云岩的特征及成因分析；（6）高五段的沉积环境。

现象描述：该处主要观察高于庄五段（Jxg_5）的典型岩性组合，由黑色细晶白云岩、灰白色波状纹层白云岩和灰白色水平纹层泥晶白云岩组成的旋回而构成（图 3-36A），其中灰白色波状纹层白云岩的纹层主要由泥晶白云石组成，纹层间充填细晶白云石，纹层常出现断裂、破碎（图 3-36B 和 C）；水平纹层泥晶白云岩主要由泥晶白云石组成，沿层有晶粒大小变化，但不明显（图 3-36B 和 D）；黑色细晶白云岩中纹层发育，常见同心球状构造或称结核状构造（应是核形石），破碎严重，裂缝中充填粗晶白云石。核形石的白云石晶粒较其间的基质晶粒粗，且更加明亮（图 3-36E 和 F）。

图 3-36 高于庄组五段（Jxg_5）地质现象及显微照片

Fig.3-36 Geological phenomena of the Jxg_5 and photomicrographs of the rocks

（A）灰黑色细晶白云岩、灰白色波状纹层白云岩和水平纹层泥晶白云岩沉积旋回，考察点 P7；（B）灰白色波状纹层白云岩和水平纹层泥晶白云岩，为图 A 中 B 框放大图，考察点 P7；（C）波状纹层状白云岩，单偏光，取样位置如图 B 中的 C 点所示，纹层为泥晶白云石为主，纹层间为细晶白云石，局部破碎，考察点 P7；（D）水平纹层泥晶白云岩，单偏光，取样位置如图 B 中 D 点所示，考察点 P7；（E）纹层状和结核状（核形石）细晶白云岩，为图 A 中 E 框放大图，考察点 P7；（F）结核状细晶白云岩，单偏光，结核的白云石晶粒较结核间基质的白云石晶粒略粗，更加明亮，结核间裂缝发育，多被粗晶白云石充填，且部分呈鞍状，考察点 P7

8. 考察点 P8：高于庄组五段（Jxg₅）上部典型岩性组合

位置：该考察点距上一考察点向东约240m处的马翟路旁，GPS点位：E 117.471185°，N 40.146267°。该点位于新开出来的路边，连续出露，现象很清楚，与考察点7属于同一层位，但出露的地层更多，现象更好。

考察内容：高五段上部岩石类型及沉积旋回的特征。

现象描述：此点主要观察高于庄组五段上部的典型岩性特征，可见具球状构造的黑色细晶白云岩、波状纹层白云岩及水平纹层白云岩频繁互层（图3-37A-D），其中具球状构造的黑色细晶白云岩中破裂最为严重，裂缝中多被乳白色粗晶白云石充填（图3-37E和F）。

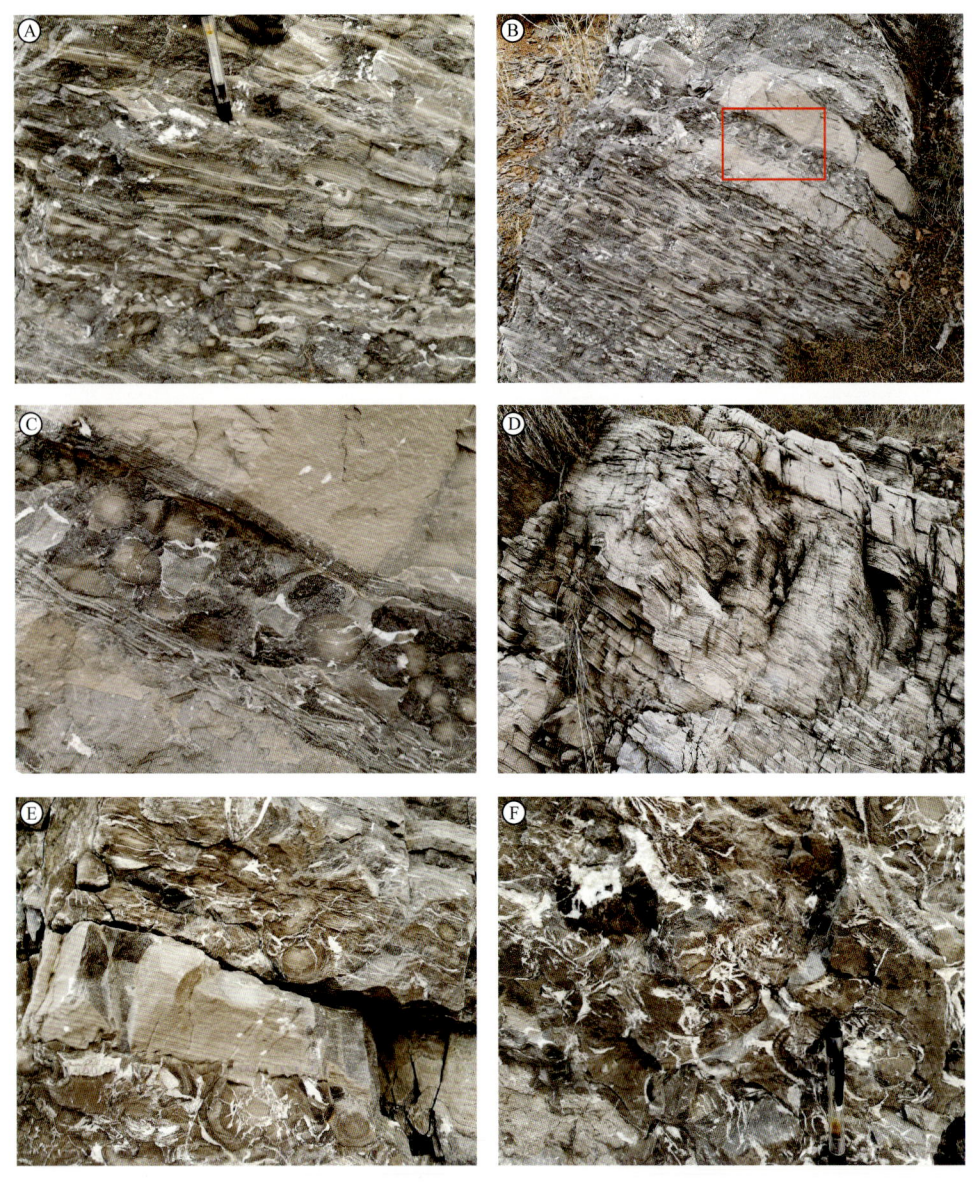

图 3-37 高于庄组五段（Jxg₅）典型岩石类型

Fig.3-37 Typical rocks of the Jxg₅

（A）灰黑色细晶白云岩，内部发育纹层，高于庄组五段，考察点P8；（B）黑色细晶白云岩（含核形状结核）、波状纹层白云岩和水平纹层白云岩旋回，高于庄组五段，考察点P8；（C）图B中红色方框放大；（D）灰白色波状纹层白云岩和水平纹层泥晶白云岩，高于庄组五段，考察点P8；（E）黑色细晶白云岩（含核形石，破裂严重）与水平纹层白云岩，高于庄组五段，考察点P8；（H）黑色细晶白云岩，内含核形石，破碎严重，裂缝内被乳白色粗晶白云石充填，高于庄组五段，考察点P8

9. 考察点 P9：高于庄组六段（Jxg$_6$）典型岩性组合

位置：沿马翟路距上一考察点向南约 80m 处，GPS 点位：E 117.470690°，N 40.145607°。

考察内容：（1）岩石类型及特征；（2）沉积旋回的特征；（3）燧石的特征及成因；（4）沉积环境分析；（5）高于庄组与杨庄组的分界。

现象描述：此处主要为高于庄组六段的含燧石结核或条带的泥粉晶白云岩，可见燧石条带顺层分布（图 3-38A）。燧石结核明显切割白云岩纹层（图 3-38B），多为成岩期形成，在镜下粉晶白云石常漂浮在燧石质基底之中（图 3-38C 和 D）。

高于庄组与杨庄组界限清晰，后者为红白相间的泥岩，为整合接触。

图 3-38 高于庄组六段（Jxg$_6$）典型岩石类型及其显微照片

Fig.3-38　Typical rocks of the Jxg$_6$ and their photomicrographs

（A）含硅质条带粉晶白云岩；（B）含硅质结核和条带粉晶白云岩；（C）硅质粉晶白云岩，高六段，单偏光，粉晶白云石漂浮于硅质基底之中；（D）图 C 正交偏光下的显微照片

第七节　罗庄子镇翟庄村北—青山村—花果峪村蓟县系杨庄组（Jxy）考察路线

中元古界蓟县系杨庄组主要为一套醒目的红白相间的泥岩（紫红色泥岩与灰白色泥岩间互），即"五花肉泥岩"，夹石灰岩和白云岩，泥岩中常含粉砂及白云石。上部见一层几十厘米厚的含鲕粒砂岩。总厚度 773m。

紫红色泥岩的颜色为强氧化色，是暴露的低能强氧化环境沉积，为广阔的潮上坪环境；灰白色泥岩的颜色为弱还原色，是经常被海水淹没的潮间泥坪或浅且低能的局限潮下带环境沉积；凝块状石灰岩或白云岩为浅的潮下带沉积，层状叠层石石灰岩或白云岩为碳酸盐潮坪；含鲕粒的砂岩为潮下高能浅滩沉积（详见第十五节）。整个杨庄组为潮上带、潮间带和浅的潮下带交替沉积。

按照岩性组合，可以划分为三段：底部杨一段（Jxy_1）"五花肉泥岩"夹含燧石的石灰岩和白云岩；中部杨二段（Jxy_2）"五花肉泥岩"；上部杨三段（Jxy_3）"五花肉泥岩"夹含燧石的石灰岩和白云岩。

杨庄组在天津市蓟州区罗庄子镇杨庄村、翟庄村、青山村及花果峪村一带广泛出露，但考虑到分布区中部存在断层及出露情况的影响，该组的考察点较为分散，主要分3条路线进行考察：第1条路线位于翟庄村北，主要考察杨一段的岩性特征；第2条路线位于青山村采摘园，主要观察杨二段和杨三段的岩性特征；第3条路线位于花果峪村东南，主要观察杨二段和杨三段的岩性特征。其中，第2条路线沿山坡上行，露头整体较为连续、齐全，并可见与上覆雾迷山组的分界。该处地层柱状图分别如图3-39所示。

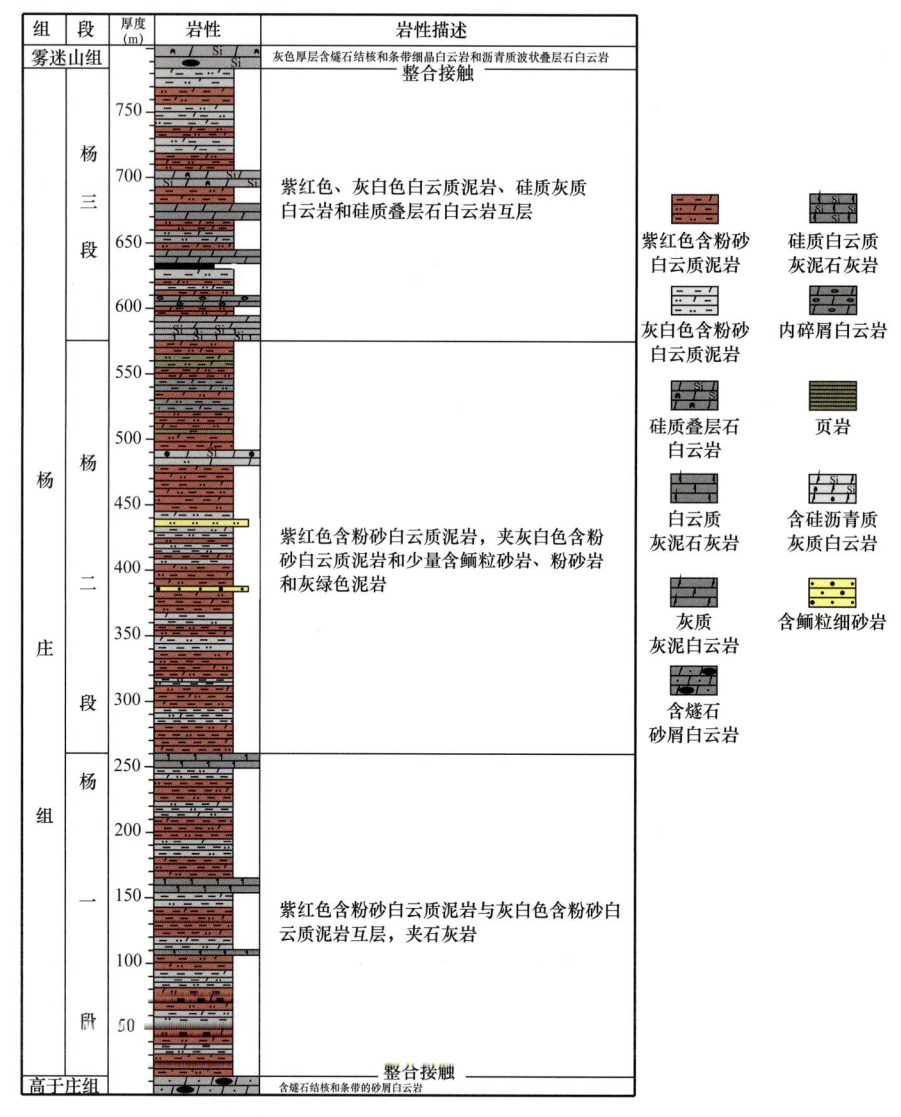

图3-39 天津市蓟州区罗庄子镇青山村采摘园中元古界蓟县系杨庄组（Jxy）地层柱状图

Fig.3-39 Stratigraphic column of the Yangzhuang Formation（Jxy）in the picking garden of Qingshan Village, Luozhuangzi Town, Jizhou District, Tianjin

一、线路位置

由于受地层出露情况和断层的影响，该路线考察点较为分散，主要分为3条路线进行考察（图3-40）：第1条路线的起点位于罗庄子镇翟庄村北（GPS坐标：E 117.467395°，N 40.141950°），终点位于马翟路与成龙路交叉口向东约125m处（GPS坐标：E 117.465003°，N 40.139757°）；第2条路线位于罗庄子镇青山村，起点位于青山采摘园大门处（GPS坐标：E 117.445361°，N 40.124436°），终点位于五名山半山坡处（GPS坐标：E 117.443030°，N 40.120781°）；第3条路线位于罗庄子镇花果峪村沿马平公路向东南方向约920m处（GPS点位：E 117.464169°，N 40.118511°）。该线路总长度大约11.2km，其3条路线皆位于主干公路旁，大小车辆通行便捷，仅第2条路线为爬坡路段，但山坡修有水泥路和步行台阶，考察点位于道旁，适于步行考察。

二、考察目的与任务

（1）观察中元古界蓟县系杨庄组岩性特征与地层序列；
（2）观察花斑状白云岩、叠层石白云岩等具有典型沉积构造的岩石；
（3）分析沉积环境及其演化。

三、考察内容与考察点

该路线沿大红峪沟至翟庄村北由底至顶依次考察高于庄组的岩性与地层特征，并在路线中选择了7个详细考察点，分别为翟庄村北的高于庄组与杨庄组分界点（图3-40中P1）、杨一段典型岩性（图3-40中P2）、青山村的杨二段典型岩性（图3-40中P3）、杨二段与杨三段分界点（图3-40中P4）、杨三段典型岩性（图3-40中P5）、杨庄组与雾迷山组分界点（图3-40中P6），及花果峪村的杨二段与杨三段分界点（图3-40中P7）等。各考察点的具体描述如下。

图3-40　天津市蓟州区中元古界蓟县系杨庄组（Jxy）考察点位置及路线

Fig.3-40　Distribution of investigation stops and route of the Yangzhuang Formation（Jxy），Jixian System，Mesoproterozoic in Jizhou District, Tianjin

1.考察点P1：高于庄组（Jxg）与杨庄组（Jxy）分界

位置：该考察点位于翟庄村北，成龙路与马翟路交叉口向东北约400m处，GPS点位：E 117.467395°，N 40.141950°。

考察内容：高于庄组与杨庄组分界。

现象描述：此处为高于庄组（Jxg）与杨庄组（Jxy）的分界（图3-41A），分界之下为高于庄组六段顶部的中层含燧石结核和条带的白云岩（图3-41B），分界之上为杨庄组底部的紫红色白云质泥岩夹灰白色含硅质结核白云质灰岩（图3-41A和C）。两个组之间为连续沉积。

图3-41 蓟县系高于庄组（Jxg）与杨庄组（Jxy）分界处地质现象及岩石显微照片

Fig.3-41 Geological phenomena near the boundary between the Gaoyuzhuang Formation（Jxg）and the Yangzhuang Formation（Jxy）of Jixian System, and photomicrographs of their rocks

（A）蓟县系高于庄组与杨庄组分界，分界面之下为高于庄组含燧石结核和条带的含粉砂砂屑白云岩，分界面之上为杨庄组紫红色白云质泥岩夹灰白色含粉砂泥晶白云岩，考察点P1；（B）含粉砂泥晶白云岩，高于庄组六段顶部，单偏光，考察点P1；（C）杨庄组一段底部灰白色含燧石结核泥晶白云岩，考察点P1；（D）杨庄组一段典型岩石类型，考察点P2

2. 考察点P2：杨庄组一段（Jxy₁）典型岩性组合

位置：该考察点位于马翟路与成龙路交叉口向东约125m处，GPS点位：E 117.465003°，N 40.139757°。

考察内容："五花肉泥岩"的特征及其沉积环境。

现象描述：此处杨庄组一段（Jxy₁）的典型岩石出露良好，岩性主要为紫红色含粉砂白云质泥岩夹灰白色含白云质泥岩，两者厚度比例约3∶1（图3-41D）。

3. 考察点P3：杨庄组二段（Jxy₂）典型岩性组合

位置：该考察点位于青山村马平线旁青山采摘园沿坡上行约250m处，GPS点位：E 117.444195°，N 40.122783°。

考察内容：（1）"五花肉泥岩"的特征；（2）偶见的一层砂岩的特征；（3）沉积环境分析。

现象描述：此处为杨庄组二段（Jxy₂）的典型岩性的分布段，主要由紫红色含粉砂白云质泥岩与灰白色含粉砂白云质泥岩互层组成，即"五花肉泥岩"（图3-42A）。中部偶见一层鲕粒质石英中砂岩（图 3-42D 和 E）。紫红色含粉砂白云质泥岩所具有的醒目的紫红色多是因其内富含 Fe^{3+} 所致，一般指示暴露的炎热干燥强氧化环境。紫红色含粉砂白云质泥岩中还常见灰绿色斑点和条带（图 3-42B），且在镜下斑点和条带相对明亮（图 3-42C）。这种现象是紫红色泥岩局部被还原形成的。还原性地下水沿高渗透性层面或裂隙渗流，使 Fe^{3+} 还原为 Fe^{2+}，颜色由红色变为灰绿色。局部层面上波痕发育（图 3-42F）。

图 3-42 杨庄组二段（Jxy_2）典型岩石类型及其显微照片

Fig.3-42 Typical rocks of the Jxy_2 and their photomicrographs

（A）杨二段典型岩石类型，紫红色与灰白色含粉砂白云质泥岩互层，考察点 P3；（B）紫红色与灰白色含粉砂白云质泥岩互层，紫红色泥岩中常见灰绿色斑点和条带，杨二段，考察点 P3；（C）紫红色含粉砂白云质泥岩，单偏光，白色虚线圈定范围代表其中的灰绿色斑点和条带，相对明亮，杨二段，考察点 P3；（D）鲕粒质石英中砂岩，平行层理发育，杨二段，考察点 P3；（E）鲕粒质石英中砂岩，正交光，为图 D 的显微照片，其中鲕粒由泥晶白云石组成，部分同心圈层明显，呈放射状，杨二段，考察点 P3；（F）紫红色夹灰白色含粉砂白云质泥岩，层面发育波痕（红色箭头所示），杨二段，考察点 P3

4. 考察点 P4：杨庄组二段（Jxy_2）与三段（Jxy_3）分界

位置：该考察点位于青山村马平线旁青山采摘园沿坡上行约 300m 处，GPS 点位：E 117.443833°，N 40.122186°。

考察内容：杨庄组二段与三段的分界及分界面上下的岩性变化。

现象描述：此处为杨庄组二段（Jxy_2）和三段（Jxy_3）的分界，以厚层碳酸盐岩出现为标志。分界面之下为杨二段（Jxy_2）紫红色与灰白色含粉砂白云质泥岩互层，界面之上为杨三段（Jxy_3）厚层灰色、灰黑色白云岩、灰质条带白云岩和叠层石白云岩夹少量灰绿色或紫红色含粉砂白云质泥岩。常见的碳酸盐岩序列由厚层凝块石（一种蓝绿藻）石灰岩（潮下带沉积）、中层含燧石条带层状叠层石石灰岩组成（潮坪沉积）。层状叠层石石灰岩中常夹薄层硅化了的鲕粒白云岩和砾屑白云岩（图3-43），为风暴流搬运到潮坪上的颗粒碳酸盐岩沉积。

图 3-43 杨庄组三段（Jxy_3）典型岩石类型及其显微照片

Fig.3-43 Typical rocks of the Jxy_3 and their photomicrographs

（A）杨三段中的碳酸盐岩沉积旋回，图中标有 B 的是含硅质凝块状石灰岩，标有 C 的是含燧石条带层状叠层石石灰岩，标有 D 的是见残余鲕粒的燧石条带，标有 E 的是砾屑白云岩；（B）含硅质凝块状石灰岩，硅质未染色，单偏光，取样位置如图 A 中 B 点所示，考察点 P4；（C）叠层石石灰岩，正交光，注意不同纹层间方解石晶体的排列方式存在一定的差异，取样位置如图 A 中 C 点所示；（D）含云鲕状硅质岩（燧石），正交光，取样位置如图 A 中 D 点所示；（E）砾屑白云岩，图 A 中 E 红色框的放大图；（F）砾屑白云岩，单偏光，为图 E 样品的显微照片，砾屑多由粉晶白云石组成

5. 考察点 P5：杨庄组三段（Jxy_3）岩性特征及典型碳酸盐岩旋回

位置：该考察点位于青山村马平线旁青山采摘园沿坡上行约 400m 处，GPS 点位：E 117.443405°，N 40.121443°。

考察内容：（1）凝块石白云岩（或石灰岩）与层状叠层石白云岩（或石灰岩）形成的沉积旋回；（2）"脉状构造"及其成因分析；（3）沉积环境分析。

现象描述：此处为杨庄组三段（Jxy_3）中典型的灰色、深灰色碳酸盐岩旋回段，从下至上依次由凝块石细晶白云岩或石灰岩（潮下带沉积）、中层波状和层状叠层石泥晶白云岩或石灰岩（潮坪沉积）等组成（图 3-44），代表由潮下带到潮坪的沉积序列。

图 3-44 杨三段（Jxy_3）典型碳酸盐岩旋回及其岩石显微照片

Fig.3-44 Typical carbonate cycle of the Jxy_3 and photomicrographs of the rocks

图①②③④分别是图Ⓐ中白色框放大图；（A）杨庄组三段典型碳酸盐岩旋回，从下至上依次为厚层—块状含粉砂泥晶白云岩（4m）、灰绿色薄层页岩（5cm）、中层层状叠层石白云岩（0.2m）、凝块状藻石灰岩层（5.1m）及层状叠层石石灰岩（3.2m），考察点 P5；（B）泥晶白云岩，单偏光；（C）层状叠层石泥晶白云岩，单偏光；（D）含硅斑状白云岩，单偏光；（E）含硅云质斑状灰岩，正交光；（F）含硅云质斑状灰岩，正交光；（G）含硅斑状白云岩，单偏光；（H）纹层状含硅云质灰岩，正交光；（I）纹层状含硅云质灰岩，单偏光；（J）纹层状含硅云质灰岩，单偏光；图（B）—（J）显微照片分别对应于图（A）中相应的★所标记的 B—J 位置

在波状和层状叠层石白云岩或石灰岩中常见灰白色脉，其内充填泥晶白云石，宽数毫米，长几到几十厘米，与层面平行、斜交或垂直。这些脉是藻席中的泥裂后期被云泥充填而形成的。

6. 考察点 P6：杨庄组（Jxy）与雾迷山组（Jxw）分界

位置：该考察点位于青山村马平线旁青山采摘园沿坡上行约500m处，GPS点位：E 117.443030°，N 40.120781°。

考察内容：杨庄组与雾迷山组分界及其上下的岩性。

现象描述：此处为杨庄组与雾迷山组的分界，二者呈整合接触关系。界面之下为杨三段顶部的灰白色含粉砂泥质白云岩；界面之上为灰色厚层含燧石结核和条带细晶白云岩和沥青质波状叠层石泥晶白云岩，在地貌上雾迷山组呈陡崖（图3-45A和B）。

图3-45 青山采摘园半山坡处杨庄组（Jxy）与雾迷山组（Jxw）分界及花果峪村东南马平公路旁杨庄组典型岩石类型

Fig.3-45 Boundary between the Yangzhuang Formation（Jxy）and Wumishan Formation（Jxw）in halfway of the mountain in the picking garden in the Qingshan Village, and typical rocks of the Yangzhuang Formation（Jxy）aside the Maping Road near the Huaguoyu Village

（A）杨庄组与雾迷山组分界，二者间呈整合过渡接触，分界面之上为雾迷山组厚层含燧石结核和条带细晶白云岩，界面之下为杨庄组顶部灰白色含粉砂质泥晶白云岩，考察点P6；（B）杨庄组与雾迷山组分界，考察点P6；（C）紫红色夹灰白色白云质泥岩，杨庄组二段，考察点P7；（D）杨庄组二段与三段分界，考察点P7

7. 考察点 P7：杨二段（Jxy₂）与杨三段（Jxy₃）分界

位置：该考察点位于花果峪村沿马平公路向东南方向约920m处，GPS点位：E 117.464169°，N 40.118511°。

考察内容：杨二段与杨三段分界及其上下的岩性。

现象描述：此处位于马平公路旁，杨庄组二段和三段在该处出露完好，现象明显（图3-45C和D）。杨二段以紫红色夹灰白色白云质泥岩（即"五花肉泥岩"）为典型特征，杨三段以"五花肉泥岩"夹灰色、深灰色含燧石碳酸盐岩为主。

第八节　罗庄子镇青山村—磨盘峪村—二十里铺村蓟县系雾迷山组（Jxw）考察路线

中元古界蓟县系雾迷山组是蓟州区元古宇中沉积厚度最大的组，最厚可达3416m。该组的岩性以碳酸盐岩占绝对优势，占雾迷山组总厚度的80%～90%；沉积韵律极为发育，巨厚的雾迷山组地层由多个沉积韵律层叠置而成，见图3-46和图3-47。

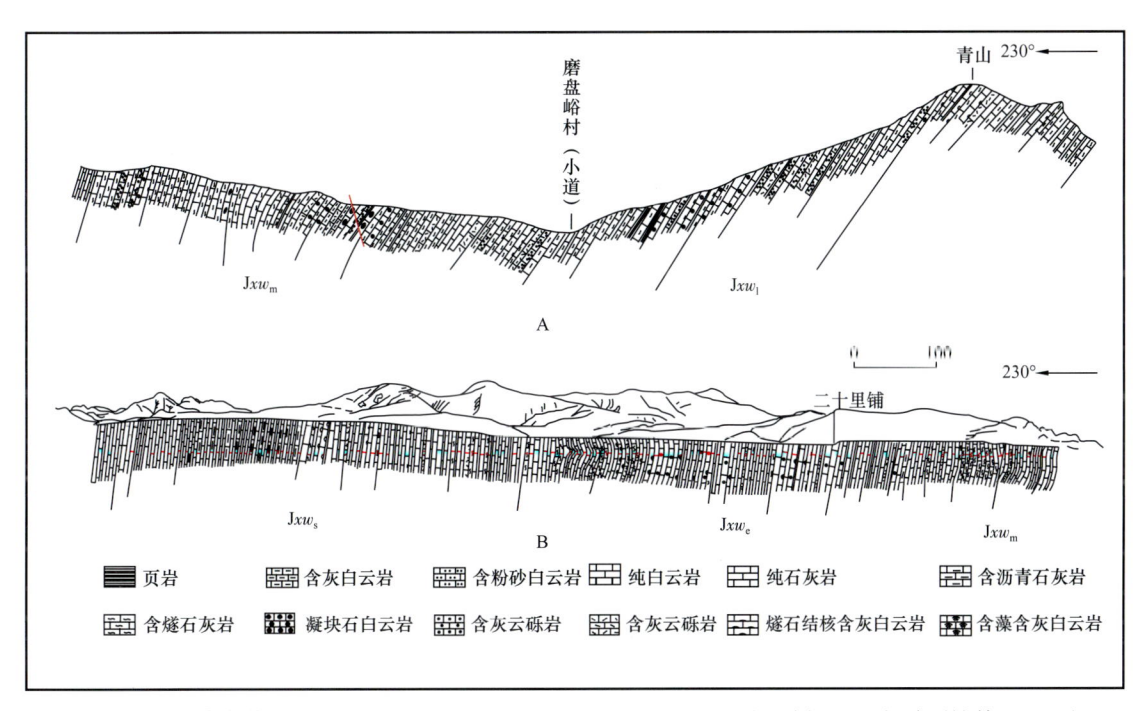

图3-46　天津市蓟州区二十里铺—洪水庄蓟县系雾迷山组（Jxw）实测剖面图（据陈晋镳等，1980）
Fig.3-46　Sketch profile of the Wumishan Formation (Jxw) from Ershilipu to Hongshuizhuang in Jizhou District, Tianjin
（A）罗庄青山—磨盘峪南山雾迷山一段（罗庄亚组）和二段（磨盘峪亚组）剖面；（B）罗庄二十里铺—闪坡岭雾迷山三段（二十里铺亚组）和四段（闪坡岭亚组）剖面

该区南北向的蓟县到黄崖关长城的S101公路横穿雾迷山组，近乎垂直地层走向，本是很好的露头剖面，但由于常年风吹雨淋，刚拓宽公路时开辟出来的地层新鲜面被浮土和植物覆盖，现象不清晰了。此外，该剖面就在路边，车辆来来往往，不安全，不适合大队伍考察。

现在的考察路线是S101路东的二十里铺村—磨盘峪村—果园西村—兴隆堡村—四洼村—S302（省道）。这条路线是弯弯曲曲的柏油小路，刚好能错车。沿这条路线，有断断续续的出露点，现象清楚，来往车辆很少。但由于小路弯曲，时而向新地层走，时而往老地层走，容易把人弄糊涂，不容易在头脑中建立一个完整的地层层序，除非时刻看着地图。

不过由于雾迷山组是由多个相似的沉积旋回叠置而成，窥一斑可以见全貌，因此解剖一处，可以基本掌握全组。沿该路线，最好的考察点是磨盘峪村附近。

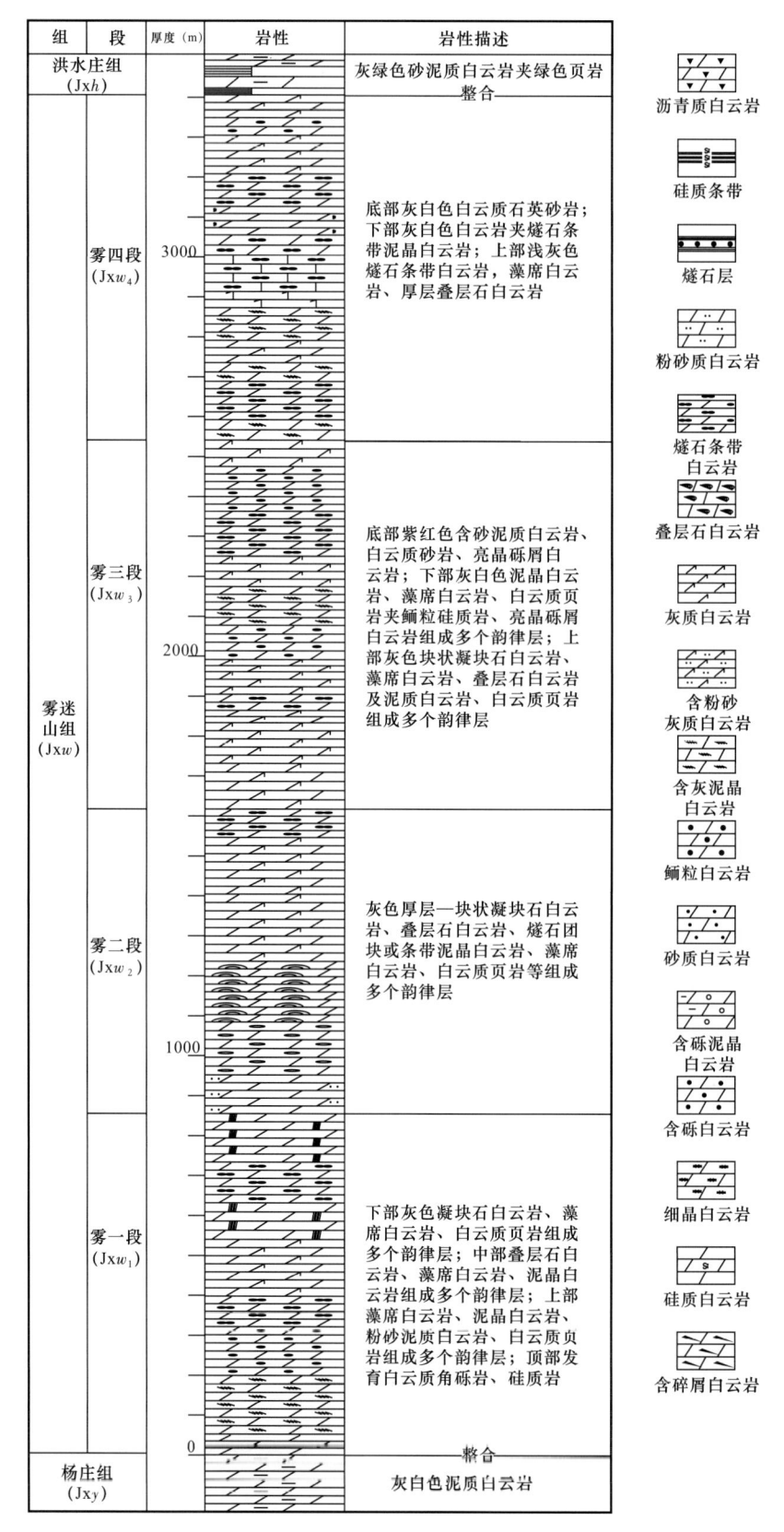

组	段	厚度(m)	岩性	岩性描述
洪水庄组 (Jxh)				灰绿色砂泥质白云岩夹绿色页岩
				整合
雾迷山组 (Jxw)	雾四段 (Jxw₄)	3000		底部灰白色白云质石英砂岩；下部灰白色白云岩夹燧石条带泥晶白云岩；上部浅灰色燧石条带白云岩，藻席白云岩、厚层叠层石白云岩
	雾三段 (Jxw₃)	2000		底部紫红色含砂泥质白云岩、白云质砂岩、亮晶砾屑白云岩；下部灰白色泥晶白云岩、藻席白云岩、白云质页岩夹鲕粒硅质岩、亮晶砾屑白云岩组成多个韵律层；上部灰色块状凝块石白云岩、藻席白云岩、叠层石白云岩及泥晶白云岩、白云质页岩组成多个韵律层
	雾二段 (Jxw₂)			灰色厚层—块状凝块石白云岩、叠层石白云岩、燧石团块或条带泥晶白云岩、藻席白云岩、白云质页岩等组成多个韵律层
	雾一段 (Jxw₁)	1000		下部灰色凝块石白云岩、藻席白云岩、白云质页岩组成多个韵律层；中部叠层石白云岩、藻席白云岩、泥晶白云岩组成多个韵律层；上部藻席白云岩、泥晶白云岩、粉砂泥质白云岩、白云质页岩组成多个韵律层；顶部发育白云质角砾岩、硅质岩
		0		整合
杨庄组 (Jxy)				灰白色泥质白云岩

图例：沥青质白云岩、硅质条带、燧石层、粉砂质白云岩、燧石条带白云岩、叠层石白云岩、灰质白云岩、含粉砂灰质白云岩、含灰泥晶白云岩、鲕粒白云岩、砂质白云岩、含砾泥晶白云岩、含砾白云岩、细晶白云岩、硅质岩、含碎屑白云岩

图3-47 天津市蓟州区二十里铺—洪水庄蓟县系雾迷山组（Jxw）地层综合柱状图

Fig.3-47 Stratigraphic column of the Wumishan Formation（Jxw）, Jixian System, Neoproterozoic from Ershilipu to Hongshuizhuang in Jizhou District, Tianjin

一、线路位置

该线路位于天津市蓟州区二十里铺村和磨盘峪村，起点位于二十里铺村东（E 117.418064°，N 40.121807°），终点位于磨盘峪村东南方向磨津路石碑处（E 117.438914°，N 40.114359°），路线总长度约4km，具体考察点和路线分布如图3-48所示。考察点P1位于二十里铺，与考察点P2至P4距离较远。建议考察点P1观察后乘车至磨盘峪村，考察点P2至P3需步行进入村北侧的山谷。随后，乘车沿磨津路向东南方向400m至考察点P4。

图3-48 天津市蓟州区中元古界蓟县系雾迷山组（Jxw）观察点位置及线路

Fig.3-48 Distribution of investigation stops and route of the Wumishan Formation (Jxw), Jixian System, Mesoproterozoic in JizhouDistrict, Tianjin

二、考察目的与任务

（1）观察中元古界蓟县系雾迷山组岩性特征及地层序列；
（2）观察雾迷山组二段下部旋回层的基本模式；
（3）分析沉积环境和旋回的成因。

三、考察内容与考察点

该路线从二十里铺至磨盘峪依次考察雾迷山组的一段及二段地层，并在路线中选择了4个详细的考察点，分别为雾迷山一段下部岩性及沉积韵律观察点（图3-48中P1）、一段中部岩性及沉积韵律观察点（图3-48中P2）、一段上部岩性及沉积韵律观察点（图3-48中P3）及二段底部连续的米级旋回沉积（图3-48中P4）。各考察点的详细描述如下：

1.考察点P1：雾迷山组一段（Jxw_1）下部岩石类型及沉积韵律

位置：该考察点位于二十里铺村东小路路口，GPS点位：E 117.418064°，N 40.121807°。

考察内容：雾一段下部典型的岩石类型及沉积韵律。

现象描述：该点观察雾迷山一段下部典型的岩石类型及沉积韵律。岩石类型主要包括灰白色泥晶白云岩、深灰色凝块石细晶白云岩、浅绿灰色白云质页岩，地层产状188°∠48°。其中，在灰白色泥晶白云岩顶面常发育波痕构造（图3-49A）。深灰色凝块状（杂斑状）细晶白云岩中常见锥状叠层石，浅色斑与深色斑混杂（图3-49B、C）。浅绿灰色白云质页岩发育水平层理。沉积旋回自下而上，依次发育浅绿灰色白云质页岩（局限碎屑岩潮下带）—凝块状白云岩（碳酸盐岩潮下带）—层状叠层石白云岩（含燧石条带和结核）（碳酸盐岩潮坪）（图3-49B）。

图3-49 蓟州区二十里铺雾迷山组一段下部典型的岩石类型及沉积构造

Fig.3-49 The lithology and sedimentary structures of the lower part of the Jxw_1 in Jizhou District, Tianjin

（A）灰白色泥晶白云岩表面的波痕构造，雾迷山组一段，考察点P1；（B）凝块状白云岩、藻席白云岩及白云质页岩序列，雾迷山组一段，考察点P1；（C）凝块状白云岩，雾迷山组一段，考察点P1；（D）层状叠层石白云岩中，雾迷山组一段，考察点P1

2. 考察点P2：雾迷山组一段（Jxw_1）中部岩石类型及沉积韵律

位置：该考察点位于磨盘峪村北侧近南北向的山谷中，从村边向东北方向走到400m左右处。GPS点位：E 117.443598°，N 40.120346°。

考察内容：雾一段下部典型的岩石类型及沉积韵律。

现象描述：此处主要观察雾迷山组一段中部的典型岩石类型及沉积韵律。该处发育的主要岩石类型包括：块状泥晶白云岩、含燧石的层状和波状叠层石白云岩，两者交互沉积。叠层石可呈波状、丘状（图3-50A），或呈层状（图3-50B）；镜下能够看出其颗粒大小变化导致明显的亮—暗层（图3-50C）；含有较多风化后突出表面的硅质结核、条带（图3-50D）。灰白色泥晶白云岩中总体具块状构造，层理模糊（图3-50E和F）。

3. 考察点P3：雾迷山一段（Jxw_1）上部岩石类型及沉积韵律

位置：该考察点仍位于山谷中，从上一考察点向南200m处。GPS点位：E 117.441724°，N 40.117973°。

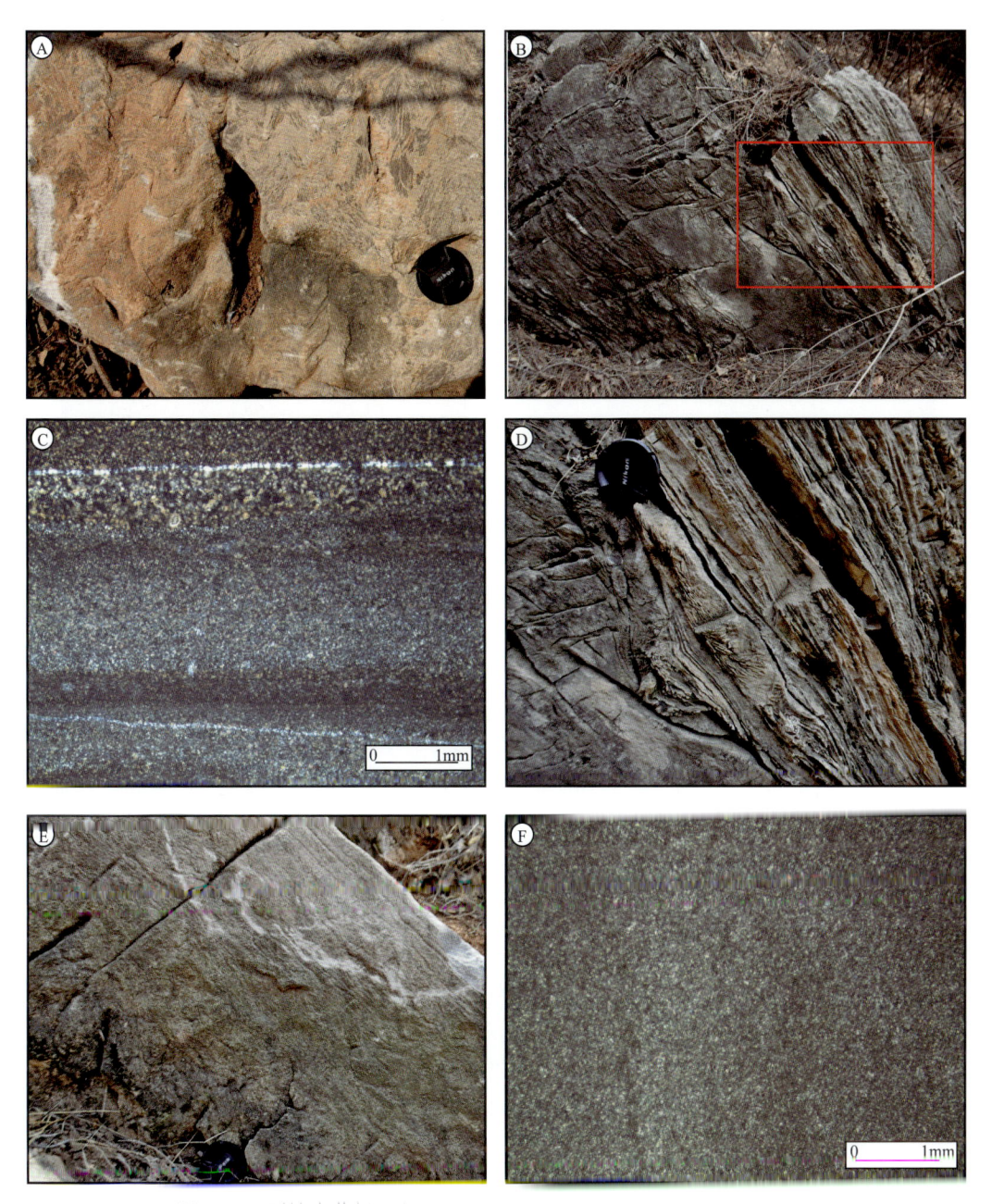

图 3-50　天津市蓟州区磨盘峪雾迷山组一段中部岩石类型及沉积构造

Fig.3-50　The lithology and sedimentary structures of the middle part of the Jxw_1 in Jizhou District, Tianjin

（A）波状叠层石白云岩，雾迷山组一段，考察点 P2；（B）含燧石条带层状叠层石白云岩，雾迷山组一段，考察点 P2；（C）雾迷山组一段层状叠层石白云岩的镜下特征，亮—暗层明显，正交光，考察点 P2；（D）燧石条带，雾迷山组一段，考察点 P2；（E）泥晶白云岩，雾迷山组一段中部，考察点 P2；（F）泥晶白云岩镜下特征，单偏光，雾迷山组一段中部，考察点 P2

考察内容：雾一段上部典型的岩石类型及沉积韵律。

现象描述：此处观察雾迷山组一段上部主要的岩石类型及沉积韵律。该处主要发育的岩石类型包括灰黑色凝块状细晶白云岩（图 3-51A）、灰色含燧石层状叠层石泥晶白云岩（图 3-51B、C）及浅绿灰色白云质页岩（图 3-51D），所形成的典型韵律自下而上依次发育浅绿灰色白云质页岩（潮下带）—灰黑色凝块状细晶白云岩（潮下带）—灰色含燧石层状叠层石泥晶白云岩（潮坪），为向上变浅的沉积旋回，与考察点 P2 类似。

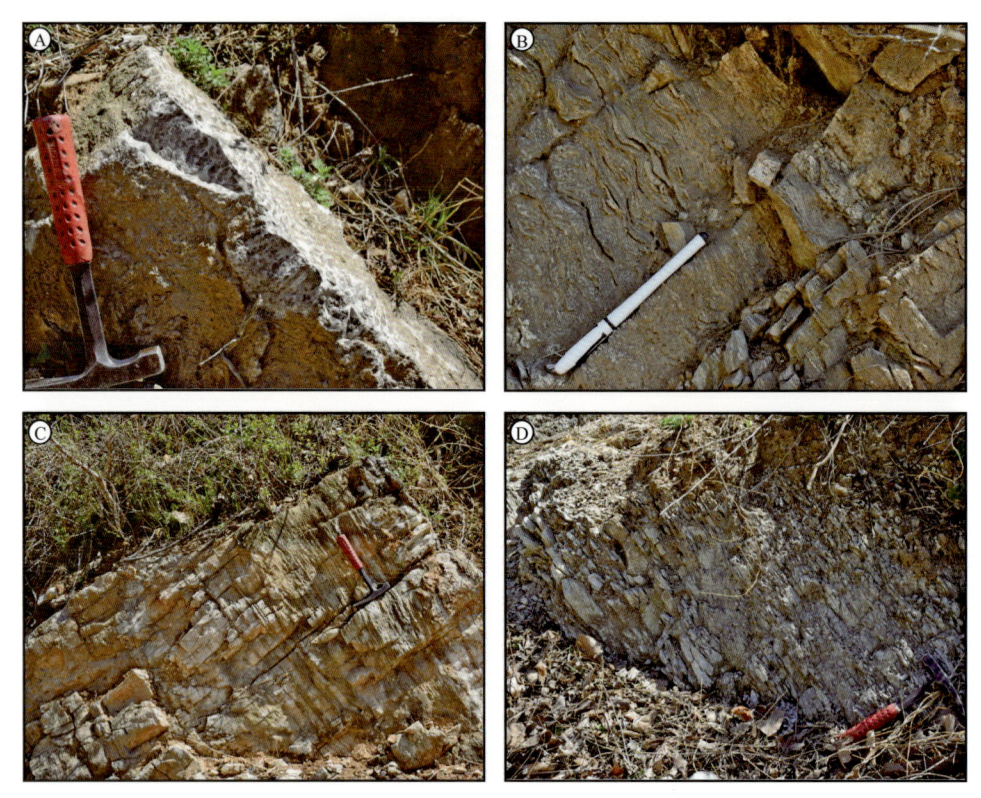

图 3-51 蓟州区磨盘峪雾迷山组一段（Jxw_1）上部岩石类型及沉积构造

Fig.3-51 The lithology and sedimentary structures of the upper part of the Jxw_1 in Jizhou District, Tianjin

（A）灰黑色凝块状白云岩，雾迷山组一段上部，考察点 P3；（B）灰色层状叠层石白云岩，雾迷山组一段上部，考察点 P3；（C）灰白色含硅质条带、结核的层状叠层石白云岩，雾迷山组一段上部，考察点 P3；（D）浅绿灰色云质页岩，雾迷山组一段上部，考察点 P3

4. 考察点 P4：雾迷山组二段（Jxw_2）沉积旋回

位置：该考察点位于磨盘峪村东南方向磨津路石碑处，GPS 点位：E 117.438914°，N 40.114359°。该点是雾迷山组露头最好的点。

考察内容：雾二段下部典型的岩石类型及沉积韵律。

现象描述：该点是雾迷山组二段（Jxw_2）旋回考察点。沿磨津路向东南方向，可以连续观察雾迷山组二段下部岩性旋回特征，沉积旋回发育很好，自下而上依次为浅绿灰色白云质页岩（潮下带）—灰黑色凝块状细晶白云岩（潮下带）—灰色含燧石层状叠层石泥晶白云岩（潮坪）（图 3-52）。

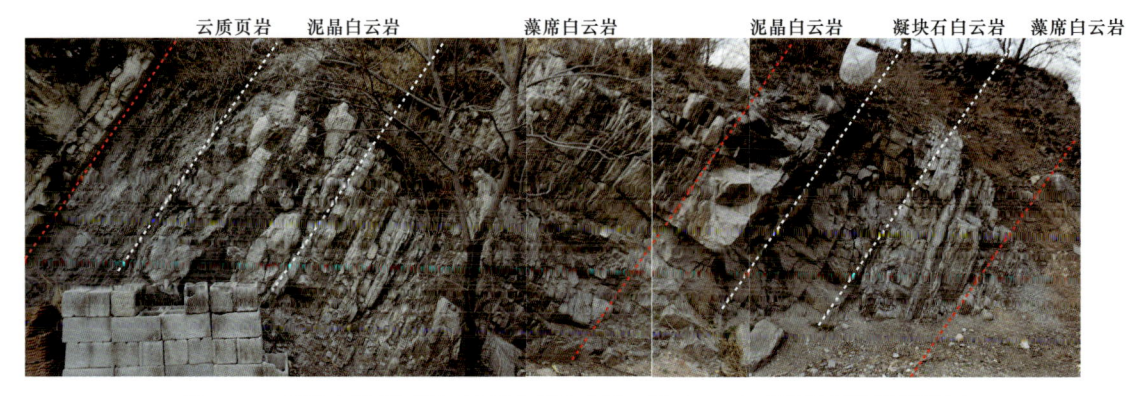

图 3-52 天津市蓟州区磨盘峪村东雾迷山组二段（Jxw_2）高频旋回的沉积特征

Fig.3-52 The sedimentary features of high frequency cyclothems of the Jxw_2 in Jizhou District, Tianjin

第九节　渔阳镇小岭子村蓟县系洪水庄组（Jxh）考察路线

中元古界蓟县系洪水庄组为一套灰黑色、灰绿色页岩，夹白云岩和薄层粉砂岩，厚度131m。该组分两段：下部的洪一段为中层白云岩夹灰绿色、蓝绿色和灰色页岩；上部的洪二段主要为灰黑色、灰绿色页岩，顶部为灰绿色页岩夹泥质泥晶白云岩透镜体（藻礁）和薄层粉砂岩。洪水庄组的页岩中赋存大量微古植物（孙淑芬，2000）。除在燕山东段有缺失或与下伏雾迷山组呈平行不整合接触外，在本区及其他地区皆与上覆铁岭组和下伏雾迷山组为整合接触关系（袁鄂荣等，1994；陈晋镳和武铁山，1997；汪凯明等，2014）。

洪水庄组在天津市蓟州区罗庄子镇洪水庄村及其东南一带出露良好。尤其沿小岭子村村口的村级石板路出露清晰，但这里出露的是洪二段，并能见到与上覆铁岭组的接触界限。洪水庄组与雾迷山组的接触界限，在府君山公园以东约2km、四洼村以南约1km处的路边能见到。小岭子村村口的洪二段信手剖面图和整个洪水庄组的地层综合柱状图分别见图3-53和图3-54。

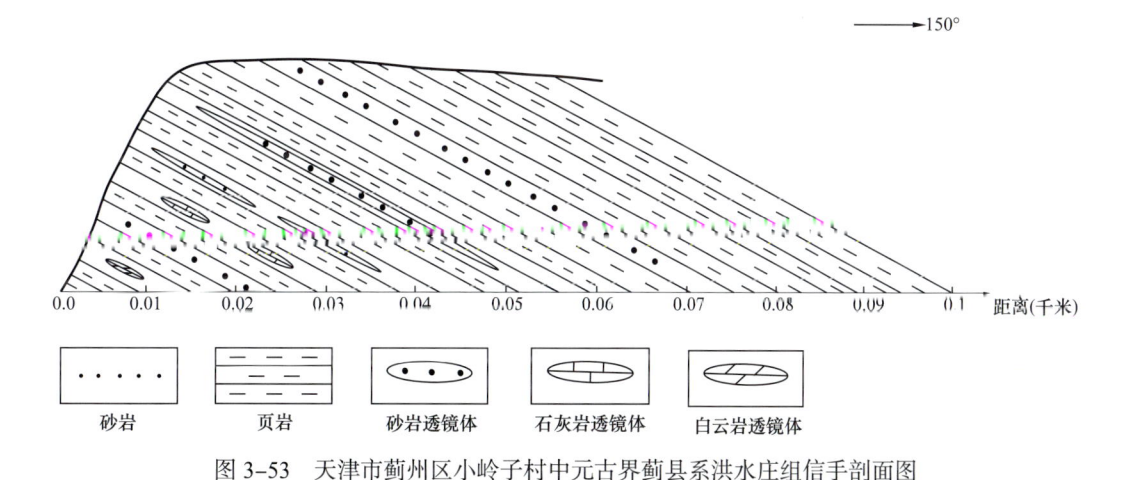

图3-53　天津市蓟州区小岭子村中元古界蓟县系洪水庄组信手剖面图
Fig.3-53　Sketch profile of the Hongshuizhuang Formation in Xiaolingzi village, Jizhou District, Tianjin

一、线路位置

该线路位于天津市蓟州区渔阳镇小岭子村村口，起点位于津围线以东，北桃园与南桃园之间分岔路（GPS坐标：E 117.388025°，N 40.084112°），终点在向东100m（GPS坐标：E 117.389207°，N 40.083879°），路线总长度约110m。该路线位于乡村公路旁，总体较为宽阔，大小型车辆皆可通行，具体路线和考察点分布如图3-55所示。

二、考察目的与任务

（1）观察中元古界蓟县系洪水庄组岩性特征与地层序列；
（2）观察黑色页岩、碳酸盐岩及石英砂岩透镜体等典型岩石特征；
（3）分析沉积环境。

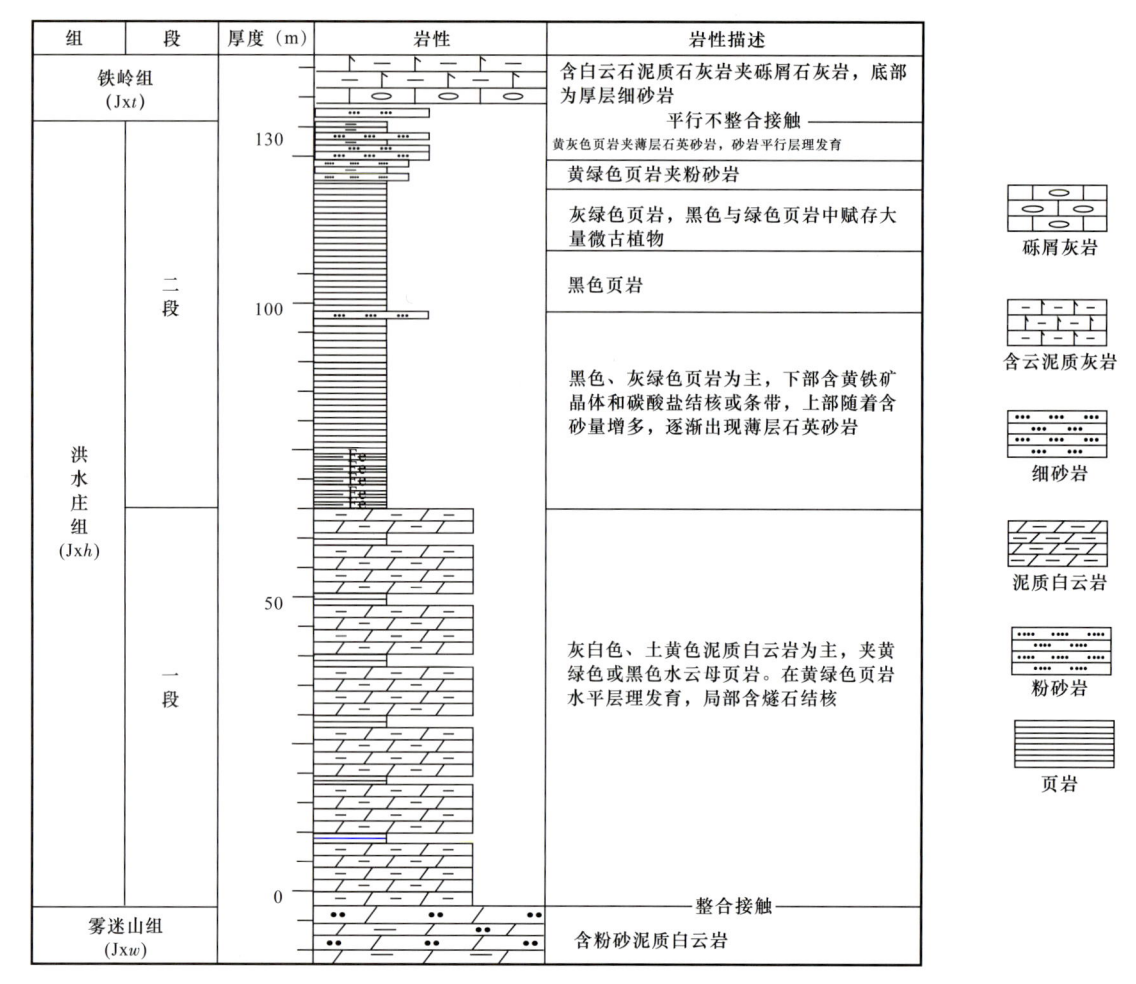

图 3-54 天津市蓟州区中元古界蓟县系洪水庄组（Jxh）地层综合柱状图

Fig.3-54 Stratigraphic column of the Hongshuizhuang Formation (Jxh) in Jixian District, Tianjin

图 3-55 天津市蓟州区中元古界蓟县系洪水庄组（Jxh）考察点位置及线路

Fig.3-55 Distribution of investigation stops and route of Hongshuizhuang Formation (Jxh), Jixian System, Mesoproterozoic in Jixian District, Tianjin

三、考察内容与考察点

该线路从津围线以东，北桃园与南桃园之间分岔路口向东考察蓟县系洪水庄组的岩性特征，由于出露条件限制，在该线路中仅选择了两个详细的考察点，分别包括洪水庄组页岩（图3-55中P1）和石英砂岩透镜体考察点（图3-55中P2）。

1. 考察点P1：蓟县系洪水庄组（Jxh）页岩

位置：该考察点位于津围线以东，北桃园与南桃园之间分岔路口以东，GPS点位：E 117.388025°，N 40.084112°。

考察内容：（1）自下而上页岩颜色的变化及其反映的环境变化；（2）白云岩透镜体（藻礁）。

现象描述：深灰、灰绿色页岩夹白云岩（或石灰岩）透镜体（图3-56）。透镜体长1.5~3m，厚0.5m。洪水庄组页岩水平纹层平直清晰，横向分布稳定，说明沉积时水介质能量很小，为页岩陆棚沉积。

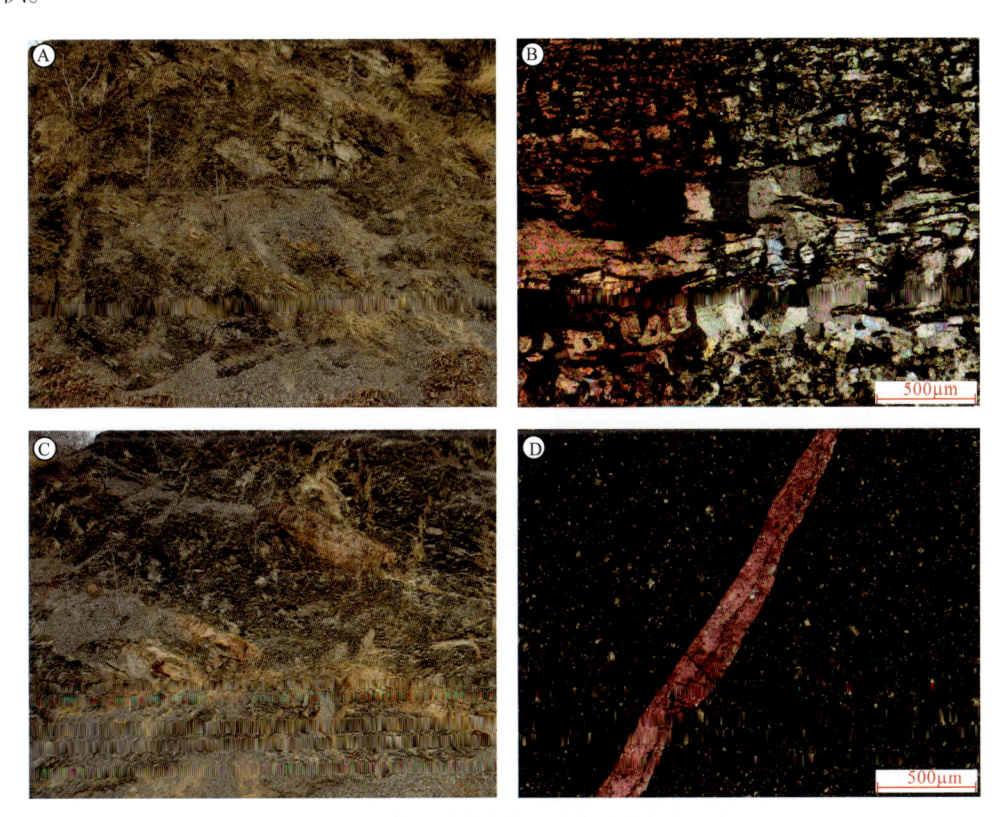

图 3-56　蓟县系洪水庄组暗色页岩与白云岩

Fig.3-56　Dark shale and dolomite of the Hongshuizhuang Formation at Jixian System

（A）蓟县系洪水庄组深灰色页岩，考察点P1；（B）灰岩透镜体显微照片，正交光，考察点P1；（C）页岩夹碳酸盐岩透镜体（藻礁），风化后呈褐色，考察点P1；（D）白云岩透镜体显微照片，正交光，考察点P1

2. 考察点P2：蓟县系洪水庄组（Jxh）石英细砂岩透镜体

位置：该考察点距考察点1以东10m处，GPS点位：E 117.408010°，N 40.084112°。

考察内容：细砂岩透镜体，分析其成因。

现象描述：暗色页岩中夹有中薄层石英细砂岩透镜体，砂岩透镜体的长1.5~2m，厚约0.1m（图3-57）。

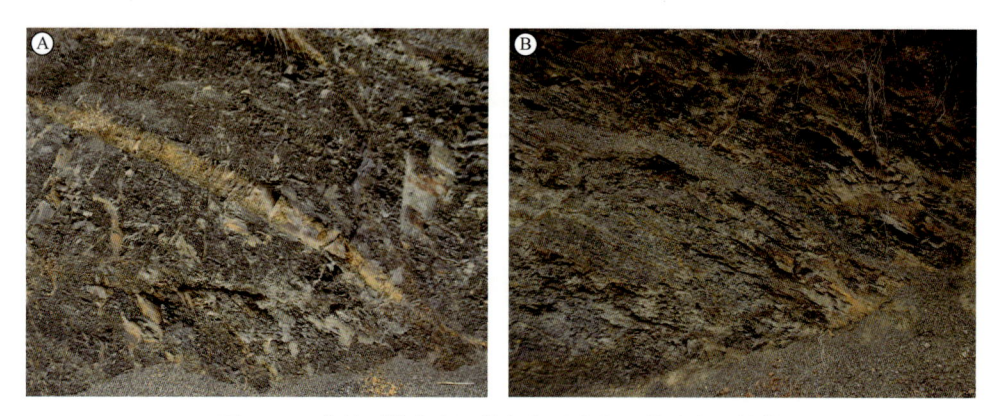

图 3-57 蓟县系洪水庄组黄灰色页岩夹石英砂岩透镜体

Fig.3-57 Yellow-gray shale with quartz sandstone lens of the Hongshuizhuang Formation at Jixian System

（A）灰绿色页岩夹砂岩透镜体，考察点 P2；（B）黄绿色页岩，视域宽 1m，考察点 P2

第十节　渔阳镇小岭子村—罗庄子镇铁岭子村南蓟县系铁岭组（Jxt）考察路线

中元古界蓟县系铁岭组为一套以碳酸盐岩为主，夹紫色和蓝绿色页岩的沉积地层，上部叠层石十分发育（图 3-58、图 3-59）。该组地层总厚度约为 303m。根据岩性组合的差异，前人将铁岭组划分为两段（代庄子亚组和老虎顶亚组）（陈晋镳等，1980），而本书将其划分为三段：

铁一段（Jxt$_1$）厚 155m，底部为灰白色厚层石英细砂岩，向上过渡为灰色含锰藻泥晶白云岩（藻礁）；主体为中厚层灰色叠层石白云岩夹灰绿色、蓝绿色页岩；顶部为蓝绿色海绿石页岩和含铁锰质暗棕色页岩。

铁二段（Jxt$_2$）厚 60m，为云泥条带灰泥石灰岩夹竹叶石灰岩和少量叠层石石灰岩，局部见燧石结核。

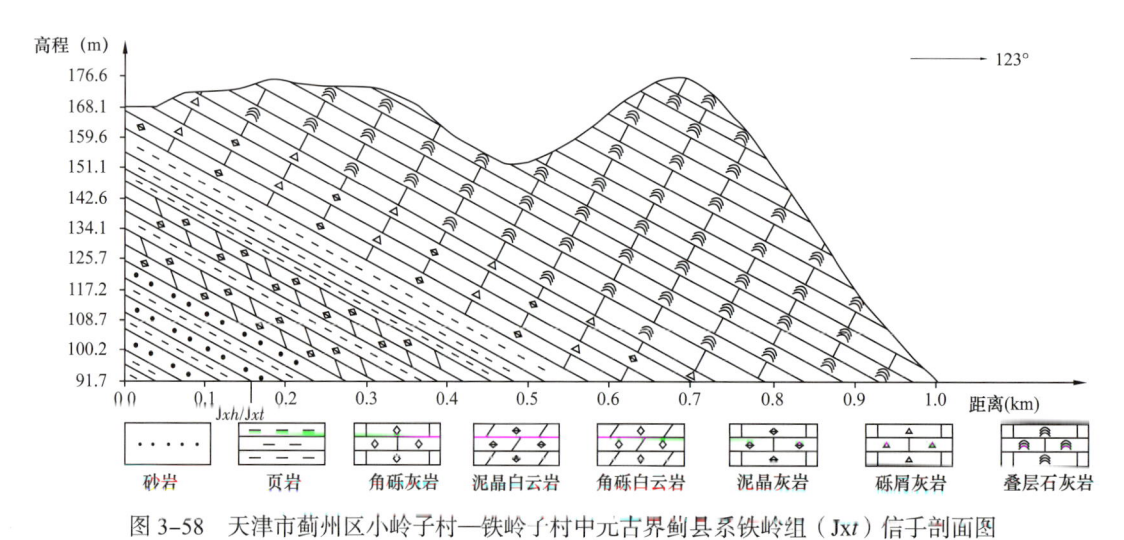

图 3-58 天津市蓟州区小岭子村—铁岭子村中元古界蓟县系铁岭组（Jxt）信手剖面图

Fig.3-58 Sketch profile of Tieling Formation (Jxt), Jixian System, Mesoproterozoic from Xiaolingzi to Tieling in Jizhou District, Tianjin

铁三段（Jxt_3）厚 88m，为叠层石石灰岩。叠层石类型多样，形态各异，包括层状、波状、柱状和墙状叠层石，是该区中元古界叠层石类型最多、发育最好的层段。该段顶部为几米厚的云泥条带灰泥石灰岩。

该组与下伏蓟县系洪水庄组地层之间为整合接触，与上覆青白口系下马岭组地层之间为平行不整合接触。

铁岭组在天津市蓟州区小岭子村—铁岭子村南一带出露较全，是该组最佳的考察线路。

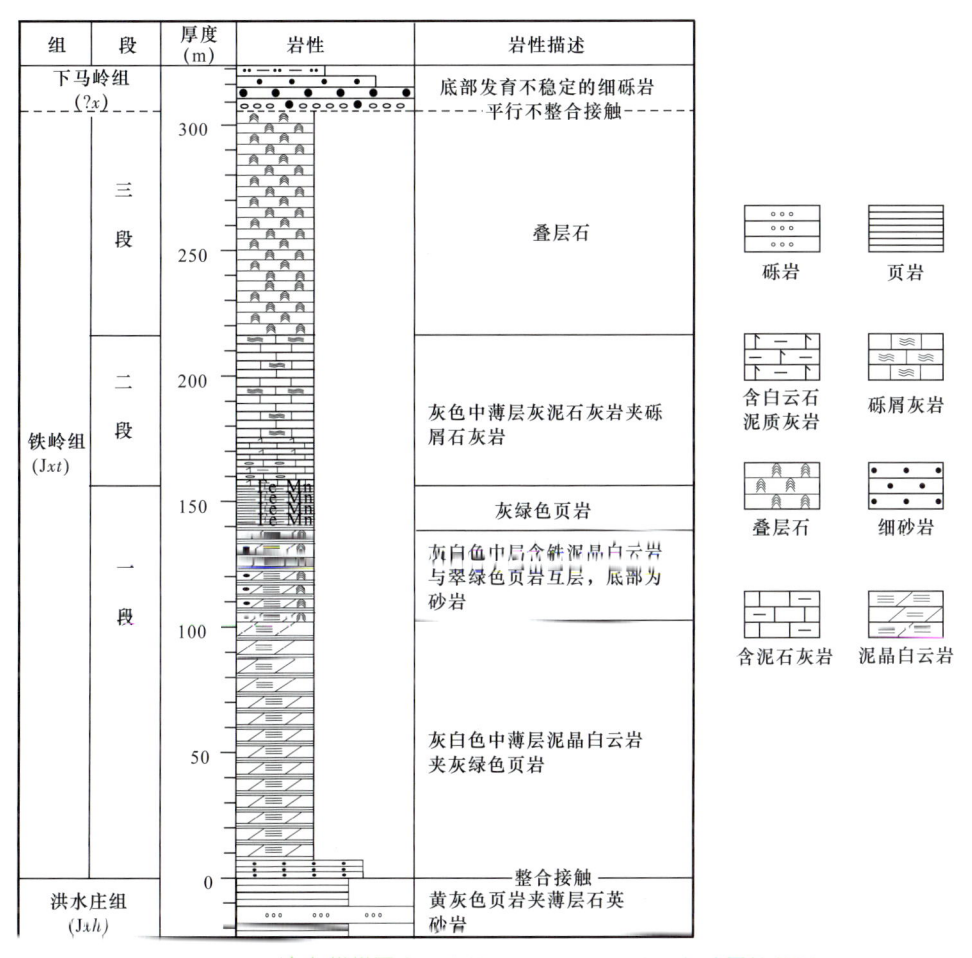

图 3-59　天津市蓟州区中元古界蓟县系铁岭组（Jxt）地层柱状图

Fig.3-59　Stratigraphic column of Tieling Formation（Jxt）from Xiaolingzi to Tieling in Jixian District, Tianjin

一、线路位置

该线路位于天津市蓟州区小岭子村—铁岭村一带，起点位于津围线以东小岭子村村口附近（GPS 坐标：E 117.389208°，N 40.083879°），终点在铁岭子村南（GPS 坐标：E 117.397112°，N 40.079933°），线路总长度约 1000m。该线路前半段为水泥路，道路较为宽阔，中小型车辆可通行，但至叠层石公园后，地形坡度较大，无车行道路，但铺设有步行台阶。考察点多沿步行台阶分布，适宜步行考察，车辆可绕行至铁岭子村南的考察路线与津燕路交叉口等待。具体线路和考察点分布如图 3-60 所示。

图 3-60　天津蓟州区中元古界蓟县系铁岭组（Jx*t*）考察点位置及路线

Fig.3-60　Distribution of investigation stops and route of Tieling Formation (Jx*t*) in Jixian District, Tianjin

二、考察目的与任务

（1）观察天津市蓟州区中元古界蓟县系铁岭组岩性特征与地层序列；
（2）观察含铁白云岩、灰岩、蓝绿色页岩、叠层石等典型岩石特征；
（3）沉积环境分析。

三、考察内容与考察点

该剖面上，铁岭组出露齐全，交通方便。在该线路中有 6 个详细的考察点，分别为铁岭组与洪水庄组（Jx*h*）的分界（图 3-60 中 P1）、白云岩透镜体考察点（图 3-60 中 P2）、灰色中层泥晶白云岩夹灰绿色页岩（图 3-60 中 P3）、蓝绿色页岩（图 3-60 中 P4）、铁岭组一段与二段的分界（图 3-60 中 P5）和铁岭组三段（Jx*t*$_3$）叠层石（图 3-60 中 P6）。各考察点的详细描述如下。

1. 考察点 P1：蓟县系铁岭组（Jx*t*）与洪水庄组（Jx*h*）分界

位置：该考察点位于津围线以东，北桃园与南桃园之间分岔路蓟县系洪水庄组（Jx*h*）与铁岭组（Jx*t*）分界牌处 GPS 点位：E 117.389207°，N 40.083879°。

考察内容：（1）铁岭组与洪水庄组的分界；（2）砂岩的沉积序列及沉积环境分析。

现象描述：该点为铁岭组与洪水庄组的分界，位于大套砂岩的底界。两组之间为整合接触关系（图 3-61A 和 B）。分界面之下为洪水庄组顶部为灰黄色页岩夹石英砂岩透镜体，分界面之上为铁岭组底部的大套灰白色石英细砂岩。砂岩中平行层理发育，并见有小型正断层。

2. 考察点 P2：白云岩透镜体

位置：该点距 P1 约 100m。

考察内容：白云岩透镜体的形态特征和成因分析。

现象描述：白云岩透镜体顶面起伏不平，风化后呈褐色，为藻丘。

3. 考察点 P3：蓟县系铁岭组一段（Jx*t*$_1$）灰白色中薄层泥晶白云岩夹灰绿色页岩

位置：该考察点距上一考察点向东南约 90m 处，GPS 点位：E 117.390059°，N 40.083879°。

图 3-61 蓟县系洪水庄组（Jxh）和铁岭组（Jxt）分界及界线附近石英砂岩镜下特征

Fig.3-61 Boundary between the Hongshuizhuang Formation (Jxh) and the Tieling Formation (Jxt), and photomicrograph of quartz sandstone near the boundary

（A）蓟县系洪水庄组（Jxh）和铁岭组（Jxt）地层分界，两者之间为整合接触，考察点 P1；（B）洪水庄组和铁岭组界面之上铁岭组灰白色石英细砂岩显微照片，正交光，考察点 P1

考察内容：（1）铁一段下部的岩石类型及特征；（2）沉积旋回的特征；（3）白云岩中叠层石的类型及形态；（4）沉积环境分析。

现象描述：该点为铁岭组一段中部灰色中厚层泥晶白云岩夹灰绿色页岩，厚约 50m，向上页岩层逐渐变薄、变少，白云岩层逐渐增多（图 3-62）。白云岩中叠层石常见，潮下带到潮坪沉积的旋回发育；页岩夹层代表短暂海侵事件，为陆棚沉积。

图 3-62 蓟县系铁岭组灰色中薄层泥晶白云岩夹灰绿色页岩

Fig.3-62 Gray medium-thin bedded microcrystalline dolomites with gray-green shales in the Tieling Formation

（A）铁岭组一段中层泥晶白云岩夹灰绿色页岩，考察点 P3；（B）铁岭组一段灰色中层泥晶白云岩显微照片，正交光，考察点 P3

4. 考察点 P4：蓟县系铁岭组一段（Jxt_1）蓝绿色页岩

位置：该考察点距上一考察点向东南约 260m 处，GPS 点位：E 117.391667°，N 40.081389°。

考察内容：蓝绿色页岩的特征与成因分析。

现象描述：该点为铁岭组一段上部的灰白色中层含铁泥晶白云岩与蓝绿色页岩互层（图 3-63）。蓝绿色乃海绿石所致，该类页岩为陆棚沉积。

5. 考察点 P5：蓟县系铁岭组一段（Jxt_1）与二段（Jxt_2）分界及铁二段的岩性特征

位置：该考察点距上一考察点向东南约 53m 处，GPS 点位：E 117.392237°，N 40.081207°。

考察内容：（1）铁岭组一段与二段的分界；（2）平行不整合分界面的特点；（3）褐色页岩的沉积环境；（4）铁二段的岩石类型及沉积环境。

图 3-63 蓟县系铁岭组剖面灰色中薄层含铁泥晶白云岩与蓝绿色页岩

Fig.3-63 Interbeddings of grayish–white medium–thin ferriferous microcrystalline dolostones and emerald shales in the Tieling Formation

（A）铁岭组一段上部灰色中层含铁泥晶白云岩与蓝绿色页岩互层，考察点 P4；（B）灰色中层含铁泥晶白云岩显微照片，正交光，考察点 P4

现象描述：该点为铁岭组一段与二段分界点，为一个平行不整合。铁一段顶部为灰绿色页岩，铁二段底部为褐色铁质细砾岩，厚约 0.2m。此界面区域分布稳定，铁岭组一段沉积后地层抬升，遭受风化剥蚀，这次构造运动称为"铁岭上升"（杜汝霖和李培菊，1980；陈晋镳和武铁山，1997）。铁二段主体为薄层云泥条带灰泥石灰岩夹竹叶状砾屑石灰岩。石灰岩中发生不同程度的白云化，为蒸发泵和回流渗透白云化（图 3-64）。

图 3-64 蓟县系铁岭组一段顶界与二段地质现象

Fig.3-64 Geological phenomena of the top of the Jxt_1 and the Jxt_2

（A）铁岭组一段与二段分界，一段顶部为蓝绿色页岩，二段底部为褐色铁质细砾岩，考察点 P5；（B）灰色砾屑灰岩与泥岩互层，考察点 P5；（C）灰色砾屑灰岩与灰白色中薄层泥晶灰岩互层，考察点 P5；（D）灰白色中薄层泥晶灰岩显微照片，正交光，可见方解石脉，考察点 P5

6. 考察点 P6：蓟县系铁岭组三段的叠层石

位置：该考察点位于距上一考察点向东南约 160m 处，GPS 点位：E 117.394179°，N 40.080387°。

考察内容：（1）叠层石的类型及特征；（2）叠层石间充填物的类型；（3）墙状叠层石不同方向切面上的形态特征；（4）叠层石形态控制因素分析；（5）叠层石沉积环境。

现象描述：该点为铁岭组三段类型多样的叠层石，厚约 80m（图 3-65）。根据叠层石纹层形态特征，可观察到柱状、波状、层状和墙状叠层石。不同形态的叠层石代表不同的沉积环境，具有沉积相指示意义。根据沉积水动力由弱变强，潮间带上部和潮上带主要发育层状叠层石，潮间带中部主要发育波状叠层石为主，潮间带下部和潮下带以柱状叠层石为主（金振奎等，2013）。墙状叠层石发育在潮间带中部和下部，其延伸方向平行潮汐水流方向。叠层石柱之间充填灰泥（常白云化）或蓝绿色泥，没有颗粒，反映水体能量不高。

图 3-65 蓟县系铁岭组三段地质现象

Fig.3-65 Geological phenomena of the the Jxt_3

（A）铁岭组三段为柱状叠层石，其上可见溶蚀孔，考察点 P6；（B）波状叠层石，考察点 P6；（C）叠层石顶部，考察点 P6；（D）柱状叠层石，表面见溶蚀孔隙，考察点 P6；（E）层状叠层石，考察点 P6；（F）叠层石显微照片，见亮暗层，正交光，考察点 P6

第十一节　罗庄子镇铁岭子村南—下庄子村待建系下马岭组（?x）考察路线

前人将下马岭组归为"待建系"，笔者认为"待建系"这个术语很别扭，应暂归蓟县系，有了新发现再说。在该区，下马岭组以灰黑色、灰绿色页岩、粉砂质页岩为主，夹泥质粉砂岩和细砂岩。该组在天津市蓟州区府君山向斜两翼近核部一带小范围出露，其中骆驼岭一带出露最佳。该组可分两段，下段夹较多细砂岩，为页岩夹细砂岩段；上段主要为页岩。此处该组的信手剖面和地层柱状图分别如图3-66和图3-67所示。

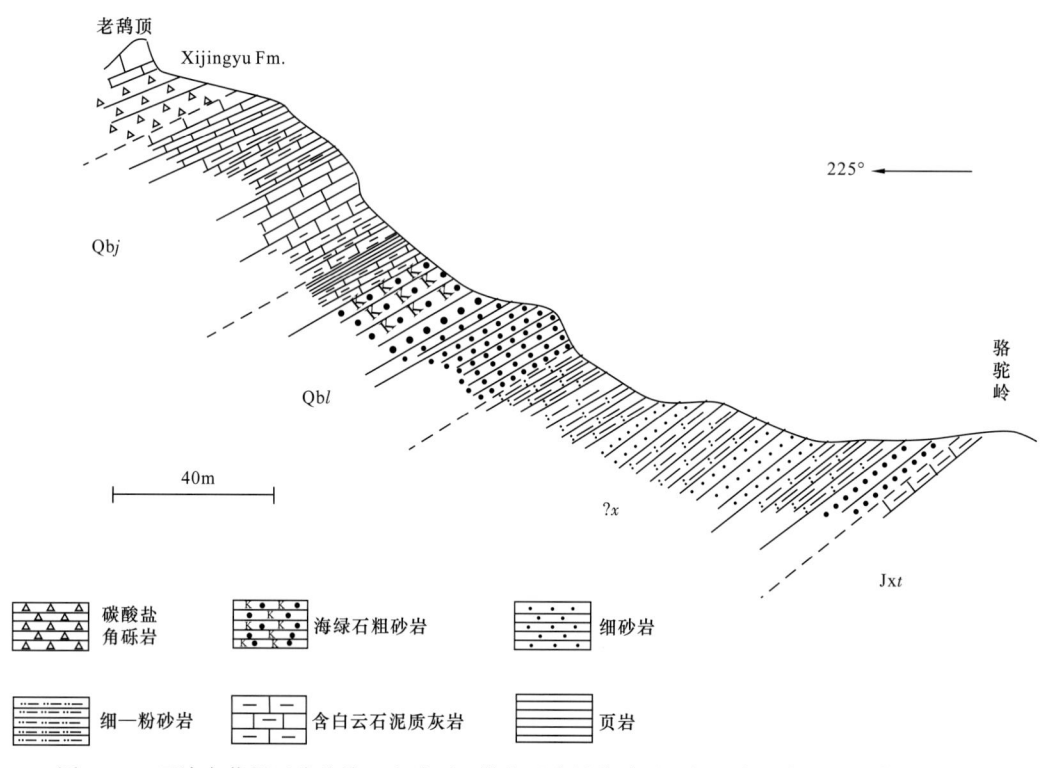

图3-66　天津市蓟州区骆驼岭—老鸹顶一带中元古界待建系和新元古界青白口系信手剖面图
（据天津地质矿产研究所，1964）

Fig.3-66　Sketch profile of the Mesoproterozoic Daijian System and Neoproterozoic Qingbaikou System from Luotuoling to Laoguading in Jizhou District, Tianjin（Tianjin Institute of Geology and Mineral Resources, 1964）

一、线路位置

该线路位于天津市蓟州区罗庄子镇铁岭子村一带，起点位于津燕路旁铁岭子村向南约650m处（GPS坐标：E 117.397112°，N 40.079933°），终点位于津燕路旁下庄子村向东南约250m处（GPS坐标：E 117.403186°，N 40.075445°），线路总长度约788m。具体线路及考察点的分布如图3-68所示。该线路主要路段沿着津燕路，道路状况良好，大小车辆皆可通行，考察点处设有步行台阶和指示牌。

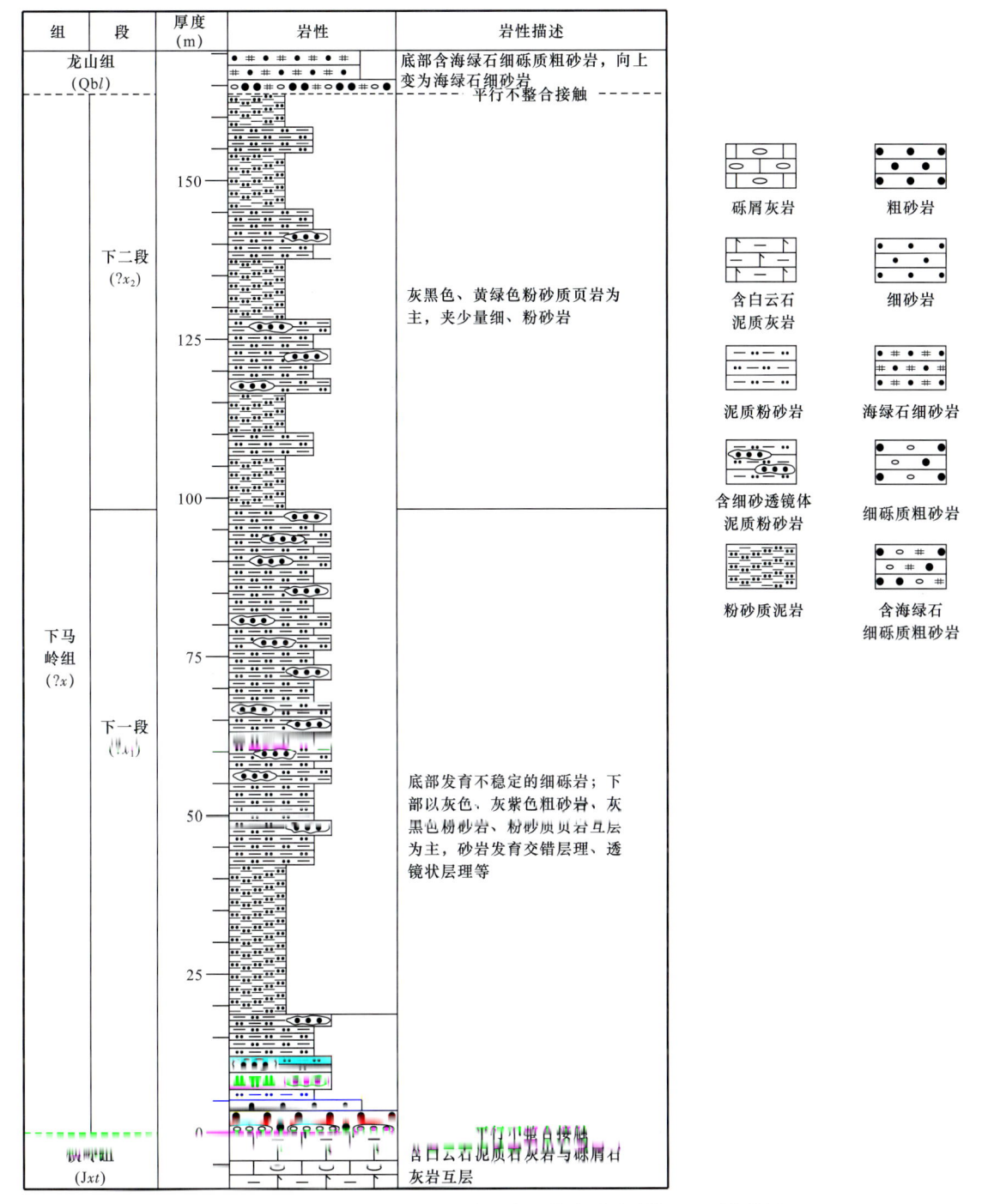

图 3-67　天津市蓟州区骆驼岭中元古界待建系下马岭组地层柱状图

Fig.3-67　Stratigraphic column of the Xiamaling Formation in Luotuoling, Jizhou District, Tianjin

二、考察目的与任务

（1）观察中元古界待建系下马岭组岩性特征与地层序列；

（2）观察待建系下马岭组与蓟县系铁岭组的平行不整合接触关系；

（3）沉积环境分析。

三、考察内容与考察点

该线路从铁岭子村至下庄子村由底至顶依次考察下马岭组的岩性变化和地层特征。由于出露条件限制，在该线路中仅选择有两个详细的考察点，分别为下马岭组与铁岭组分界点（图3-68中P1）和下一段典型岩性考察点（图3-68中P2）。

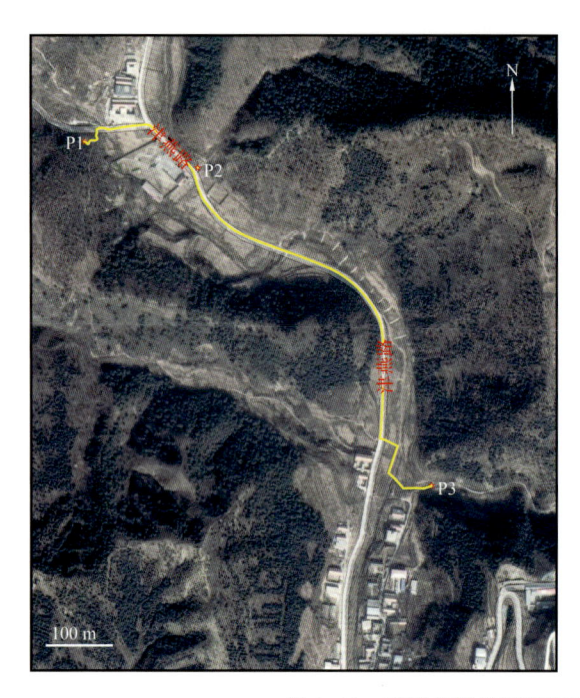

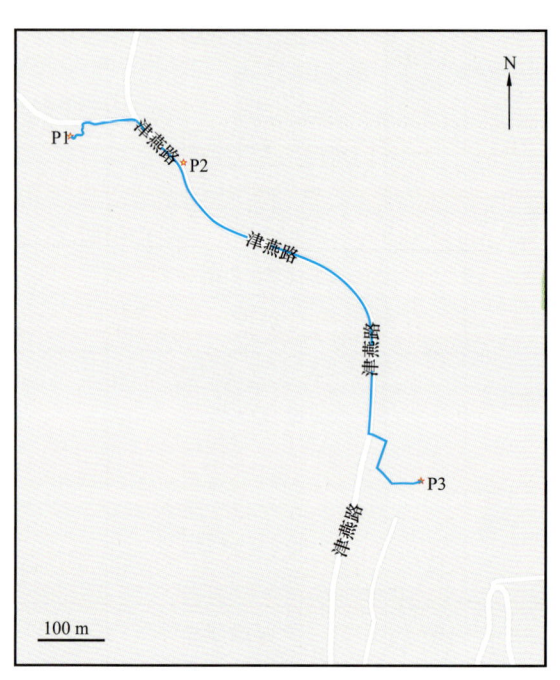

图 3-68 天津市蓟州区待建系下马岭组考察点位置及路线

Fig.3-68 Distribution of investigation stops and route of the Daijian System in Jizhou District, Tianjin

1. 考察点 P1：待建系下马岭组（?x）与蓟县系铁岭组（Jxt）的分界

位置：该考察点位于津燕路旁铁岭子村向南约650m处下马岭组与铁岭组分界碑处，GPS点位：E 117.397112°，N 40.079933°。

考察内容：（1）下马岭组和铁岭组分界；（2）界面上、下的岩石类型及特征；（3）风化壳。

现象描述：该点为下马岭组和铁岭组分界，二者之间呈平行不整合接触（图3-69A）。该不整合面所代表的构造运动被称之为"芹峪抬升"（陈晋镳等，1980）。分界面之下是铁岭组顶部的薄层云泥条带灰泥石灰岩夹竹叶状石灰岩（图3-69B），为潮坪沉积。竹叶状石灰岩剖面呈透镜状，竹叶状砾屑呈放射状和倒"小"状（图3-69C），为风暴沉积。分界面之上为下马岭组底部的铁锈色铁质细砂岩，底部有底砾岩，局部含赤铁矿扁豆体（图3-69D和E），为海侵初期沉积。

2. 考察点 P2：下马岭组一段岩性特征

位置：该考察点位于铁岭子村茂源祥农家院沿津燕路向东南约140m处，GPS点位：E 117.399128°，N 40.079632°。

考察内容：（1）下一段的岩石类型及特征；（2）沉积环境分析。

现象描述：该点为下马岭组一段典型岩性的出露点，由底部铁锈色的细砂岩、粉砂岩逐渐过渡为该处的薄层灰色、灰绿色页岩夹薄层泥质粉砂岩（图3-69F）。粉砂岩在层内断续分布，部分呈透镜状。总体为页岩陆棚环境。

图 3-69 蓟县系铁岭组和待建系下马岭组分界处地质现象

Fig.3-69 Geological phenomena near the boundary between the Jixian System Tieling Formation and the Daijian System Xiamaling Formation

(A) 铁岭组与下马岭组分界，二者呈平行不整合接触，考察点 P1；(B) 薄层云泥条带灰泥石灰岩，铁岭组顶部，考察点 P1；(C) 竹叶石灰岩透镜体，铁岭组顶部，砾屑呈放射状和倒"小"状排列，考察点 P1；(D) 块状鱼扑泥质细砂岩，下马岭组底部，考察点 P1；(E) 钙质泥质细砂岩，偏光，考察点 P1；(F) 薄层砂岩出灰泥质泥质细砂岩，下马岭组二段，考察点 P2

第十二节　渔阳镇后寺沟青白口系龙山组（Qbl）考察路线

新元古界青白口系龙山组为一套碎屑岩沉积，主要岩性包括含砾长石砂岩、石英砂岩、海绿石砂岩及杂色砂岩等。该组自下而上粒度由粗变细，总厚度约 118m。根据岩性组合变化，龙山组划分为两段：下部的龙一段为砂岩段，底部为灰绿色、灰黄色含海绿石细砾质长石石英砂岩夹灰黄色泥质粉砂岩，砂岩层内发育板状、楔状等交错层理和脉状层理；上部的龙二段以灰绿色和灰黑色页岩为主，夹薄层海绿石细砂岩，顶部为紫红色泥岩。该组与下伏待建系下马岭组（?x）和上覆青白口系景儿峪组皆呈平行不整合接触关系。

龙山组在蓟州区骆驼岭及后寺沟一带广泛出露，且在后寺沟一带出露连续、完好，且界线明显，是该组最佳的考察路线。该处沿线露头的信手剖面图和综合柱状图分别如图3-66和图3-70所示。

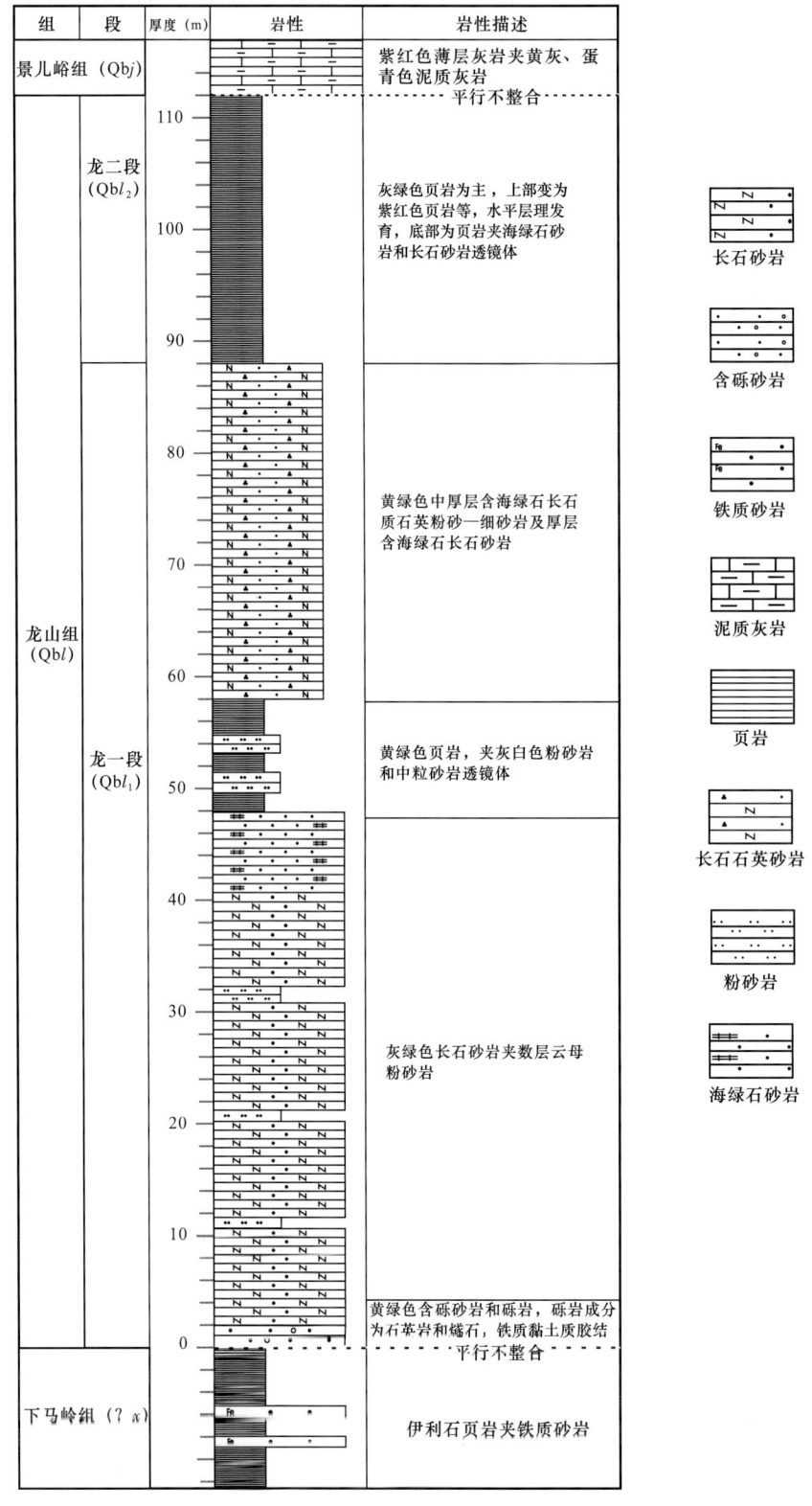

图3-70 天津市蓟州区城北骆驼岭青白口系龙山组（Qb*l*）地层柱状图

Fig.3-70 Stratigraphic column of the Longshan Formation (Qb*l*) in Jizhou District, Tianjin

一、线路位置

该线路位于天津市蓟州区罗庄子镇后寺沟一带，起点位于津燕路旁下庄子村向北约120m处（GPS坐标：E 117.409593°，N 40.076888°），终点位于后寺村向南约350m处（GPS坐标：E 117.416372°，N 40.075982°），线路总长度约703m。具体线路及考察点的分布如图3-71所示。该考察线路位于山谷处，为石阶路，建议步行考察，考察点处设有步行台阶和指示牌。

二、考察目的与任务

（1）观察新元古界青白口系龙山组岩性特征及地层序列；
（2）观察下马岭组和龙山组间的平行不整合以及波痕、交错层理等沉积构造；
（3）沉积环境分析。

三、考察内容与考察点

该线路从下庄子村至后寺村南由底至顶依次考察青白口系龙山组的岩性变化和地层特征，选择4个详细的考察点，包括龙山组与下马岭组分界点（图3-71中P1）、海绿石砂岩考察点（图3-71中P2）、海绿石页岩考察点（图3-71中P3）以及脉状层理、交错层理沉积构造考察点（图3-71中P4）。各考察点的详细描述如下。

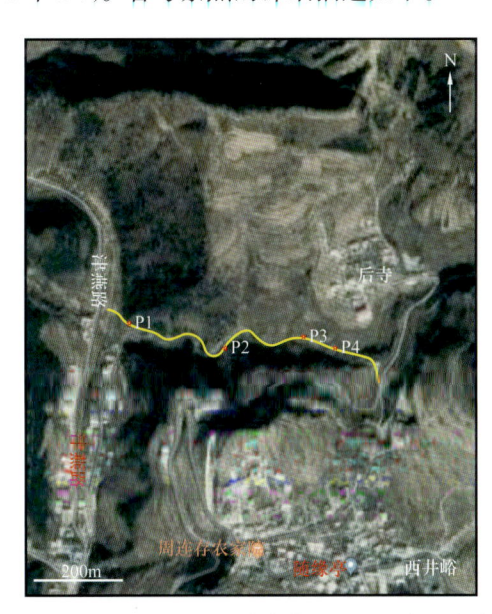

图3-71 天津市蓟州区新元古界青白口系龙山组（Qbl）考察点位置及路线

Fig.3-71 Distribution of investigation stops and route of the Longshan Formation（Qbl）in Jizhou District, Tianjin

1. 考察点P1：青白口系龙山组（Qbl）与待建系下马岭组（?x）分界

位置：该考察点位于津燕路旁下庄子村北120m处下马岭组与龙山组分界碑处，GPS点位：E 117.409593°，N 40.076888°。

考察内容：（1）龙山组与下马岭组界限的特征；（2）界限上、下的岩石类型、特征及沉积环境分析。

现象描述：该点观察待建系下马岭组（?x）与青白口系龙山组（Qbl）的分界，即待建系与青白口系分界。下马岭组顶部蓝绿色海绿石页岩之上为龙山组黄灰色（风化色）底砾岩，二者间呈突变接触，为侵蚀面（图 3-72A）。下马岭组沉积后期，地壳抬升遭受剥蚀。与北京西山青白口村一带对比，在蓟州区下马岭组缺失中、上部地层，二者间存在较大沉积间断，呈平行不整合接触，对应的构造运动被称为"蔚县上升"（陈晋镳等，1980）。

龙山组底部砾质粗砂岩颗粒以石英、燧石及少量长石为主，含海绿石，填隙物为泥质—铁质，内部具凹凸不平的侵蚀面（图 3-72C 和 D）；颗粒磨圆较好，为次棱角状到次圆状，分选较差；发育槽状交错层理和冲刷面构造（图 3-72B），为辫状河河道沉积。

图 3-72 下马岭组（?x）与龙山组（Qbl）分界处地质现象

Fig.3-72 Geological phenomena near the boundary between the Xiamaling Formation（? x）and the Longshan Formation（Qbl）

（A）下马岭组与龙山组之间为平行不整合，界面之上为龙山组厚层砾质粗砂岩，界面之下为下马岭组蓝绿色页岩沉积，考察点 P1；（B）龙山组底部黄灰色砾质粗砂岩的交错层理及冲刷面，考察点 P1；（C）黄灰色砾质粗砂岩，单偏光，龙山组底部；（D）图 C 的正交偏光显微照片

2. 考察点 P2：龙山组一段（Qbl₁）岩性特征及沉积构造

位置：该考察点距上一考察点向东上坡约 200m 处，GPS 点位：E 117.412086°，N 40.076710°。

考察内容：（1）龙一段岩石类型及特征；（2）砂岩中的构造类型及古水流方向；（3）沉积环境分析。

现象描述：该点观察龙山组一段的岩性及沉积构造。龙山组一段主要发育灰绿色长石质石英砂岩，常含海绿石，细到粗砂岩都有，局部含细砾；颗粒磨圆普遍好，为次圆状；分选普遍好；填隙物为硅质胶结物（图 3-73C 和 D）；层面发育波痕构造（图 3-73A），内部交错层理发育（图 3-73B）。砂岩为典型的海洋滨岸沉积。

图 3-73 龙山组一段（Qbl_1）长石石英砂岩及其沉积构造

Fig.3-73 Arkose sandstone and its sedimentary structure in the Qbl_1

（A）砂岩的波痕构造，龙一段，考察点 P2；（B）板状交错层理，龙一段，考察点 P2；（C）长石质石英砂岩，硅质胶结，龙一段，单偏光；（D）图 C 的正交光显微照片

3. 考察点 P3：龙山组一段（Qbl_1）中部海绿石页岩及其沉积构造

位置：该考察点位于海绿石页岩指示牌处，GPS 点位：N 40.076815°，E 117.414159°。

考察内容：海绿石页岩的特征。

现象描述：该点观察龙山组一段中部海绿石页岩及其沉积构造。海绿石页岩新鲜面呈蓝绿色，具页理（图 3-74A）；常见海绿石页岩与灰白色粉砂岩互层发育（图 3-74B）；为陆棚环境。

图 3-74 龙山组一段（Qbl_1）中部海绿石页岩及其沉积构造

Fig.3-74 Glauconitic shale and its sedimentary structure in the the Qbl_1

（A）海绿石页岩，考察点 P3；（B）蓝绿色海绿石页岩与粉砂岩薄互层，考察点 P3

4. 考察点 P4：龙山组一段（Qbl_1）顶部的海绿石砂岩及沉积构造和龙山组二段（Qbl_2）的紫红色页岩

位置：该考察点距上一考察点向东约 85m 处，GPS 点位：N 40.076499°，E 117.415360°。

考察内容：（1）海绿石砂岩中的脉状层理；（2）紫红色、杂色页岩。

现象描述：该点观察龙山组一段顶部的海绿石砂岩及沉积构造和龙山组二段的杂色页岩。海绿石细砂岩风化后呈黄绿色，主要矿物包含石英和海绿石。海绿石呈定向排列，石英颗粒分选好，磨圆次圆状（图 3-75C）。发育双向交错层理（图 3-75A）以及脉状层理（图 3-75B），指示潮间带的沙坪沉积环境。龙二段上部为紫红色页岩（图 3-75D），指示暴露的氧化环境—潮上带环境。

图 3-75 龙山组一段（Qbl_1）海绿石砂岩及沉积构造和龙山组二段（Qbl_2）杂色页岩

Fig.3-75 Glauconitic sandstone and its sedimentary structure in the Qbl_1 and Mottled shale in the Qbl_2

（A）龙山组一段海绿石砂岩中的双向交错层理，考察点 P4；（B）脉状层理，龙山组一段，考察点 P4；（C）海绿石砂岩，单偏光，龙山组一段；（D）龙山组二段紫红色页岩，考察点 P4

第十三节　渔阳镇西井峪村北青白口系景儿峪组（Qbj）考察路线

景儿峪组主要为一套中—薄层碳酸盐岩夹薄层海绿石砂岩和页岩的沉积，大致可细分为三个段：下部的景一段以灰色、灰紫色中—薄层含泥灰岩为主，底部为几十厘米厚的含海绿石含细砾粗砂岩；中部的景二段主要为灰色、蛋青色中层灰泥石灰岩夹泥质石灰岩；上部景三段主要为灰色薄层泥质含灰白云岩、白云质灰岩夹灰绿色页岩。

该组在天津市蓟州区府君山向斜的两翼近核部区域小范围出露，尤其在老鸹顶一带出露良好，此处该组的信手剖面和地层柱状图分别如图 3-66 和图 3-76 所示。

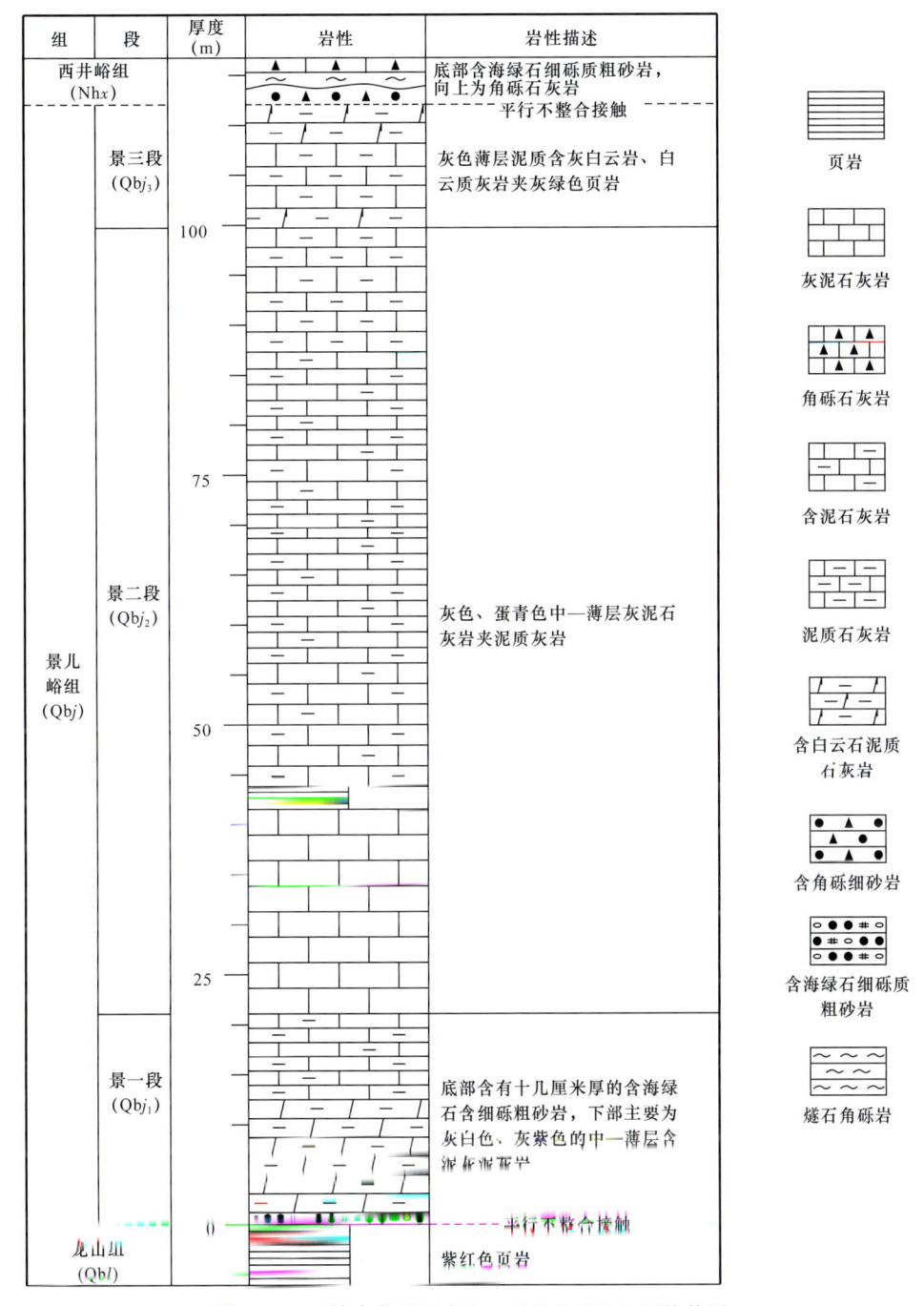

图 3-76 天津市蓟州区青白口系景儿峪组地层柱状图

Fig.3-76 Stratigraphic column of the Jingeryu Formation in Jizhou District, Tianjin

一、线路位置

该线路位于天津市蓟州区城关镇西井峪村北岭一带，起点位于西井峪村至后寺村的公路旁（GPS 坐标：E 117.409733°，N 40.074470°），终点位于向南上坡拐弯处的采石坑内（GPS 坐标：E 117.409780°，N 40.073561°），线路总长度约 110m。具体线路及考察点的分布如图 3-77 所示。该线路主要路段位于通村公路附近，通村公路状况良好，大小车辆皆可通行，可乘车至考察点附近，考察线路较短，但需沿坡上行，坡道为水泥路面，适宜步行前往。

- 89 -

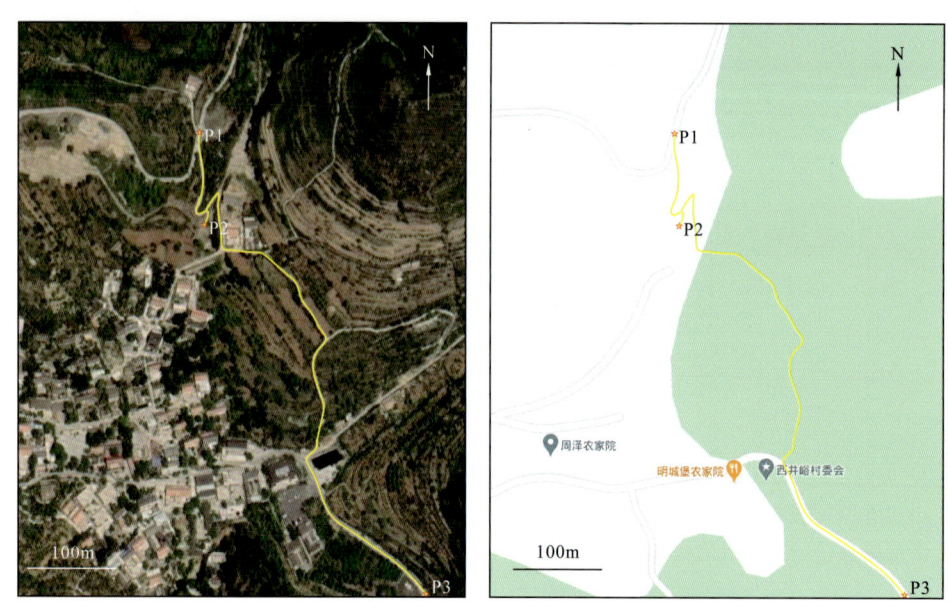

图 3-77 天津市蓟州区青白口系景儿峪组考察点位置及线路

Fig.3-77 Distribution of investigation stops and route of the Jingeryu Formation in Jizhou District, Tianjin

二、考察目的与任务

（1）观察新元古界青白口系景儿峪组岩性特征与地层序列；
（2）观察青白口系龙山组与景儿峪组及西井峪组的分界；
（3）沉积环境分析。

三、考察内容与考察点

该路线沿坡向上由底至顶依次考察青白口系景儿峪组的岩性变化及地层特征，并选择了 3 个详细考察点，分别为龙山组与景儿峪组分界点（图 3-77 中 P1）、景二段典型岩性（图 3-77 中 P2）和景儿峪组与西井峪组的分界点（图 3-77 中 P3）。

1. 考察点 P1：青白口系龙山组（Qbl）与景儿峪组（Qbj）分界

位置：该考察点位于西井峪村至后寺村的公路旁，GPS 点位：E 117.409733°，N 40.074470°。

现象描述：该点为青白口系龙山组与景儿峪组分界（图 3-78A 和 B），分界面之下由龙山组顶部的紫红色页岩夹灰绿色页岩组成；分界面之上为景儿峪组的灰色、紫红色中层—薄层泥质灰泥灰岩互层，底部局部发育含砾的海绿石粗砂岩（图 3-78B）。两者间常被认为是整合接触，但因景儿峪组底部存在砾石，也不排除沉积期发生过短暂间断的可能（朱士兴等，2016）。

2. 考察点 P2：景二段典型岩性

位置：该考察点距上一考察点向南上坡拐弯处的采石坑，GPS 点位：E 117.409780°，N 40.073561°。

考察内容：（1）景儿峪组二段下部的中薄层灰泥石灰岩夹灰绿色页岩薄层；（2）沉积环境分析。

现象描述：该点位于采石坑内，景儿峪组二段典型岩性段出露良好，主要为蛋青色中薄层灰泥石灰岩夹灰绿色页岩薄层（图 3-78C），层面平整，为水体较深的碳酸盐缓斜坡沉积。

图 3-78 青白口系景儿峪组（Qbj）与下伏龙山组（Qbl）及上覆西井峪组分界处地质现象

Fig.3-78 Geological phenomena near the boundaries among Qingbaikou System Jingeryu Formation and underlying Daijian System Xiamaling Formation, overlying Xijingyu Formation

(A) 青白口系景儿峪组与龙山组分界，分界面之下为紫红色页岩夹灰黑色、灰绿色页岩，考察点 P1；(D) 青白口系景儿峪组与龙山组分界，景儿峪组底部为一层含砾海绿石砂岩（红色箭头指示砾石），考察点 P1；(C) 景儿峪二段中薄层灰泥质灰岩，考察点 P2；(D) 青白口系景儿峪组与寒武系府君山组下部（本书新建的西井峪组）分界，分界面之下为景儿峪组薄层灰色含灰泥质白云岩、白云质灰岩，分界面之上为府君山组下部（西井峪组）角砾灰岩，考察点 P3

3. 考察点 P3：青白口系景儿峪组（Qbj）与南华系西井峪组（Nhx）分界

位置：该考察点位于景儿峪组与西井峪组（原府君山组下部）分界碑处。

考察内容：（1）景儿峪组与西井峪组分界；（2）界面上、下岩性特征。

现象描述：该点为青白口系景儿峪组与西井峪组的分界点，两者之间呈微角度不整合接触（图 3-78D），代表著名的"蓟县运动"（孙六铸，1957；陈晋镳等，1980）。分界面之下为景儿峪组薄层灰色含灰泥质白云岩、白云质灰岩，分界面之上西井峪组碳酸盐岩角砾岩（冰川沉积）等。

第十四节　渔阳镇府君山公园西侧南华系西井峪组（Nhx）考察线路

西井峪组仅出露于蓟州区北岭一带。此次考察西井峪组剖面位于府君山公园西侧，岩性以角砾碳酸盐岩为主，总厚 155m，总体呈块状，无层理，无层面。角砾成分多样，多呈灰白色，成分多为白云岩，少部分为灰岩。向上，燧石角砾常见。角砾多为棱角状，大小混杂，杂乱排列，分选差，为冰川沉积的冰碛岩（史书婷等，2019）。

该组底部的信手剖面和地层柱状图分别如图 3-79 所示。

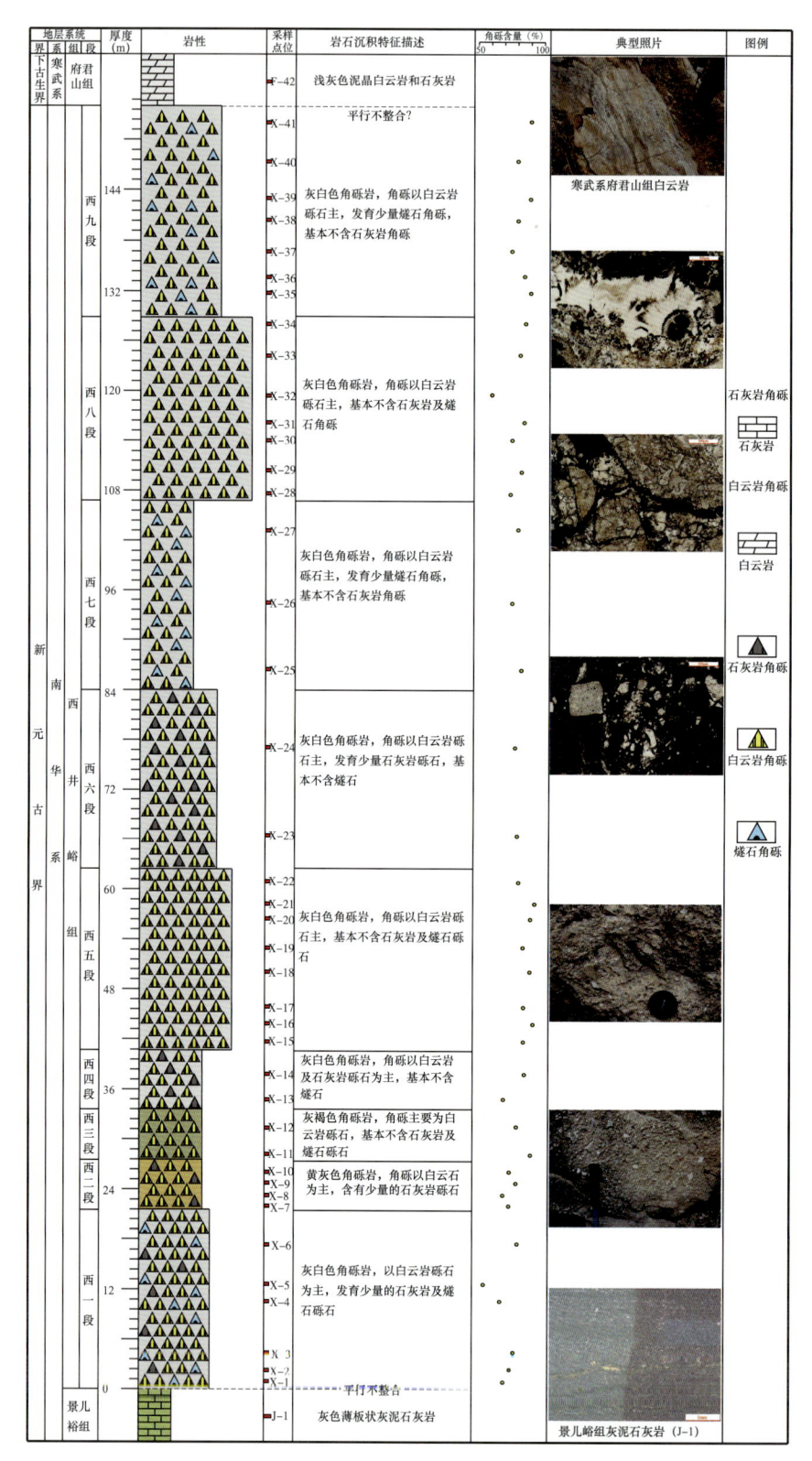

图 3-79 天津市蓟州区西井峪组综合柱状图

Fig.3-79 Stratigraphic column of the Xijingyu Formation in Jixian District, Tianjin

一、线路位置

该线路位于天津市蓟州区北部府君山公园西侧,起点位于西井峪村南(GPS 坐标: E 117.411278°, N 40.069987°),终点在桃园村北(GPS 坐标:117.418635°,40.062835°),线路总长度约 1000m。该线路道路总体较为宽阔,大小型车辆皆可通行,停车方便。具体线路和考察点分布如图 3-80 所示。

图 3-80 天津市蓟州区西井峪组考察点位置及路线

Fig.3-80 Distribution of investigation stops and route of the Xijingyu Formation in Jixian District, Tianjin

二、考察目的与任务

(1)观察天津市蓟州区西井峪组岩性特征与地层序列;
(2)观察不同类型角砾岩沉积特征;
(3)角砾岩的成因。

三、考察内容与考察点

该路线从西井峪村南至桃园村北,路况较好,便于考察。选择 4 个详细考察点,分别为下青白口系景儿峪组与南华系西井峪组分界点(图 3-80 中 P1)、西井峪组角砾岩(图 3-80 中 P2)、角砾碳酸盐岩和辉绿岩(图 3-80 中 P3)及上覆寒武系府君山组碳酸盐岩(图 3-80 中 P4)。各考察点的详细描述如下:

1.考察点 P1:青白口系景儿峪组(Qbj)与南华系西井峪组(Nhx)分界

位置:该考察点位于君山公园西侧,西井峪村南,GPS 点位:E 117.412785°,N 40.069987°。

图 3-81　青白口系景儿峪组与西井峪组的典型岩性

Fig.3-81　Typical rocks of the Jingeryu Formation and the Xijingyu Formation

（A）青白口系景儿峪组顶部灰色灰泥灰岩，考察点 P1；（B）西井峪组底部的角砾岩，表面可见白云岩碎屑零星分布，大小为 2～4cm，考察点 P1

考察内容：景儿峪组与西井峪组的分界面。

现象描述：该点为景儿峪组与西井峪组的分界，两组之间为平行不整合接触关系（图 3-81A 和 B）。分界面之下为景儿峪组顶部的中薄层灰色灰泥灰岩，分界面之上为西井峪组底部的角砾岩。

2. 考察点 P2：西井峪组下部角砾岩的特征

位置：该考察点距上一考察点向东南约 150m 处，GPS 点位：E 117.413223°，N 40.069136°。

考察内容：西井峪组角砾岩的特征（颜色、成分、大小、磨圆、分选等）。

现象描述：该点到西景峪村公交站牌之间地层发育大量角砾岩，角砾成分以白云岩为主，还有石灰岩、燧石等（图 3-82）。

3. 考察点 P3：西井峪组中部角砾岩的特征

位置：该考察点位于西井峪村公交站牌以南洞天福地牌楼处。GPS 点位：E 117.415321°，N 40.064639°。

考察内容：西井峪组中部角砾岩的特征。

现象描述："洞天福地"牌楼处地层发育大量角砾岩，有的角砾为砂屑白云岩和燧石（图 3-83）。

4. 考察点 P4：南华系西井峪组（Nhx）与府君山组（$\epsilon_1 f$）分界

位置：该考察点位于洞天福地牌楼向南下坡至坡底道路拐弯处（有一座房子），GPS 点位：E 117.415430°，N 40.062772°。

考察内容：西井峪组与府君山组分界及其上下的岩性。

现象描述：该点可见西井峪组与府君山组分界，为突变接触。界面之上为府君山组成层性很好的中薄层石灰岩和白云岩，之下为西井峪组的角砾岩（图 3-84）。

图 3-82　西井峪组角砾岩及其显微照片

Fig.3-82　Breccia of the Xijingyu Formation and their photomicrographs

（A）西井峪组角砾白云岩，表面可见燧石结核，长 2~6cm，考察点 P2；（B）西井峪组角砾灰岩，以方解石为主要矿物组成成分，考察点 P2；（C）西井峪组的角砾岩，考察点 P2；（D）角砾岩显微照片，正交光，考察点 P2

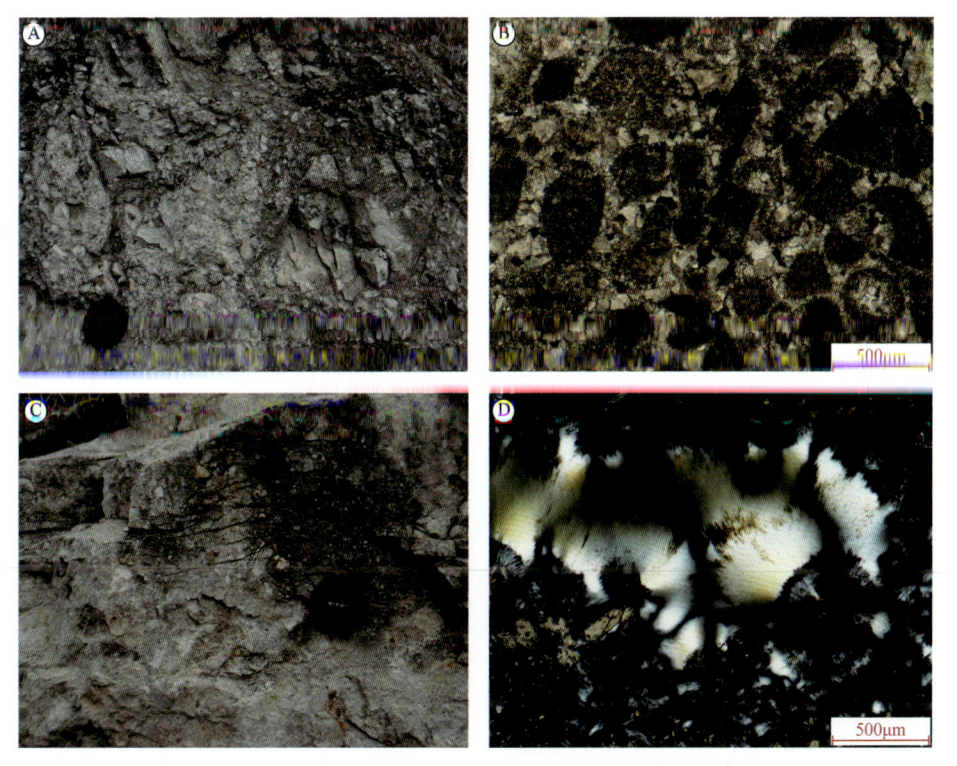

图 3-83　西井峪组白云质角砾岩及其显微照片

Fig.3-83　Dolomitic breccia of the Xijingyu Formation and their photomicrographs

（A）角砾岩，其中角砾多为白云岩，直径为 1~6 cm，考察点 P3；（B）砂屑白云岩显微照片，正交光，考察点 P3；（C）含燧石角砾的角砾白云岩，燧石角砾可达 3cm，考察点 P3；（D）燧石显微照片，正交光，考察点 P3

图 3-84　寒武系府君山组碳酸盐岩及其显微照片

Fig.3-84　Carbonate of the Cambrian Fujunshan Formation and their photomicrographs

（A）府君山组底部的灰泥灰岩，考察点 P4；（B）灰泥灰岩显微照片，正交光，考察点 P4；（C）府君山组白云岩，考察点 P4；
（D）含方解石脉白云岩显微照片，正交光，考察点 P4

第十五节　沉积相分析

一、概述

元古宙蓟州一带的沉积背景是一个北东—南西向的裂陷槽（图 3-85）。该裂陷槽在长城系沉积时期是狭长的，长近 1000km，宽仅 100km 左右，被认为是拗拉槽的一个分支，在蓟县系和青白口系沉积时期，有所变宽。

根据野外露头观察描述和薄片观察等资料，笔者对蓟县地区元古宇各组的沉积相进行了分析。本次进行相分析所采用的相标志包括：岩石类型、颜色、矿物成分、结构类（粒度、磨圆、分选、填隙物类型）、沉积构造类（交错层理、平行层理、波状层理、脉状层理、水平层理、波痕、泥裂、叠层石）、沉积序列、共生组合。元古宇没有钙质古生物化石。

元古宇的沉积类型可分 4 种：碎屑岩沉积、碳酸盐岩沉积、碎屑岩—碳酸盐岩混合沉积、冰川沉积。

以碎屑岩沉积为主的层位包括常州沟组、串岭沟组、下马岭组、洪水庄组、龙山组。这些组中几乎见不到碳酸盐岩，其沉积模式见图 3-86。这些组只是沉积于模式图中某一个或几个相带。常州沟组第一段沉积于辫状平原，第二段沉积于滨岸上部，第三段沉积于滨岸下部。串岭沟组第一

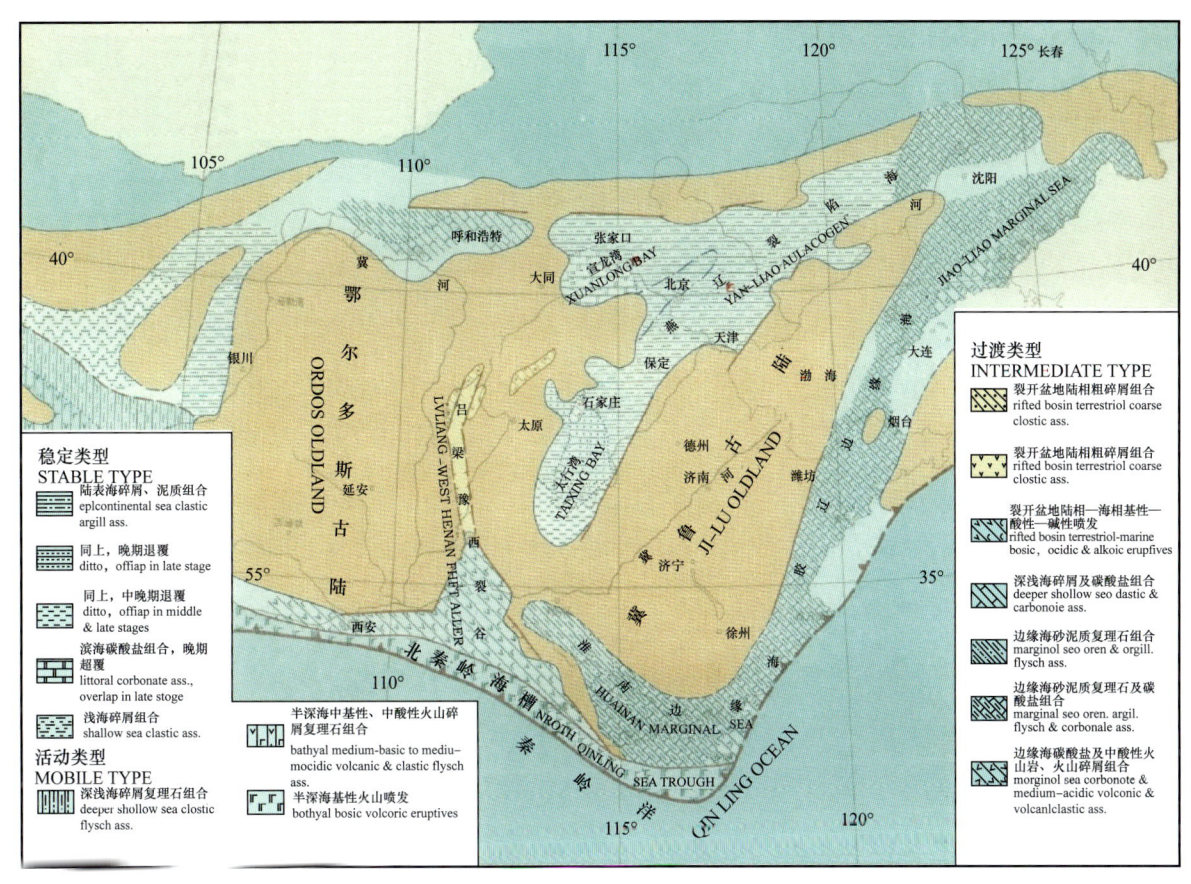

图 3-85 长城纪岩相古地理（据王鸿祯等，1985，略有修改）

Fig.3-85 Paleogeographic map of the Mesoproterozoic (now Paleoproterozoic) Changcheng period (modified from Wang et al., 1985). Red triangle represents Jixian area

红色三角代表蓟县地区

段沉积于页岩陆棚，第二和第三段沉积于缓斜坡—盆地。洪水庄组、下马岭组和龙山组主要沉积于滨岸和页岩陆棚。

高于庄组、雾迷山组、铁岭组和景儿峪组以碳酸盐岩占绝对优势，但也有一些碎屑岩。发育碎屑岩—碳酸盐岩混合沉积的层位主要包括团山子组、大红峪组、杨庄组、洪水庄组。碎屑岩-碳酸盐岩混合沉积模式见图 3-87，其中团山子组沉积相带比较齐全，从碎屑岩潮上泥坪、潮间混合坪、潮下砂坪、潮下混合坪、页岩局限浅海、碳酸盐潮坪（潮上坪和潮间坪）、开阔台地、碳酸盐缓斜坡和盆地都有。

对于海岸砂岩与碳酸盐岩互层的层段，碳酸盐台地与砂质海岸直接接触，其间未发育页岩局限浅海。

对于潮坪泥岩与碳酸盐岩互层的层段，泥质潮坪与碳酸盐台地直接接触，其间普遍缺乏海岸高能砂质相带和页岩局限浅海（图 3-88）。杨庄组和雾迷山组主要属于这种模式的沉积。杨庄组红色泥岩、白云质泥岩代表强氧化的潮上泥坪，浅绿灰色、灰白色泥岩为潮间泥坪。尽管潮间泥坪频繁暴露，但由于频繁被潮水淹没，沉积物处于潮湿状态，其内部为弱还原成岩环境，使红色泥岩还原为浅绿灰色、灰白色泥岩。这种泥岩与灰绿色页岩不同，后者是陆棚沉积的。

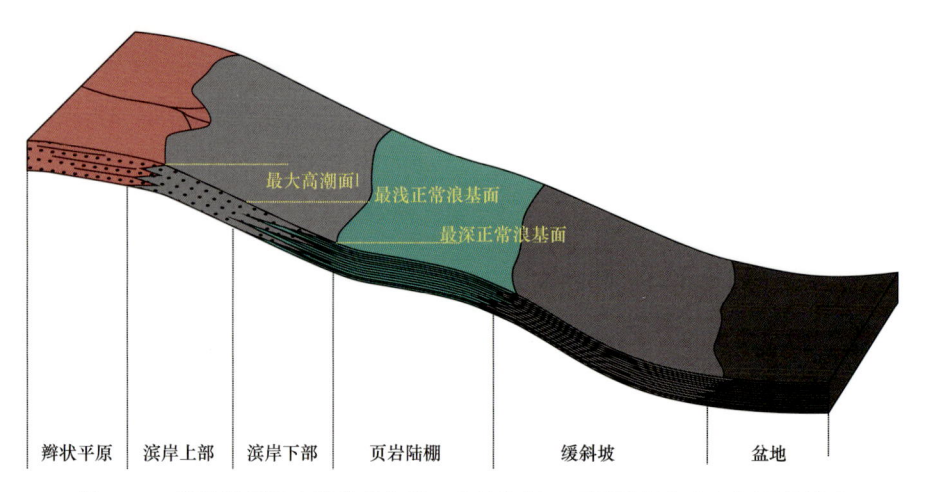

图 3-86　蓟县地区元古宇常州沟组、串岭沟组、下马岭组和龙山组沉积模式

Fig.3-86　Depositional model of the Proterozoic Changzhougou Fm., Chuanlinggou Fm., Xiamaling Fm.and Longshan Fm.in Jixian area

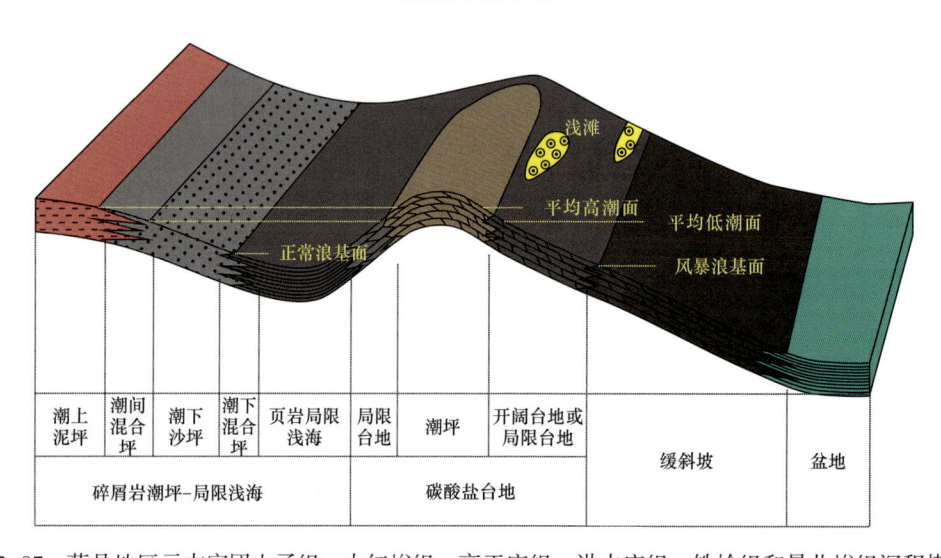

图 3-87　蓟县地区元古宇团山子组、大红峪组、高于庄组、洪水庄组、铁岭组和景儿峪组沉积模式

Fig.3-87　Depositional model of the Proterozoic Tuanshanzi Fm., Dahongyu Fm., Gaoyuzhuang Fm., Hongshuizhuang Fm., Tieling Fm.and Jingeryu Fm.in Jixian area

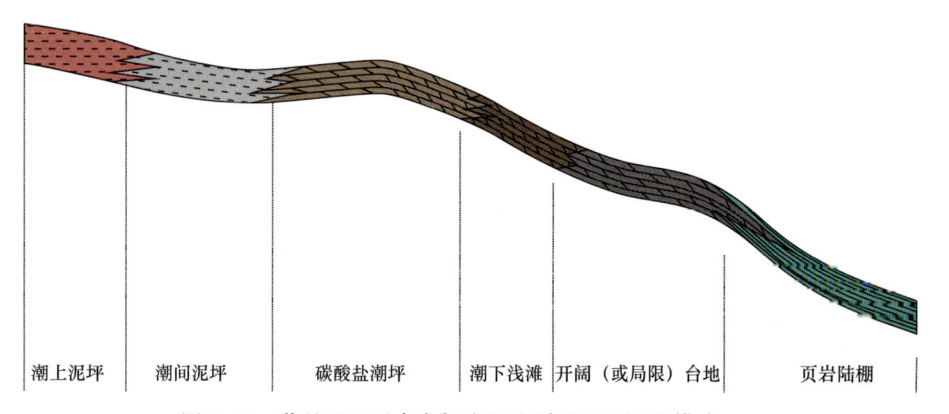

图 3-88　蓟县地区元古宇杨庄组和雾迷山组沉积模式

Fig.3-88　Depositional model of the Proterozoic Changzhougou Fm., Chuanlinggou Fm., Xiamaling Fm.and Longshan Fm.in Jixian area

二、常州沟组和串岭沟组沉积相分析

1. 常州沟组

常州沟组为一套连续沉积的、巨厚（859m）的以砂岩为主的地层，自下而上分三段，即常一段、常二段、常三段。

1）常一段

常一段的沉积相究竟是河流相还是海洋滨岸相？由于缺乏化石，只能靠特征的颜色、矿物成分、结构、构造、沉积序列和共生组合等来区分。

海洋滨岸相砂岩多为弱还原色，因为多在水下；常含海绿石（常州沟组无海绿石，但龙山组中常见）。由于水动力较稳定，其分选普遍好；由于砂以滚动搬运为主（尤其是冲洗带），因此磨圆普遍好，通常以次圆状为主。由于多处于水下，波痕多为对称波痕；由于波浪方向多变，形成的交错层理倾向多变。此外，可发育特征的冲洗交错层理。由于滨岸滩坝无侵蚀能力，因此冲刷面罕见（除非风暴流），而且滩坝多为均质序列或向上变粗的沉积序列。海侵虽可形成总体向上变细的沉积序列，但对单个滩坝来说，这种情况罕见。

河流砂岩多为氧化色，因为多在水上，无海绿石。由于水动力在洪水期和平水期波动较大，其分选普遍不太好，为中等；由于河流中砂的搬运以跳跃为主（洪水期可悬浮搬运），其磨圆普遍较差，以次棱角状为主。由于河流方向总体稳定，形成的交错层理倾向较稳定。此外，可发育特征的冲刷面、向上变细的沉积序列。

根据上述分析，该区的常一段砂岩、含砾砂岩、砾质砂岩和细砾岩为砂质辫状河沉积，而且是深大裂谷中的辫状河。证据如下：

（1）颜色红，为氧化色（图3-5和图3-7）。该段普遍为紫红色，包括偶见的泥砾和泥质夹层，属于氧化色，反映为长期暴露环境。

（2）磨圆较差，分选中等（图3-8）。砂岩中的砂多为次棱角状，磨圆较差。即使细砾，其磨圆也不太好。分选中等。这些都是河流沉积的特点。

（3）层理倾向稳定。从下到上厚260m左右的砂岩、含砾砂岩、砾质砂岩和细砾岩，见到的层理的倾向均为北北西向，十分稳定，该段上部尤为明显，这反映了水流方向的稳定性。

（4）冲刷面和向上变细的沉积序列常见（图3-6和图3-7），是河流沉积的特征。

（5）见撕裂泥砾（图3-5）。这种棱角状撕裂泥砾不可能出现在滨岸沉积中，因为反复的簸洗会使泥砾消失，保存不下来。而这种泥砾在河流沉积中十分常见。

（6）厚260m左右的地层全为砂岩、含砾砂岩、砾质砂岩和细砾岩，仅偶见侵蚀残余的泥岩薄夹层，这反映该时期的河流环境中河漫滩不发育。为什么不发育？应是河道左右频繁摆动所致，使河漫滩沉积难以保存下来。根据现代沉积观察，只有深大裂谷中的辫状河具有这种特点，能形成砂岩巨厚而无泥岩夹层的沉积。

综上所述，该区常一段砂岩、含砾砂岩、砾质砂岩和细砾岩为深大裂谷中的砂质辫状河沉积，不是海洋滨岸沉积，与上覆的常二段、常三段不同。

2）常二段和常三段

常二段和常三段以乳白色细砂岩为主。与常一段砂岩相比，常二段和常三段的砂岩颜色偏还原色，尤其是其磨圆和分选普遍好。综合分析认为，常二段为滨岸上部沉积，常三段为滨岸下部沉积。证据如下：

（1）砂岩磨圆好，分选好（图3-8）。石英颗粒多呈次圆状，能够使石英磨圆的环境，只有稳定的被动大陆边缘的滨岸环境，因为这里在波浪、冲洗流和沿岸流的作用下，砂主要以滚动方式搬运，而且由于沉降慢，砂能够长期遭受簸洗和磨蚀，因此其磨圆度普遍高。在河流中，砂无论搬运多远，都很难磨圆，因为以跳跃甚至短暂悬浮搬运为主。长江口的砂磨圆仍较差，尽管搬运了数千千米。滨浅湖中，砂虽然可以滚动搬运，但由于湖平面不稳定，变动频繁，难以持续磨圆。因此，凡是颗粒磨圆好（次圆或圆状）的砂岩，都是海相的。尽管海相砂岩风化后，可以向河流提供磨圆好的砂，但一定有其他母岩提供的磨圆不好的砂。因此河流沉积的砂岩中总是有磨圆不好的砂，而且通常占多数，尽管可以有磨圆好的。磨圆度是区分海相和陆相的重要标志。分选好也是海相砂岩的重要特征，因为正常天气的海浪、冲洗流和沿岸流都相对较稳定，搬运的颗粒大小相近。

（2）填隙物为胶结物。常二段砂岩的填隙物主要是硅质胶结物，而不是杂基。这反映沉积环境的水体是持续动荡的，泥质沉积不下来。其水深应该在最浅的正常浪基面之上，因为这样才能保证不是间歇动荡，保证没有泥质沉积的安静阶段。

（3）颜色偏还原色。为什么常一段的砂岩为紫红色，而常二段和常三段的砂岩主要为乳白色（灰白色）？笔者认为，这是因为常二段和常三段的砂岩为水下沉积，其沉积环境为弱还原，因此其颜色为弱还原。

（4）砂岩多呈块状，无层理。笔者认为，这是生物扰动导致的。虽然目前在元古宇中尚未发现生物遗体化石，但不意味着一定没有生物。很可能当时的生物是软体的，未留下化石。从沉积构造看，如果没有生物扰动破坏，滨岸砂岩中是应该发育层理的。反之，如果没有层理，就可以推测有生物存在。

（5）常三段与常二段相比，灰绿色泥质薄夹层变多，层变薄，而且向上逐渐过渡为串岭沟组滨外陆棚沉积的灰绿色页岩，反映从常二段到常三段，水体是逐渐加深的。泥质薄层代表海水安静期沉积。常二段中无泥质薄层，说明海水浅，是持续动荡环境，泥质沉积不下来，属于滨岸相的上部，水深在最浅正常浪基面之上。常三段中有，表明是间歇动荡环境。动荡时沉积砂，安静时沉积泥，有时候波浪能影响到海底，有时候不能。浪基面波动带就具有这种特征。由于其沉积仍以砂为主，说明以频繁动荡为主，安静期短暂且不频繁，属于滨岸相的下部，水深在最浅与最深正常浪基面之间。如果在最深正常浪基面之下，就以泥沉积为主了。

（6）层面波状起伏，是风暴浪形成的波痕。常二段和常三段的砂岩层面不平整，波状起伏，实际上是大型波痕。波痕大致对称，波长数十厘米，波高数厘米到十几厘米，是风暴浪这种大型波浪形成的。

总之，从常一段到常三段是一个完整的海侵序列，由陆相变为海相，水深逐渐变大。

2. 串岭沟组

串岭沟组总体为一套巨厚的页岩（889m），与下伏常州沟组为典型的连续、渐变型接触，无沉积间断。

该组自下而上分三段：串一段、串二段和串三段。

串一段页岩呈灰绿色下部夹一些薄层细砂岩和粉砂岩，顶部夹泥晶白云岩透镜体（风化后呈灰白色），见图3-13。串二段为大套灰黑色页岩，中部有辉绿岩侵入体。串三段为灰黑色页岩夹薄层（单层厚度多为1cm左右）泥质—粉砂质泥晶白云岩（图3-13和图3-16D）。白云岩含铁，风化后呈褐色。该段顶部有辉绿岩侵入体。

整个组的岩性为页岩，为安静低能环境沉积；其颜色为还原色，反映为水下沉积，水深在正常浪基面之下。页岩是湖湘还是海相？由于那个时代没有硬体化石，并不能根据是否有海相化石来判定。但由于其上为团山子组大套海相碳酸盐岩，与之伴生，内部也有泥晶白云岩透镜体，同时考虑区域沉积背景，因此认为串岭沟组应为海相页岩，而不是湖湘页岩。

海有多深呢？笔者提出，串一段为滨外陆棚，串二段和串三段为深水缓斜坡—盆地。证据如下：

（1）串一段页岩颜色较浅，为灰绿色（图3-15C和E），反映弱还原环境，在氧化还原界面附近；其内夹有具波痕和小型交错层理的薄层细砂岩和粉砂岩（图3-15D和E）。这些薄层细砂岩和粉砂岩应为风暴流沉积，说明在风暴浪基面之上。在该段中部的灰绿色页岩夹薄层粉砂岩处（串岭沟组考察点3），局部见对称波痕和疑似泥裂（图3-15F）。有些学者认为是泥裂，但笔者认为不是泥裂，因为其形状和裂缝内"充填物"（为粉砂质）都不像典型泥裂，而且是出现在水体较深的陆棚沉积中。因此，笔者认为，这些"泥裂"实际上是下伏具薄层粉砂岩中波痕的波峰和波谷不均匀风化形成的。粉砂岩薄层具有鱼鳞状小型波痕（为多方向的干涉波痕），其上覆盖页岩薄层。风化时，波谷之上的泥质披覆层被保存，而波峰之上的则被剥蚀掉并裸露，貌似泥裂的充填物。如果真是泥裂，就需海平面快速大幅度下降，使海底暴露，但这种可能性不大。

（2）串二段页岩呈灰黑色，比串一段的深，在氧化还原界面之下；而且无细砂岩和粉砂岩夹层，页岩也更纯，说明水体比串一段更深，远在风暴浪基面之下。尤其是，在该段下部（串岭沟组考察点4）见小型褶皱，可能是准同生滑塌褶皱，代表斜坡。但未见浊积岩，说明斜坡为缓斜坡，而非陡斜坡。因此认为，该段为深水缓斜坡—盆地沉积。

（3）串三段的页岩仍呈灰黑色，其沉积环境总体与串二段相同，为深水缓斜坡—盆地沉积。但该段夹有薄层泥质、粉砂质泥晶白云岩。这些白云岩是由低密度浊流从邻近的碳酸盐台地机械搬运到深水环境中沉积的白泥和一些陆源泥和粉砂形成的（另有论文发表）。

总之，串岭沟组自下而上，页岩颜色由灰绿色逐渐变为灰黑色，粉砂岩和细砂岩薄夹层逐渐消失，水体逐渐加深，由滨外陆棚变为缓斜坡—盆地。其中，串二段是整个元古宇海水最深的时期（比洪水庄组还要深），是元古宇的最大海泛面。

三、团山子组沉积相分析

团山子组厚518m（图3-17和图3-18），自下而上分四段，即团一段、团二段、团三段、团四段。

1. 团一段

团一段为深灰色中—薄层泥质泥晶白云岩和粉砂质泥晶白云岩与深灰色白云质页岩互层。白云岩风化后呈黄褐色，反映含铁。

该段最大的特征是各类岩石层面平整，水平层理极其发育（图3-20）。其颜色暗、水平层理发育，说明为水下还原环境沉积，水体较深，而且远在风暴浪基面之下，水底安静。由于覆于串三段缓斜坡—盆地沉积的灰黑色页岩之上，根据沃尔瑟相律，团一段应为深水沉积。鉴于该段以白云岩为主，其水深应比串三段浅，为斜坡环境。由于缺乏高密度浊流和滑塌构造，该斜坡应是十分平缓，即缓斜坡。

至于该段的白云岩，笔者认为是机械搬积的，不是原生或次生的。在邻近的碳酸盐台地上原生沉淀或蒸发泵白云化形成的泥晶白云石被风暴浪搅起后，以低密度浊流和远洋悬浮的方式搬运到深水环境中沉积下来。证据如下：

（1）这些泥晶白云岩是在深水环境中沉积的，无论目前哪种白云岩形成机理，都不能解释其成因。蒸发泵白云化不能，因为不是潮坪环境；回流渗透白云化不能，因为全是泥晶白云石，而不是粗粉晶或更粗；混合水白云化、埋藏白云化和热液白云化形成的白云岩都是砂糖状的。热水沉积不可能，因为水平层理极其发育。

（2）白云岩中普遍含泥质和粉砂（图3-20），说明是混浊的水流沉积。尤其粉砂的存在，更是机械搬运的证据。

如果这些白云岩是机械沉积的，就应称"云泥白云岩"，与灰泥石灰岩类似。

2. 团二段

团二段以灰色中—厚层泥晶白云岩夹薄层泥质泥晶白云岩为主，由于白云岩含铁量较高，其风化面上多呈黄褐色。

与团一段相比，团二段常见中—厚层泥晶白云岩，其顶面不太平整，有对称波痕（图3-89）；内部呈块状构造或模糊的水平层理，局部可见厚度只有几毫米到2cm左右正递变层，主要由粉砂和粉屑组成，顶面波状起伏。

图 3-89 团二段白云岩顶面的对称波痕（蓟县团山子村西）

Fig.3-89 Symmetric ripple marks on the top of dolostones in the Cht_2 at the west of Tuanshanzi village, Jizhou

层理不明显，可能是生物扰动所致。至于何种生物，目前不清楚，可能是藻类植物，也可能是软体动物，均未留下化石。泥质泥晶白云岩水平层理发育，与团一段的相似。

总体上，团二段沉积的水体比团一段浅，应为缓斜坡上部，在正常浪基面之下，但在风暴浪基面附近。正递变层应为低密度浊流沉积。

3. 团三段

团三段明显分下、中、上三部分。下部主要为深灰色薄层泥晶白云岩（风化后黄褐色）夹灰色薄层白云质石英细砂岩，砂岩中小型交错层理和波痕常见；中部为深灰色页岩与细砂岩薄互层，加数层具气孔构造的玄武岩（单层厚度几十厘米），砂岩中小型交错层理和波痕常见；上部为中—薄层泥晶白云岩夹中层白云质石英砂岩。白云岩中层状和波状叠层石、泥裂常见；砂岩中交错层理常见。

团三段为碳酸盐岩与碎屑岩型混积台地沉积。白云岩中普遍发育层状和波状叠层石，常见泥裂，说明为潮间坪和潮上坪沉积。白云岩为泥晶，应为蒸发泵白云化成因。

砂岩普遍为中、薄层，与潮坪白云岩互层，小型交错层理发育，上、下层的层理倾向常相反，反映双向水流，因此也是潮坪沉积。

中部为暗色页岩与细砂岩薄互层。暗色页岩代表水下还原环境，为安静低能期沉积；砂岩为动荡期沉积。砂岩中波痕和交错层理发育，说明其沉积水体不深。因此，团三段中部应为局限潮下沉积。

4. 团四段

团四段为薄层紫红色粉砂质页岩夹薄层细砂岩，砂岩层表面具不对称波痕，其内常见小型交错层理。自下而上，由以紫红色页岩为主到以薄层细砂岩为主，并过渡为上覆大红峪组的灰白色厚层石英细砂岩。

该段页岩具氧化色，为潮上泥坪沉积。上、下砂岩层小型交错层理的倾向相反，反映双向水流，因此砂岩与页岩薄互层为潮间混合坪沉积。

该段自下而上，由潮上泥坪变为潮间混合坪，再变为大红峪组的潮下浅滩，为一个海侵序列。

总体上，团山子组为混积台地沉积。

四、大红峪组沉积相分析

大红峪组自下而上细分为三段：大一段、大二段、大三段。

大一段以中—厚层乳白色石英砂岩为主，夹紫红色粉砂岩、含燧石砂质白云岩或白云质砂岩，以及蓝绿色凝灰岩，与下伏团山子组连续沉积。石英砂岩层面上波痕常见，平行层理和交错层理发育，分选好，磨圆好，硅质胶结，为高能动荡的潮下浅滩沉积。

大二段主要是一套火山熔岩和火山角砾岩，夹少量石英砂岩和凝灰岩。根据伴生的地层分析，其喷发环境应为滨浅海。

大三段以灰白色中厚层含燧白云岩为主，层面波状起伏。白云岩中常见锥状叠层石，叠层石是潮坪的标志，因此该段的沉积环境主要是碳酸盐台地上的潮坪。燧石交代了白云岩，是次生的。交代作用发生的较早，在埋藏后不久，即准同生期。丰富的藻席腐烂形成了酸性微环境，利于白云石溶解，硅质沉淀，硅质优先交代叠层石表明了这一点。

五、高于庄组沉积相分析

高于庄组自下而上细分为六个岩性段：高一段、高二段、高三段、高四段、高五段、高六段。

1. 高一段

高一段为灰色中—厚层含燧石条带和结核叠层石泥晶白云岩夹灰绿色薄层页岩，底部有一套厚约3m的石英砂岩，与下伏大红峪组平行不整合接触。砂岩中小型对称波痕发育。

叠层石主要是层状、波状和半球状，是潮坪环境的标志。白云岩层面波状起伏，有些是波痕，有些是波状和半球状叠层石导致的。因此，中—厚层含燧石条带和结核叠层石泥晶白云岩主要为碳酸盐台地上的潮坪沉积。燧石的成因与大红峪组的相同。

底部的石英砂岩分选好，磨圆好，硅质胶结，具有对称波痕，表明是潮下浅滩沉积。砂质浅滩之上直接覆盖碳酸盐潮坪，说明海岸有沙滩，向海很快就过渡为碳酸盐台地，与今天澳大利亚东海岸类似。

灰绿色页岩夹层是页岩陆棚沉积，代表短暂的海侵事件，说明碳酸盐台地经常被淹没于较深的水下。

2. 高二段

高二段为灰褐色含锰粉砂质页岩和含锰粉砂质白云岩、含锰粉砂岩互层，水平层理发育。中下部含有"蓟县式锰（硼）矿"。

该段为较深的陆棚沉积。泥质沉积为主，说明水体安静低能；水平层理发育，说明水深在风暴浪基面之下。颜色为灰褐色，乃含较多锰质所致。锰质是水下还原环境下形成的，与菱铁矿类似。

3. 高三段

高三段为灰色块状中—厚层泥晶白云岩与水平层理发育的泥晶白云岩互层，夹少量灰绿色页岩。白云岩含铁，风化后呈褐色。燧石结核和条带较稀少，集中在水平层理发育的泥晶白云岩中。叠层石不发育。

该段层面平整，水平层理常见，叠层石少见，说明水体比高一段白云岩的深，但比高二段页岩的浅，应为正常浪基面之下的开阔（或局限）台地沉积，水平层理发育的白云岩比块状白云岩更深些。页岩夹层代表海侵事件，与高一段的类似。白云岩晶粒很细，为泥晶，且含Fe^{2+}，很可能为准同生白云化或微生物诱发的原生沉淀成因。

4. 高四段

高四段为深灰色中层灰质白云岩、白云质灰岩夹薄层含泥的灰质白云岩、白云质灰岩。燧石结核和条带虽有，但少见。

中层灰质泥晶白云岩、白云质灰泥灰岩具有块状构造，无层理，无颗粒，无叠层石，其层面较平整（图3-90）。这些特征表明其沉积环境安静低能，水深在正常浪基面之下，可能为开阔台地。缺乏层理可能是生物扰动所致。

深灰色含泥的灰质白云岩、白云质灰岩水平层理发育（图3-90），但有的具有小型对称波痕，使薄层呈豆荚状或链条状。泥质含量较高，水平层理发育，说明其沉积水体应在风暴浪基面之下，为缓斜坡沉积。而具有波痕的含泥的灰质白云岩和白云质灰岩则在风暴浪基面附近。

总体上，该段为开阔台地与缓斜坡交替沉积。

本段底部的灰泥石灰岩瘤，可能是微生物成因。顶部的"臼齿灰岩"或称"鸡爪灰岩"，应是发育不完全的泥裂充填形成的。这说明该段在沉积末期，水体变浅，向上覆第五段潮坪沉积过渡。

图 3-90　高四段含燧石结核具块状构造的白云质灰泥石灰岩和水平层理发育的含泥白云质灰泥石灰岩

Fig.3-90　Massive dolomitic micritic limestones with chert nodules and muddy domitic micritic limestones with horizontal beddings in the Jxg$_4$

5. 高五段

高五段下部为纹层状沥青质细晶白云岩和细晶灰岩，上部为核形石状粗晶白云岩（又称"虾米石"）与纹层状中细晶白云岩、泥晶白云岩互层。纹层不平整，为层状叠层石。

大量层状叠层石和核形石的存在，说明其沉积环境为碳酸盐台地上的潮坪。细晶白云岩为热液白云岩，与附近的深断裂有关。石灰岩普遍重结晶，也是与热液水岩反应的结果。

6. 高六段

高六段为灰色中厚层含燧石结核和条带的泥晶白云岩，层状和波状叠层石常见，为碳酸盐潮坪沉积。

总之，整个高于庄组厚度巨大，经历了多次环境变迁。高一段、高五段和高六段主要为碳酸盐台地上的潮坪沉积；高二段为深水页岩陆棚沉积；高三段主要为开阔台地沉积；高四段主要为开阔台地和缓斜坡沉积。从高一段到高二段为海侵，从高二段到高三段为海退，从高三段到高四段为海侵，从高四段到高六段为海退。高二段是整个高于庄组的最大海泛面。

八、杨庄组沉积相分析

杨庄组以醒目的紫红色与灰白色相间（即红白相间）的泥岩为特征，像"五花肉"。泥岩常含粉砂和泥晶白云石。与下伏高于庄组白云岩为突变接触，可能为平行不整合。

该组自下而上可以划分为三段：杨一段、杨二段和杨三段。杨一段（Jxy$_1$）为"红白相间"的泥岩，泥岩中常含粉砂和泥晶白云石。厚209m。杨二段（Jxy$_2$）为"红夹白"泥岩，泥岩中常含粉砂和泥晶白云石。上部见一层几十厘米厚的粗砂岩。厚434 m。杨三段（Jxy$_3$）为"红白相间"的泥岩夹灰色块状藻白云岩、石灰岩以及叠层石白云岩。厚130m。

紫红色泥岩代表暴露的氧化环境，有的地方可见泥裂，为干旱气候下的潮坪。灰白色泥岩代表低能的水下还原环境，为局限海或当潟湖。偶见的砂岩为潮下浅滩沉积。灰色块状藻白云岩和石灰岩为潮下带沉积，层状叠层石泥晶白云岩为潮坪沉积。

总体上，杨庄组以泥质潮坪为主，与浅水局限海或潟湖频繁交替，乃海平面频繁波动所致。

七、雾迷山组沉积相分析

雾迷山组厚度巨大，但岩性单调，米级旋回极其发育，与下伏杨庄组连续沉积。典型的旋回是自下而上依次由绿灰色白云质页岩—灰色凝块状细晶藻白云岩—含燧石层状叠层石泥粉晶白云岩。

页岩呈弱还原色，为水下低能沉积，代表浅水局限海，位于碳酸盐台地与陆地之间。

灰色凝块状细晶藻白云岩为潮下带沉积，层状叠层石泥粉晶白云岩为潮坪沉积。

从绿灰色白云质页岩依次到灰色凝块状细晶藻白云岩、含燧石层状叠层石泥粉晶白云岩，水体依次变浅。

雾迷山组上部见少量紫红色泥岩，为泥质潮坪沉积（潮上带）。

雾迷山组中偶见灰白色石英细砂岩，分选好，磨圆好，为潮下浅滩沉积。

八、洪水庄组沉积相分析

洪水庄组主体为一套页岩，与雾迷山组连续沉积。该组自下而上可分洪一段和洪二段。洪一段为中层泥晶白云岩夹灰绿色、蓝绿色和灰色页岩。洪二段主要为灰黑色、灰绿色页岩，顶部为灰绿色页岩夹泥质泥晶白云岩透镜体（藻礁）和薄层粉砂岩。

洪一段为中层白云岩夹灰绿色、蓝绿色和灰色页岩。白云岩中常见叠层石，为碳酸盐台地沉积，页岩为陆棚沉积，两者频繁交互，说明海平面振荡频繁，时浅时深。

洪二段页岩颜色属还原色，水平层理仍发育，代表安静低能、还原性环境，为页岩陆棚沉积。向上，页岩陆棚沉积过渡为铁岭组的三角洲前缘河口坝砂岩或滨岸砂岩。

总之，雾迷山组沉积之后，发生海侵，海水逐渐加深。在经历了洪一段台地白云岩与陆棚页岩交互沉积的过渡阶段后，完全变为洪二段的页岩陆棚，末期过渡为铁岭组的三角洲前缘河口坝砂岩或滨岸砂岩。

九、铁岭组沉积相分析

铁岭组与下伏洪水庄组为连续沉积，自下而上分三段：铁一段、铁二段和铁三段。

1. 铁一段

铁一段厚155m，底部为灰白色中—薄层石英细砂岩，向上过渡为灰色含锰藻泥晶白云岩；主体为中厚层灰色叠层石白云岩夹灰绿色、蓝绿色页岩；顶部为蓝绿色海绿石页岩和含铁锰质暗棕色页岩。

在蓟县府君山公园西北方向的小岭子村的露头上，石英细砂岩分选较好，见平行层理，与下伏洪水庄组陆棚灰绿色页岩渐变接触，并显示向上变粗的沉积序列，应为三角洲前缘的河口坝沉积。在府君山公园以东约2km、四洼村以南约1km的露头上，见石英砂岩底面发育浪成对称波痕，且与下伏灰黑色页岩突变接触，应为滨岸砂岩。

主体部分中厚层灰色叠层石白云岩为碳酸盐台地上的潮坪沉积，所夹的灰绿色、蓝绿色页岩代表短暂海泛。

该组顶部较薄的蓝绿色海绿石页岩和含铁锰质暗棕色页岩代表短暂的页岩陆棚沉积。

铁一段沉积后发生构造抬升，遭受风化剥蚀，形成了与铁二段之间的平行不整合。

2. 铁二段

铁二段沉积时，该区又沉降，接受沉积，主要为云泥条带灰泥石灰岩夹竹叶石灰岩和少量叠层石石灰岩，局部见燧石结核。叠层石和竹叶石灰岩的出现说明水体不深，主要为碳酸盐台地上的潮坪沉积。竹叶石灰岩为风暴流在潮坪上的沉积。局部可见小型潮汐水道。风化后呈褐色的泥质泥晶白云岩薄层应为潮坪上的蒸发泵白云化所致。

3. 铁三段

铁三段为叠层石石灰岩段。叠层石类型多样，形态各异，包括层状、波状、柱状和墙状叠层石，是该区古元古界叠层石类型最多、发育最好的层段。

大量叠层石表明，铁三段沉积为潮坪和浅水潮下带环境。柱状叠层石之间充填灰泥或灰绿色泥，没有颗粒，说明水体能量不高；填隙物常白云化。其中墙状叠层石可指示潮汐水流方向，其延伸方向与潮汐流平行。

铁三段的叠层石体为一个矩形叠层石礁。在蓟州区一带最厚，在其他地区很薄或消失。

铁三段沉积后发生构造抬升，遭受风化剥蚀，形成了与下马岭组之间的平行不整合。

十、下马岭组沉积相分析

下马岭组与下伏铁岭组为平行不整合接触，主体为灰绿色页岩，底部为几米厚的灰白色砂岩夹页岩，局部见砾石。

砂岩为滨岸沉积，页岩为陆棚沉积，分析的依据同上。

整个下马岭组为一个海侵序列。

十一、龙山组沉积相分析

龙山组与下伏下马岭组页岩呈突变接触，为凹凸不平的侵蚀面，可能为平行不整合。

龙山组分两段：龙一段和龙二段。

龙一段为中厚层含砾粗石砂岩段，层面常有波痕，槽状交错层理发育，局部有灰白色页岩薄层，应为典型的滨岸沉积。

龙二段以灰绿色和灰黑色的页岩为主，夹薄层海绿石细砂岩，主要为页岩陆棚沉积。顶部数米为紫红色泥岩，紫红色为强氧化色，代表暴露的氧化环境，因此应为潮上带泥坪沉积。

整个龙山组为一个海侵又海退的序列。

十二、景儿峪组沉积相分析

该组底部含有十几厘米厚的含海绿石含细砾粗砂岩，下部主要为灰白色、灰紫色的中—薄层灰泥石灰岩；中部为灰色、蛋青色中—薄层灰泥石灰岩夹泥质灰泥灰岩；上部为灰色薄层泥质灰泥灰岩夹灰绿色页岩。

该组总体为较深水碳酸盐开阔台地沉积，因为层薄，层面较平整，且夹灰绿色页岩。

十三、西井峪组冰川沉积

西井峪组是金振奎新建的组，为一套大小混杂的块状碳酸盐角砾岩，厚155m。金振奎及其学生研究后认为是冰川沉积。

该组原来被划归下寒武统府君山组，但其岩性特征与上覆下寒武统府君山组的中薄层泥晶和细晶白云岩完全不同，有天壤之别，因此笔者认为有必要新建一个组。由于这套地层的剖面位于西井峪村附近，故名"西井峪组"。

西井峪组与下伏景儿峪组和上覆府君山组之间均呈显著的突变接触。与下伏景儿峪组之间被认为是平行不整合或微角度不整合，并被认为是"蓟县运动"所致。

西井峪组与上覆府君山组之间呈突变接触，很可能是平行不整合接触。但因缺乏化石和其他定年矿物，两组之间是否缺失地层并不确定。

西井峪组的年代应是景儿峪组沉积之后、府君山组沉积之前，但究竟是属于新元古代还是早寒武世，并没有确切证据，但很可能属于新元古代。我们在这套角砾岩中并未找到任何三叶虫或其他化石。如果属于寒武系，应该有化石。而且角砾岩中常见燧石角砾，但在华北寒武系中并没有燧石结核或条带，这些证据表明其母岩不可能是寒武系。从角砾的成分看，其母岩主要是中元古界蓟县系雾迷山组，少量来自铁岭组。据此，推测其年代属于新元古代而不是寒武纪。由于南方南华系是冰川沉积活跃时期，尤其是南沱组，因此推测西井峪组与上南华统南沱组相当。

1. 角砾岩结构特征

1）角砾的粒度、分选及磨圆

西井峪组角砾岩中角砾大小悬殊，最大的角砾直径在1m左右，整个剖面以直径在1~5cm的角砾为主。粒径较大的角砾在剖面横向和纵向上分布不均匀，常见局部集中的现象（图3-91A）。角砾的分选极差，大小角砾混杂（图3-91B），毫无分选性可言。角砾的磨圆极差，普遍呈棱角状和次棱状为主（图3-91C和D）。角砾形态多样，呈菱形、正方形、三角形、长条状、楔形或不规则形态（图3-91F）。燧石在野外剖面上主要为长条状，相比碳酸盐岩角砾抗风化能力极强，普遍突出角砾岩表面。白云岩角砾、石灰岩角砾以菱形、正方形、三角形为主，其中白云岩角砾风化面刀砍纹发育（图3-91E）。从角砾岩的分选和磨圆看，其结构成熟度极低。

角砾岩的填隙物为基质，其含量为10%~20%。基质以陆源白云质（少量灰质）泥和粉砂为主（图3-92A、B、D、F），基本不染色（茜素红），镜下呈暗色；局部含有粉砂级（少量为细砂级）石英颗粒（图3-92C）。

2）角砾的成分

根据野外观察及镜下鉴定，研究区角砾岩的成分主要为白云岩，部分为石灰岩和燧石，也见极少量的火成岩岩屑。白云岩角砾平均含量在80%左右（图3-93）。石灰岩及燧石角砾在垂向上分布不均匀。

西井峪组白云岩角砾的类型极为丰富。按照晶粒大小划分，从泥粉晶到粗晶白云岩角砾均有发育，但以泥晶和细粉晶白云岩角砾为主（图3-93A-E）。泥晶白云岩角砾按照沉积构造可以划分为两种，即具水平层理的泥晶白云岩角砾（图3-93A）和均质泥晶白云岩角砾（图3-93）。除此之外还发育两种颗粒白云岩，分别是亮晶砂屑白云岩角砾（图 3-93F）和亮晶鲕粒白云岩角砾（图 3-93G）。这些发育特殊颗粒和沉积构造的白云岩角砾为进一步分析其物源类型提供了依据。

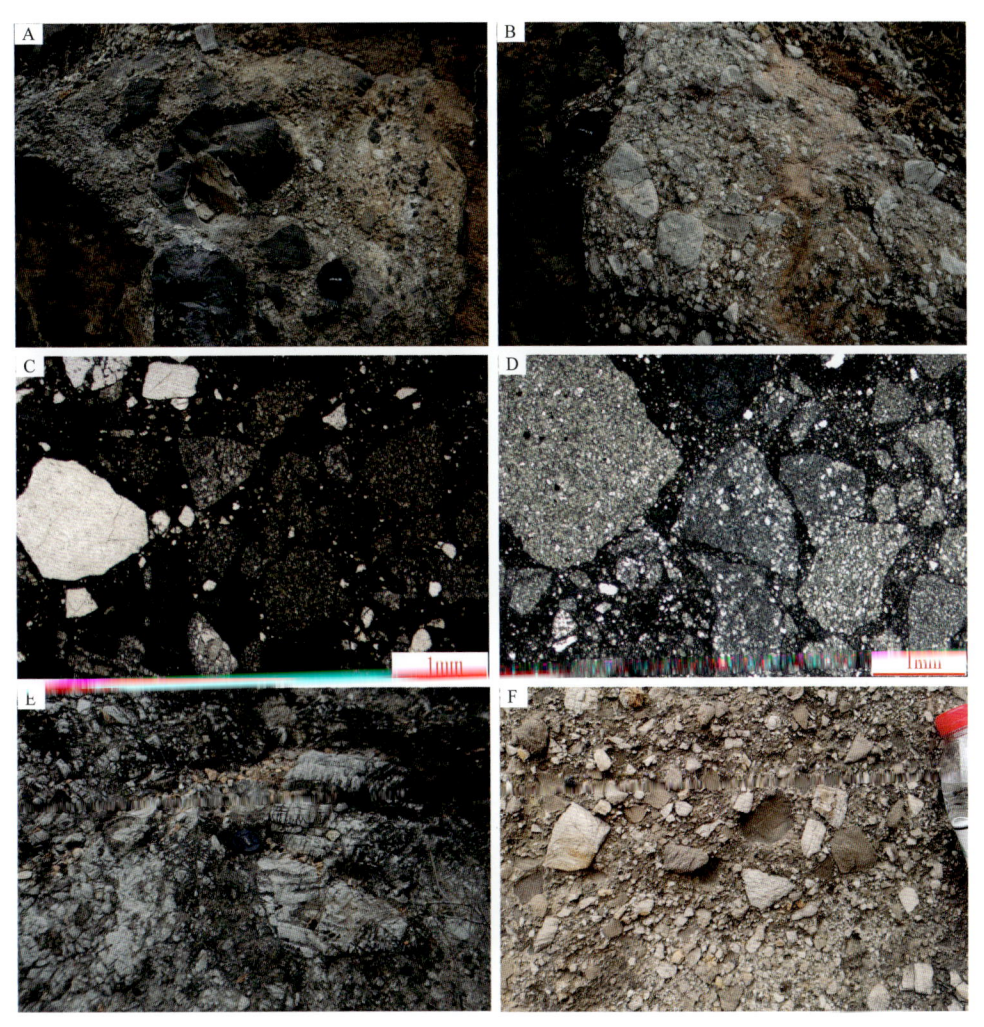

图 3-91 天津蓟州西井峪组角砾岩的大小、磨圆、分选及形态特征

Fig.3-91 Characteristic of size, rounding, sorting and form of breccia of the Xijingyu Formation in Jizhou, Tianjin

(A) 西井峪组大块角砾集中带；(B) 西井峪组大小混杂的角砾特征；(C) 角砾磨圆特征，普遍为次圆、次棱、棱角状，中晶 x1；(D) 角砾磨圆特征，普遍为次棱角状，分选适度，单偏光，样品 A-33；(E) 角砾岩中大的白云石角砾，为次圆状；(F) 形态丰富的角砾，以正方形、菱形、长方形为主

西井峪组石灰岩角砾相对较少，以粗晶和粉晶石灰岩角砾为主（图 3-93H 和 I）。由于西井峪组下伏中—新元古代碳酸盐岩地层岩性普遍为白云岩，石灰岩地层较少，造成西井峪组角砾岩中石灰岩角砾稀少，但局部存在高含量的石灰岩角砾集中带。

西井峪组角砾岩中常见燧石角砾（图 3-94）。燧石在野外主要为土黄色长条状。燧石相比碳酸盐岩角砾抗风化能力强，突出角砾岩表面，造成角砾岩表面凹凸不平。单片光下燧石普遍为白色（图 3-94A），部分燧石中含有一些粉晶白云石晶体，表现为斑点状（图 3-94C）。有些燧石具粉晶结构（图 3-94D），而有些则为隐晶质结构（图 3-94E 和 F）。碳酸盐岩发生燧石过程中往往不破坏母岩的结构，因此在研究区常见燧石化的鲕粒白云岩及水平层理发育的燧石角砾（图 3-94G）。鲕粒白云岩角砾中鲕粒圈层结构明显，填隙物普遍发生燧石化，鲕粒普遍为白云质；水平层理发育的燧石角砾是水平层理泥晶白云石被燧石交代的结果。由于交代不彻底，形成了白色燧石纹层与残余泥晶白云石纹层交互的现象（图 3-94H）。

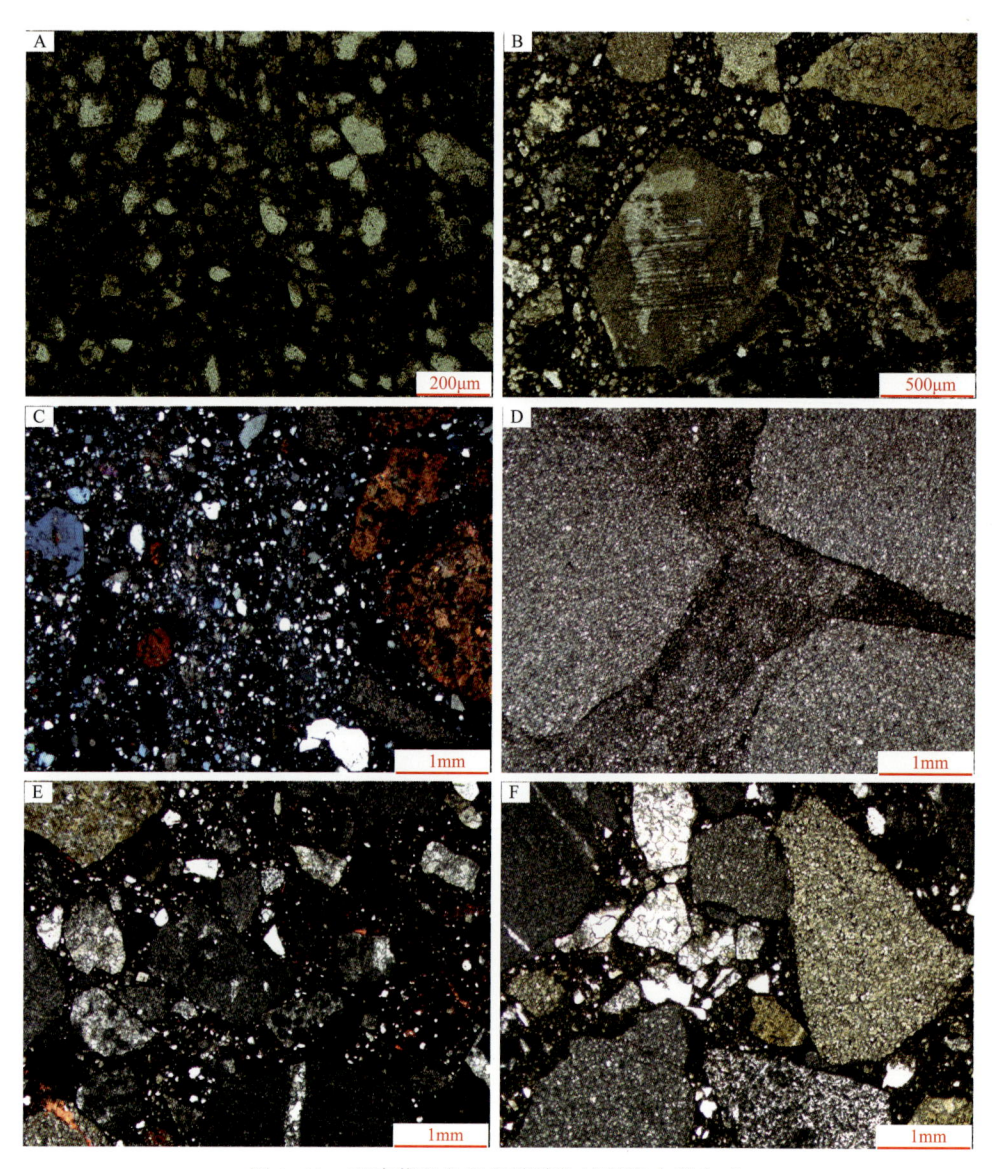

图 3-92 天津蓟县角砾岩填隙物特征及支撑方式

Fig.3-92 Characteristics of fillings and supporting type of breccia of the Xijingyu Formation in Jizhou, Tianjin

（A）细粉砂级别的白云石填隙物，单偏光，样品 X-4；（B）白云岩角砾岩，填隙物为白云石泥和粉砂，少量黏土，单偏光，样品 X-6；（C）石灰岩角砾（染红色）和粒径不一的石英颗粒填隙物，正交光，样品 X-6；（D）白云岩角砾间的白云石泥基质，单偏光，样品 X-15；（E）角砾岩的暗色白云石泥和粉砂基质，含少量灰质，单偏光，样品 X-12；（F）颗粒支撑结构的白云岩角砾岩，基质为暗色白云石泥和粉砂，单偏光，样品 X-14

西井峪组角砾岩中见有极少的火成岩角砾，其类型单一，为辉绿岩角砾，具有典型的辉绿结构（图 3-94I），其主要来源于下伏中—新元古界的辉绿岩侵入体。

2. 讨论

1）角砾岩的物源

根据成分分析，西井峪组角砾岩的母岩是其之下的中—新元古界。角砾岩成分主要为碳酸盐岩，基本不含碎屑岩，这主要是由于其下伏地层中—新元古界巨厚层状的碳酸盐岩地层决定的。虽然中—新元古界普遍为碳酸盐岩，但不同地层形成的岩石特征也十分明显，对比角砾岩中角砾的类型与下伏各组地层岩石特征，可为角砾岩的物源提供依据。

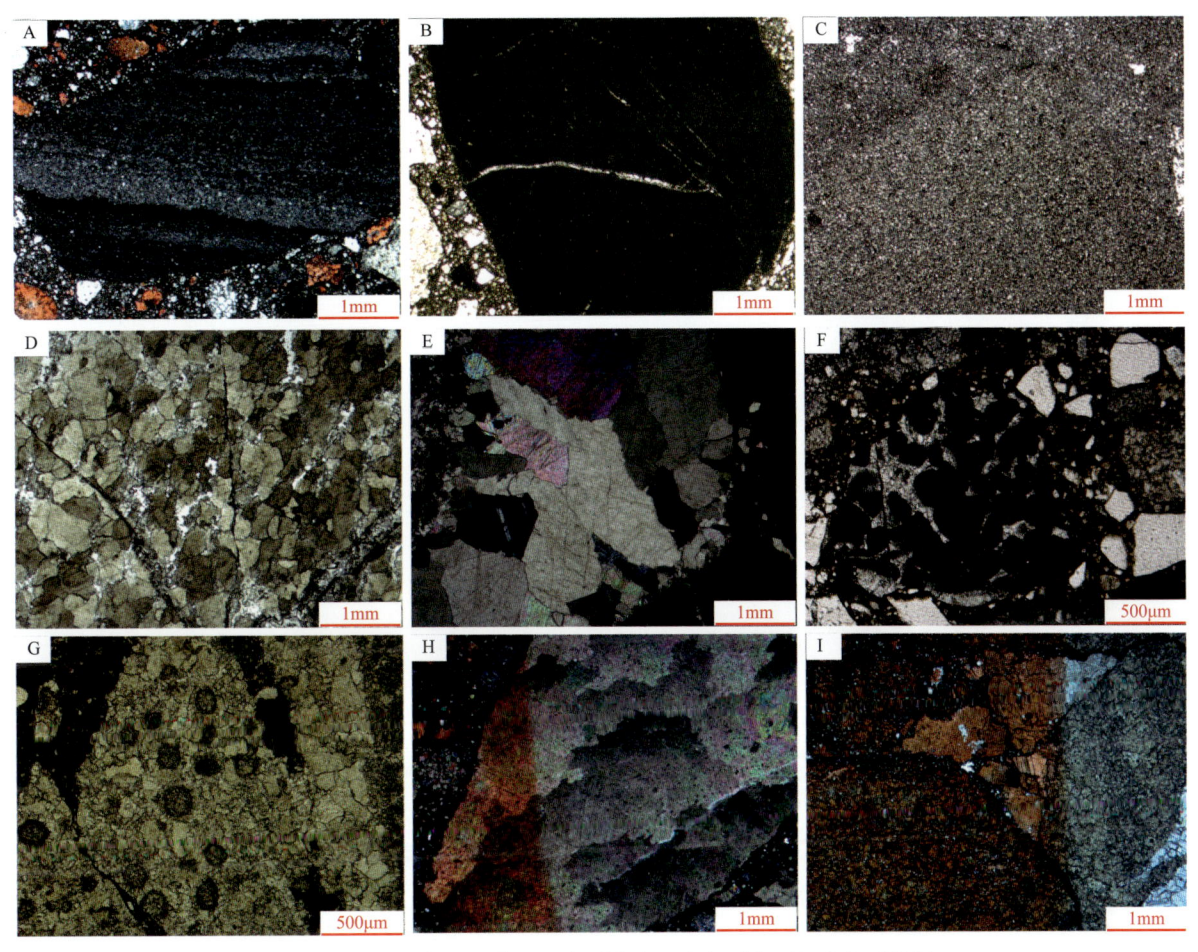

图 3-93 天津蓟州西井峪组角砾岩白云岩角砾和石灰岩角砾特征

Fig.3-93 Characteristics of dolomite breccia and limestone breccia in the Xijingyu Formation, Jizhou, Tianjin

（A）水平层理发育的泥晶白云岩角砾，单偏光，样品 X-6；（B）泥晶白云岩角砾，单偏光，样品 X-6；（C）粉晶白云岩角砾，单偏光，样品 X-19；（D）中细晶白云岩角砾，单偏光，样品 X-21；（E）粗晶白云岩角砾，正交光，样品 X-8；（F）砂屑白云岩角砾，单偏光，样品 X-1；（G）鲕粒白云岩角砾，单偏光，样品 X-40；（H）粗晶石灰岩角砾（染红色），正交光，样品 X-0；（I）粉晶石灰岩角砾（染红色），单偏光，样品 X-1

新元古界青白口系下龙山组以灰绿色海绿石泥砂岩沉积为主，龙山组以海绿石石英砂岩和页岩沉积为主。西井峪组与青白口系直接接触，但角砾中除了发育少量的灰泥石灰岩角砾，基本不含有石英砂岩岩屑。灰泥石灰岩角砾呈灰色，而不是灰绿色，这说明青白口系不是角砾岩的物源。

中元古界待建系下马岭组是一套以灰绿色页岩为主的地层，底部有少量砂岩，但通过野外观察及镜下鉴定，西井峪组角砾岩不含页岩角砾，甚至角砾岩的填隙物也以云泥为主，黏土矿物极少。这些特征表明，待建系下马岭组不是砾岩的物源。

中元古界蓟县系铁岭组以石灰岩为主，且发育多种类型的叠层，而西井峪组角砾岩中见到一些直径在 1m 左右的巨大的叠层石灰岩角砾，叠层石的特征与铁岭组的一致，说明铁岭组是西井峪组角砾岩的物源之一，但不是主要物源，因为角砾岩的绝大部分角砾是白云岩，而且铁岭组中燧石条带仅偶见。

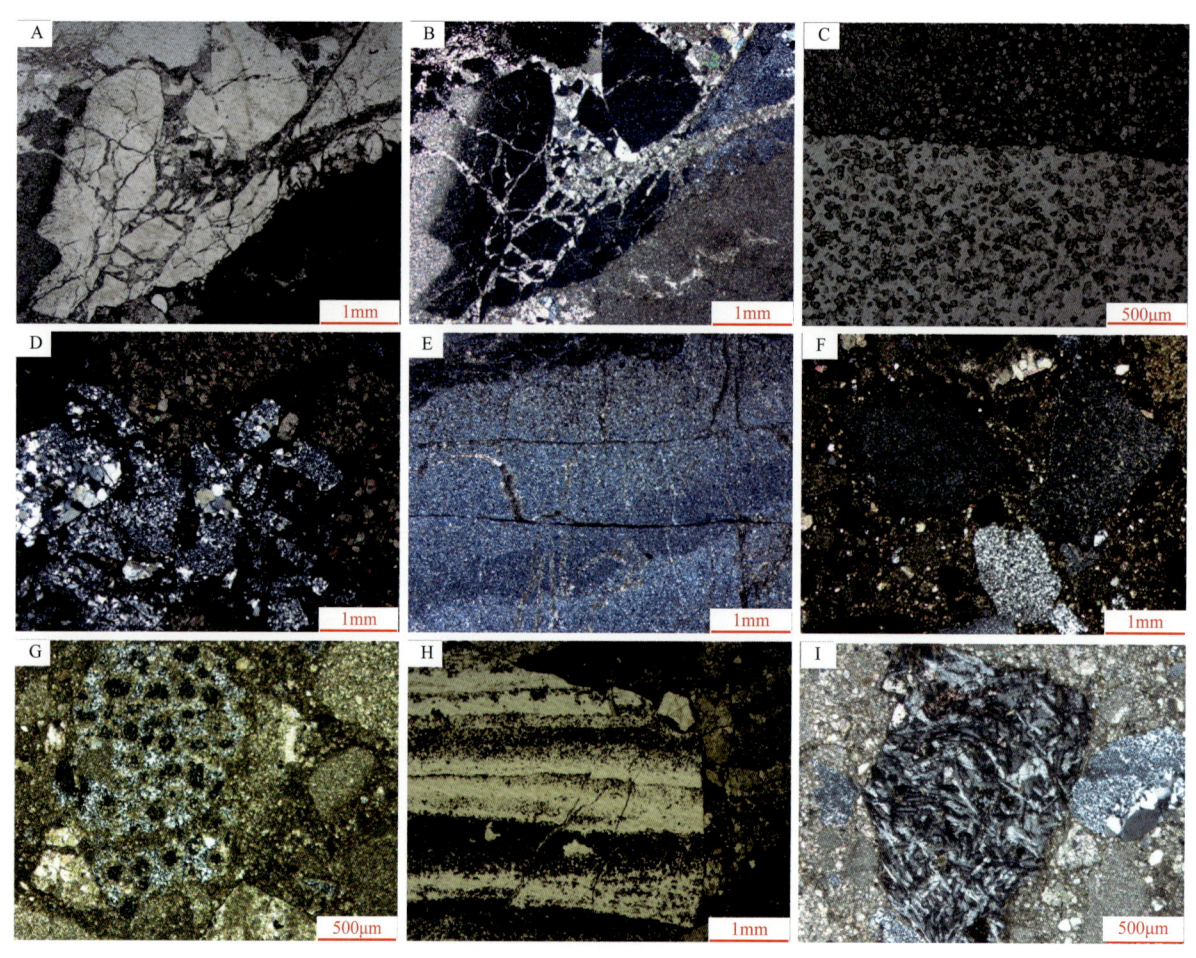

图 3-94 天津蓟州西井峪组燧石及火成岩角砾特征

Fig.3-94 Characteristics of chert and pyrolith breccia in theXijingyu Formation, Jizhou, Tianjin

（A）破碎的燧石角砾，单偏光下为白色，单偏光，样品 X-30；（B）破碎的燧石角砾，正交光下为隐晶质，与 A 为同一视域，正交光，样品 X-30；（C）燧石角砾表明漂浮粉晶白云石颗粒，单偏光，样品 X-13；（D）小米粒结构的燧石，正交光，样品 X-1；（E）隐晶质的燧石角砾，正交光，样品 X-23；（F）隐晶质与小米粒结构的燧石对比，正交光，样品 X-2；（G）燧石化的云质鲕粒白云岩角砾，正交光，样品 X-29；（H）水平层理泥晶白云岩燧石化的角砾，单偏光，样品 X-40；（I）辉绿岩角砾，正交光，样品 X-3

中元古界蓟县系洪水庄组以灰绿色和黑色页岩为主，显然不是西井峪组角砾岩的物源。

中元古界蓟县系雾迷山组发育数千米巨厚的燧石条带白云岩，而且白云岩从泥晶到粗晶都有，包括鲕粒白云岩、砂屑白云岩等，局部层位偶见少量砂岩。西井峪组角砾岩的白云岩角砾及燧石角砾的特征与雾迷山组的燧石条带白云岩相似，因此雾迷山组是重要物源之一。

中元古界蓟县系杨庄组主要为红色泥质泥晶白云岩夹浅灰色泥质泥晶白云岩，但西井峪组角砾岩中罕见这样的角砾，因此该组成为主要物源的可能性很小。

中元古界长城系高于庄组下部和上部发育燧石条带白云岩，中部为石灰岩和灰质白云岩，岩性特征与西井峪组角砾岩的相似，但由于高于庄组在杨庄组之下，而西井峪组角砾岩中又缺乏来自杨庄组的角砾，因此推测高于庄组成为主要物源的可能性很小。

长城系大红峪组下部为石英砂岩，中部为火山角砾岩和火山岩，上部为燧石条带白云岩；团山子组下部为白云岩，中部为白云岩夹砂岩，上部为紫红色白云质泥岩夹砂岩；串岭沟组为大套页岩；常州沟组为大套砂岩。西井峪组角砾岩中既无砂岩角砾，也无火山岩角砾，而且这些组均在杨庄组之下，因此都不可能成为其物源。

综上所述，西井峪组角砾岩的白云岩和燧石角砾主要来自中元古界蓟县系雾迷山组的燧石条带白云岩，石灰岩角砾来自铁岭组的石灰岩。这意味着物源区出露的地层主要是雾迷山组和铁岭组。由于不整合为平行不整合，说明物源区没有褶皱，主要是受断层控制的大幅抬升。物源区沟谷纵横，切穿了铁岭组，切入了雾迷山组。

2）角砾岩的成因

关于西井峪组角砾岩的成因，有两种可能，即冰川成因和泥石流成因。笔者认为，本区的角砾岩不是泥石流沉积，而是冰川沉积，是冰碛岩。

泥石流沉积的砾岩虽然与冰碛岩相似，如砾石杂乱排列，分选差，磨圆差，但本区的一些证据表明西井峪组的角砾岩不是泥石流沉积。证据如下：

（1）填隙物不是泥质。泥石流沉积可以形成砾岩，但其填隙物通常为泥质。泥质与水混合后形成"泥浆"，其润滑作用，使泥石流在重力作用下向前流动。但本区角砾岩的填隙物主要为白云质泥和粉砂。这种基质靠碳酸盐岩风化是难以形成的，因为风化到这么细的程度，早就溶解掉了。这种基质只能靠碳酸盐岩磨蚀，而冰川流动可以将碳酸盐岩基岩磨蚀成白云质的泥和粉砂。

（2）所有砾石的磨圆度都很差。泥石流沉积虽然磨圆差，多呈棱角或次棱角状，但也常见磨圆较好的砾石，这些磨圆好的砾石是在山区河流中磨圆的。而本区的角砾岩中就没有磨圆好的砾石（如次圆状）。

（3）无递变层理。泥石流沉积中常见逆粒序层理。而本区厚达155m的角砾岩全为块状，没有任何逆粒序层理。

（4）无清晰层面。一期泥石流沉积的厚度多为几米。不同期的泥石流沉积叠置时，会产生明显的层面。而本区厚达155m的角砾岩全为块状，内部找不到层面，即使不同粒度的角砾岩接触处，根本看不到清晰的层面。

（5）未与其他水成沉积物互层共生。泥石流沉积常与洪积扇或其他水流沉积互层共生。但本区达155m的地层全部为杂乱排列的角砾岩。

（6）坡度太小。在角砾岩沉积前，本区的地形应是十分平坦的，地层为平行不整合。在北京西山一带，景儿峪组与府君山组之间也为平行不整合，间接佐证了此观点。而且，西井峪组之下、之上的地层均为浅水碳酸盐岩地沉积，不存在陡斜坡。在平坦的地形上泥石流流动困难，很难形成巨厚沉积。

根据以上证据，笔者认为本区的角砾岩不是泥石流沉积。认为是冰川沉积的依据如下：

（1）岩石类型为角砾岩，角砾大小混杂（大者直径可达1m左右），杂乱排列，分选很差，磨圆很差，填隙物为白云质的泥和粉砂。这些特征与冰川沉积符合。角砾主要是物理风化和冰川侵蚀形成的，并被冻结在冰川中搬运。冰川消融后，大大小小的角砾被释放、堆积下来，导致角砾大小混杂，杂乱排列，分选很差，磨圆很差。作为填隙物的基质主要是冻结在冰川底部的角砾与基岩磨蚀形成的。

（2）整体呈块状，155m厚的地层无层理，无层面。虽然冰川可以多期沉积，但各期沉积之间不会有层面，这与正常的水成、风成或事件沉积是显著不同的。但由于母岩区被侵蚀的地层可以随时间而变化，因此各期沉积的粒度和成分可以有差异。西井峪组下部的角砾岩中石灰岩角砾较多，可能是母岩区有较多的铁岭组出露。随着铁岭组被剥蚀殆尽，石灰岩角砾基本消失。

（3）本次未发现砾石表面的冰川擦痕，但这并不奇怪，因为大部分砾石不会有擦痕。只有冻结在冰川底部的砾石表面会有，而其数量很少，不容易发现。而且角砾成分以碳酸盐岩为主，容易受到风化作用的影响，因此冰川擦痕难以保存。

总之，从角砾岩的结构、构造和岩石共生组合看，西井峪组的角砾岩为典型的冰碛岩。

这些冰碛岩的发现对恢复新元古代地球气候特征具有重要意义，因为据古地磁资料，在该时期华北板块处于低纬度热带、亚热带地区，景儿峪组和府君山组大量碳酸盐岩沉积也说明了这一点。但低纬度地区为什么会出现冰碛岩，或许真与"雪球地球"有关。

第四章 野外地质工作基本技能和方法

第一节 野外地质工作的基本原则

由于研究目的、时间安排、地层出露条件、人员安排、天气状况等方面的差异，野外地质工作的详尽程度不一。但是，无论是区域地质调查，还是剖面的实测和系统采样，都是对观察到的原始地质现象进行科学性的记录，因而必须遵守"系统、客观、细致"的原则。

野外地质记录是最为宝贵的第一手资料，也是后期进行研究和分析的重要依据之一，因此必须注意以下几点：（1）野外记录的内容必须系统、全面，记录的内容主要包括观察到的地质现象、联想到的地质问题、做出的判断和分析、观察点位置及标志性地物、天气等一系列信息，系统和全面的记录方便于后期室内的及时回忆、整理和分析；（2）野外记录的内容必须客观、真实，如实反映所观察的地质现象，不能凭主观随意夸大、缩小或扭曲客观的地质情况；（3）野外记录的内容应该图文并茂，尤其对一些很难通过简短文字加以说明的特殊地质现象（如断层、褶皱、沉积构造、化石、岩性变化、沉积体侧向变化、接触关系和典型相标志等），可以通过精炼、细致的手绘图或照片来进行生动、全面而详细的记录；（4）野外记录的内容应该简洁美观、条理清晰，记录的简洁性和条理性是衡量记录质量的重要标准，是避免后期整理、分析出现差错和争议的前提。

第二节 野外地质工作常用物品及其使用方法

充分的前期准备是野外地质工作顺利开展的重要保障，筹备充足的野外常用物品并熟知它们的使用方法至关重要。野外地质工作的常用物品主要如表4-1所示，下面就部分常用物品的用途和使用方法进行详细介绍。

表4-1 野外地质工作常用物品
Table 4-1 Common Equipment for field geological work

野外常用物品	用途	备注
地质包	装纳和拌扣野外地质物品和样品等	面料结实、耐磨且轻便（如帆布），多收纳口袋、多挂扣
地质锤	采集样品、用作比例尺等	按研究地层的岩性差异，选择扁头（沉积岩）或尖头（岩浆岩和变质岩）标准地质锤
手持放大镜	放大观察岩石	×10放大倍数为最佳
地质罗盘	测量方位、倾角等	需要预先设置研究区磁偏角，天津蓟州区约为6°53′（W）
卷尺	实测剖面长度	一般长度以50 m为佳
野外记录簿	记录	硬皮、防水、页面带网格、尺寸不小于180mm×220mm（方便野外露头素描）为最佳

续表

野外常用物品	用途	备注
袖珍折叠刀	测试矿物硬度、清理植物、水果除皮等	配备刀片、剪刀、螺丝刀、开瓶器等多种工具的折叠式瑞士军刀最佳
稀盐酸	鉴别白云石和方解石	稀盐酸浓度一般为 10%
笔	记录、标记或用作比例尺	记录一般用铅笔，样品标注一般用记号笔
直尺	辅助绘图和做比例尺	一般长度不宜超过 20cm
样品袋	装样品	不同尺寸的、结实的、耐磨的布袋为佳
手持 GPS 仪	记录位置信息	无
数码相机	采集照片	无
地质图、地形图等相关资料	定点、定方位	搜集地质图件相关网站或书籍、软件（谷歌、百度、高德地图等）获取或向地质资料服务部门购买
指导书	指导野外地质工作	无
衣物、鞋子、帽子	保暖、防滑、遮阳	最好穿戴透气的长袖上衣和长裤、专业登山鞋及遮阳帽
药物	急救和防护	常见感冒药、消炎药、降暑药、防晒霜、驱蚊剂、驱蛇剂等

一、地质罗盘

地质罗盘，又称为"袖珍经纬仪"，是野外地质工作必不可少的工具之一。地质罗盘虽然种类繁多，但是基本构造相似，一般由指向磁针、水准器、刻度盘及瞄准设备等，大致如图 4-1 所示。地质罗盘的工作原理是：具有磁性的指向磁针靠顶针支撑于罗盘中心，且可以自由转动，当罗盘水平且磁针停止转动时，磁针所指示方向即为磁子午线的方向，因而可以确定目标相对于磁子午线的方位角。

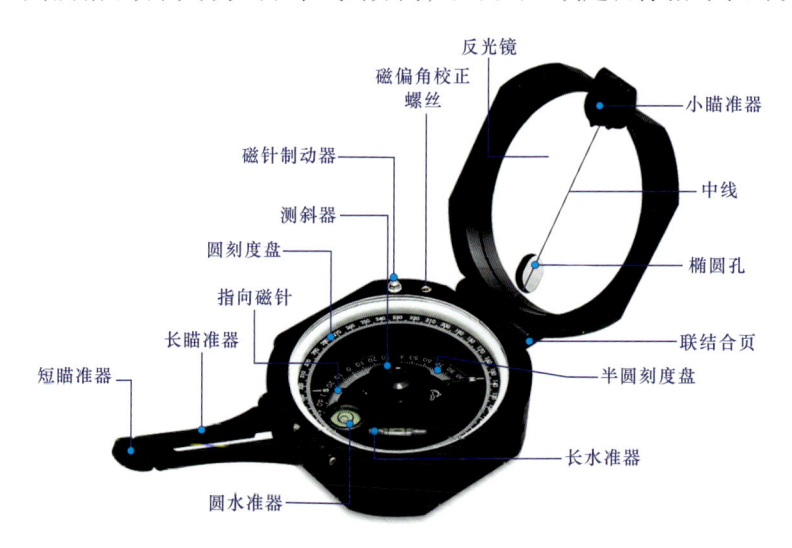

图 4-1 地质罗盘的构造

Fig.4-1 Details of geological compass

1. 地质罗盘的功能

地质罗盘主要是用于测量所观察的点、线或面的空间方位（例如点的方位，褶皱轴部、线理和波痕等的延伸方向，岩层面、断层面、节理面等的产状）。地质罗盘主要包括两套测量系统，即由圆水准器、圆刻度盘和指向磁针组成的水平测量系统和由长水准器、半圆刻度盘和测斜器组成的垂直测量系统；前者用于测量走向、倾向等水平方位，后者用于测量倾角、坡角等。

2. 地质罗盘的使用方法

1）校正地磁偏角

由于地球的地磁南北极（磁子午线）与地理南北极（真子午线）不重合，因而为了获取所测对象的真正方位，在野外地质工作开展前必须预先对罗盘进行地磁偏角校正。通常，人们将地磁偏角定义为地磁北极与地理北极间的交角，且以地磁北极向东偏移为正，地磁北极向西偏移为负。我国大部分地区的地磁偏角都是向西偏，只有少数地区（如新疆等）是向东的。地磁偏角的校正方法如图4-2所示，如果地磁偏角为西偏时，校正时用罗盘配备的小钥匙转动罗盘侧面的地磁偏角校正螺丝，使圆刻度盘沿逆时针方向转动相应的偏转角度；相反，东偏时，使圆刻度盘顺时针转动相应的偏转角度。天津市蓟州区的地磁偏角约为西偏7°17′，因而需要使圆刻度盘沿逆时针方向旋转7°17′即可（图4-2C）。

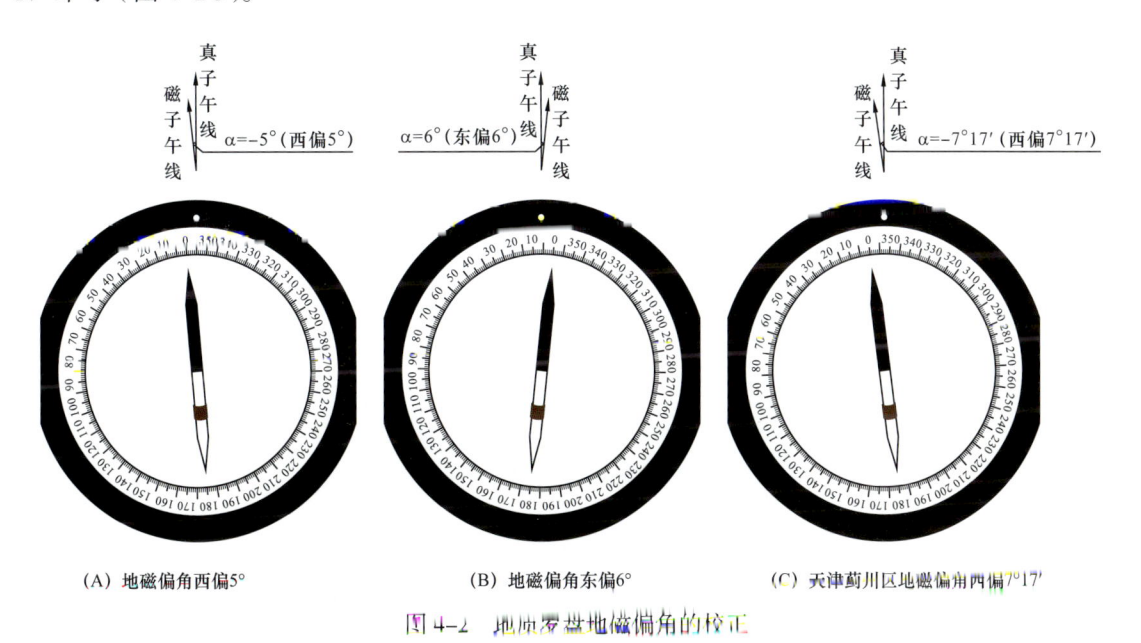

（A）地磁偏角西偏5°　　　（B）地磁偏角东偏6°　　　（C）天津蓟川区地磁偏角西偏7°17′

图4-2 地质罗盘地磁偏角的校正

Fig.4-2 Calibration of magnetic declination for geological compass

2）测量方位角

方位角测量的关键是保证测量者、罗盘和目标物保持在一条线上，即"三点一线"。由于目标物和测量者的相对位置的限制，一般采用以下两种方式进行测量：（1）当目标物高程较测量者高或略低时，可将罗盘置于胸前，保持罗盘圆水准器气泡居中（即保持水平），将长瞄准器适度抬起，并对准目标物，调节反光镜，使长瞄准器的短瞄准器和目标物同时出现在反光镜中，调整方位，使得短瞄准器尖端和目标物中轴线与反光镜中线重合，稳定姿势，并松动磁针制动器，待磁针稳定，方可按下磁针制动器并读数，此时指北针读数即为目标物相对测量者的方位（图4-3A）；（2）当目标物高程明显较测量者低时，在罗盘水平情况下，目标物和瞄准器很难同时出现在反光镜中，因而可将罗盘转动180°，使得反光镜面朝测量者，并保持圆水准器气泡居中，适度调节方位，

— 117 —

使得长瞄准器、反光镜的椭圆孔和目标物的中轴线位于同一直线，松动磁针制动器，待磁针稳定，方可按下磁针制动器，此时指南针读数即为目标物相对于测量者的方位（图4–3B）。

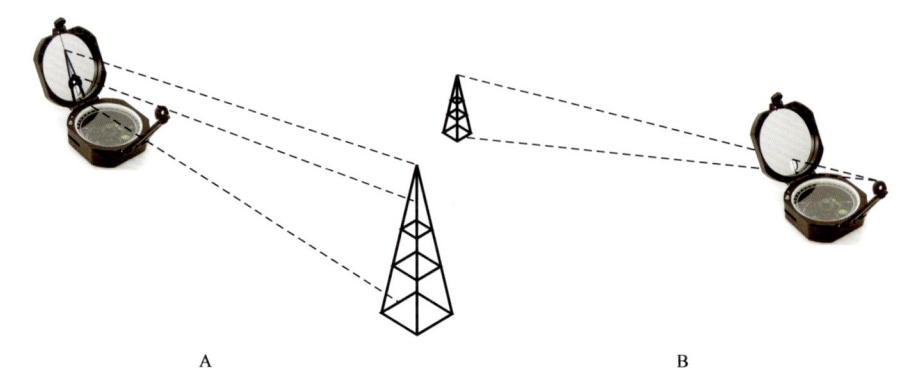

图4-3　利用地质罗盘测量方位角

Fig.4-3　Azimuth measurement by geological compass

3）测量线状和面状构造的产状

（1）线状构造（如波痕波脊、褶皱轴部和线理等）的产状主要包括倾伏向和倾伏角。其中，线状构造的倾伏向为其在水平面上的垂直投影的沿下倾方向的方位，可以通过将罗盘的长瞄准器与线状构造的平面垂直投影的下倾方向重合，并调节圆水准器的气泡居中，磁针自由摆动至稳定时，指北针对应的圆刻度盘读数即为其倾伏向；线状构造的倾伏角为线状构造与其在水平面上的垂直投影间的夹角，可以通过将罗盘长边贴在线状构造上，并调节长水准器使气泡居中，此时半圆刻度盘的读数即为其倾伏角大小。具体的线状构造的产状测量实例如图4-4中所示的波痕产状的测量。

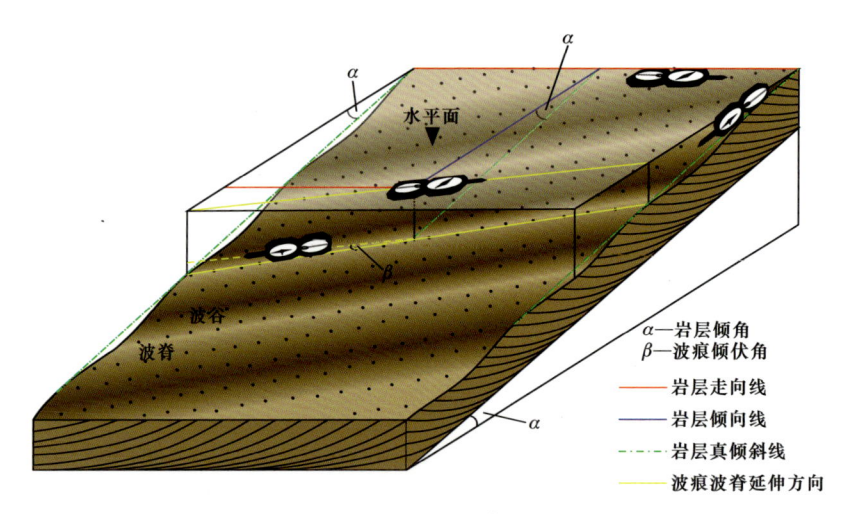

图4-4　利用地质罗盘测量岩层产状及其层面上波痕产状

Fig.4-4　Occurrences measurement of strata and ripple marks on its bedding by geological compass

（2）面状构造（如岩层面、断层面和节理面等）的产状主要包括走向、倾向和倾角。其中，面状构造的走向为其与任意水平面的交线的延伸方向，可以通过将罗盘的长边紧贴在测量面上（可将记录簿平铺于测量面上形成平面），并调节罗盘角度使得圆水准器的气泡居中，磁针自由摆动至稳定时，圆刻度盘读数即为其走向，可以用两个相差180°的方位角进行表示。倾向一般与走向垂直，且指向测量面的倾斜方向，可通过走向计算获得，同时也可以通过地质罗盘直接测量得到，具体测量方法如下：将罗盘盖或者短边紧贴在测量面的顶面或底面之上，并调节罗盘角度，使得圆水准器的气泡居中，磁针自由摆动至稳定，此时若紧贴的为测量面的顶面，则指北针在圆刻度盘上的读数即为其倾向；若紧贴的为测量面的底面，则指南针在圆刻度盘上的读数即为其倾向，注意倾向与走向不同，只有唯一的方向。面状构造的倾角是在测量倾向后，直接将罗盘垂直转动至长边紧贴测量面，并转动测斜手柄至长水准器的气泡居中，此时半圆刻度盘对应的读数即为其倾角。具体的面状构造的产状测量的实例如图 4-4 中所示的岩层面产状的测量。

4）测量坡角

在野外地质工作中（尤其是剖面实测）常涉及地形坡度的测量，利用地质罗盘测量坡角的方法是：一般是由身高相同的前后测手各手持一个罗盘，或在可忽略测量者身高影响时，也可由一个测量者单独完成。首先，将长瞄准器调整至与罗盘盘面平齐，并将前端的短瞄准器扳至垂直，使罗盘盖与盘面呈约 45°；然后，将罗盘垂直，并通过短瞄准器的小孔和反光镜下方的椭圆孔仰视或俯视观察另一测手或目标物，调整测斜器，直至长水准器的气泡居中，此时半圆刻度盘对应的读数即为坡度。通常，记录上坡时坡角为"＋"，下坡时坡角为"－"（图 4-5）。

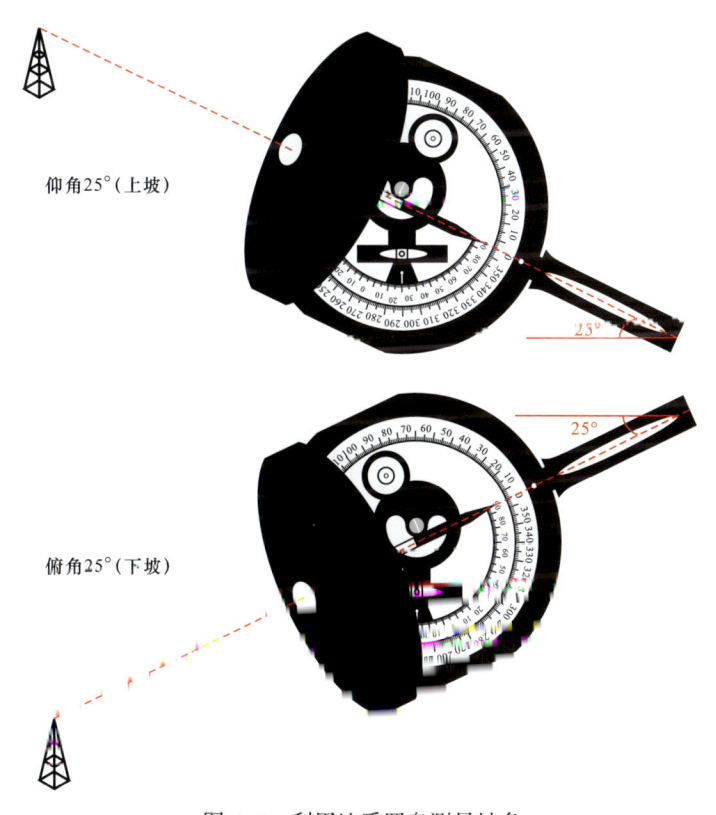

图 4-5　利用地质罗盘测量坡角

Fig.4-5　Slope angle measurement by geological compass

地质罗盘除了以上功能外，还有很多其他用途，例如通过测量相对于地图上两个已标记物之间的相对方位，进而确定测量者的位置。当然，随着目前智能手机功能的日益强大，手机完全可以替代地质罗盘的部分功能，但手机可能存在信号弱或电量不足等潜在的缺陷，因而，虽然地质罗盘操作较为复杂，但仍是可靠性很高的野外地质工具。

二、手持放大镜

手持放大镜是野外地质工作中不可或缺的工具之一，一般用于观察岩石的矿物成分、颗粒类型和大小、生物化石等，一般放大 10 倍类型的放大镜基本满足野外地质工作的需求。手持放大镜的正确使用方法是：左手拿捏需要观察的手标本，右手的大拇指和食指夹持放大镜，右手中指弯曲，无名指轻压于手标本表面上，放大镜靠近眼睛，并同时调节左右手及放大镜与手标本距离，直至看到稳定、清晰的放大现象为止（图 4-6）。

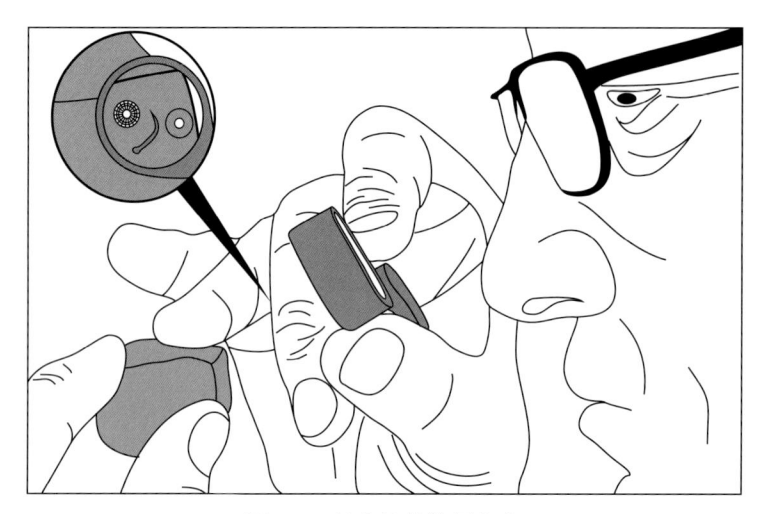

图 4-6 放大镜的使用方法

Fig.4-6 Usage of magnifying glass

三、野外记录簿

野外记录簿是野外地质工作中记录地质现象观察结果的最重要载体，其重要性不言而喻，在这里我们推荐使用尺寸不小于 180mm×220mm（方便野外露头素描）、页面带有网格的硬皮野外记录簿，且有防水功能最佳。野外记录的系统、客观和细致直接关系到野外地质成果的质量，因此野外记录簿的使用也有一定的规范。

野外记录簿一般分为四个功能区（图 4-7）：

（1）页眉区：位于页面的顶部，主要用于记录工作日期、星期及天气情况。

（2）左批注栏：位于页面左侧，由竖线间隔，主要用于编写野外记录的题录（如线路、参与人员、考察目的、点号、点位、位置、GPS 点位、点性、现象描述、其他及问题和想法等），以便记录条理清晰。

（3）文字记录栏：位于页面中部，主要用于记录各题录下的详细内容，或者简略的示意图，是地质记录最为重要的部分。记录的主要内容可以大致总结为以下几个方面：① 考察点的位置信息；② 地质现象的描述（详见后续章节有关地质现象描述的内容）；③ 样品和照片采集情况、目的（其他题录下可以记录采样目的和预计的实验分析计划等）；④ 观察时的结论、问题和想法。

（4）右批注栏：位于页面右侧，由竖线间隔，主要用于标注、修正和补充记录。

	2019 年 04 月 28 日　星期日　天气晴
线路	天津市蓟州区下营镇大红峪沟—罗庄子镇翟庄村北高于庄组（Jxg）考察路线
参与人员	×××、××、×××、××、×××、××
考察目的	（1）观察中元古界蓟县系高于庄组（Jxg）的岩性特征及地层序列； （2）观察古元古界长城系大红峪组（Chd）与中元古界蓟县系高于庄组（Jxg）的平行不整合接触关系； （3）观察中元古界蓟县系高于庄组（Jxg）中发育的叠层石、圆脊波痕、瘤状和臼齿等典型构造。
点号	No.1
位置	营树路岔口处长城系大红峪组与蓟县系高于庄组分界牌处
GPS点位	E 117.476749°；N 40.164697°
点性	古元古界长城系大红峪组与中元古界蓟县系高于庄组地层界线点
现象描述	该点为大红峪组（Chd）与高于庄组（Jxg）的分界，即中元古界长城系与蓟县系的分界，两组之间为平行不整合接触关系。分界面之上为高于庄组底部的厚约3 m的中—厚层含长石石英中砂岩（样品Jxg-1），砂岩表面发育对称浪成波痕、平行层理，波痕较为连续，波高约3cm，波脊的延伸方向大致为290°，砂岩层的产状为：205°∠38°；分界面之下为厚层大红峪组含锥状叠层石硅质白云岩（样品Chd-25），且叠层石多被硅化。界线处大型的锥状叠层石 Conophyton daoyuense 的顶部被高于庄组底部砂岩截切的现象（照片P003），局部出现铝土矿风化壳代表高于庄组沉积前该处曾遭受暴露，并发生侵蚀作用。
其他	样品Chd-25主要观察溶蚀孔洞发育情况，样品Jxg-1主要观察砂岩成分和结构特征，为沉积环境的确定提供依据。
问题和想法	大红峪组由下至上从火山岩和火山碎屑岩沉积逐渐过渡为叠层石白云岩，代表构造背景由强烈断陷、火山活动强烈，逐渐过渡为稳定的浅水碳酸盐潮坪；高于庄组底部砂岩层表面发育对称浪成波痕，表明其形成于浅水碎屑岩滨岸环境。高于庄组碎屑岩的物源有待研究，且两组之间平行不整合与构造活动的对应关系如何？构造活动对其物源是否存在控制作用？

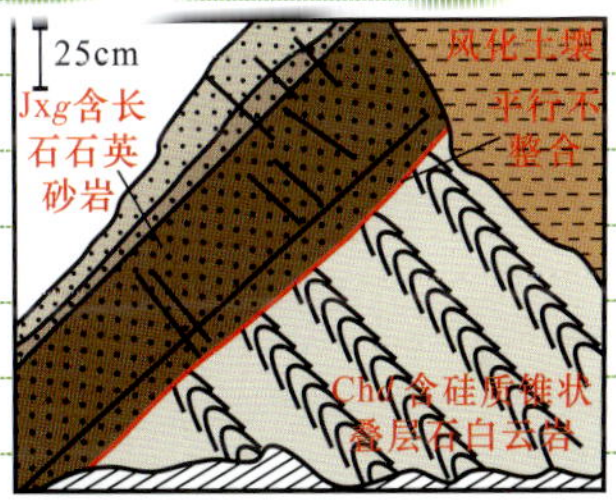

图 4-7　野外地质记录实例

Fig.4-7　Example of field geological record

除了上述页外，一般野外记录簿都配有网格纸页，该页主要是便于在野外绘制相关图件，以配合和补充文字的说明，进而更加全面、客观、形象地反映所观察的地质现象。野外记录过程中要求使用2H型号的铅笔书写，在记录前应仔细观察地质现象，观察、测量和记录同时进行，少记部分应该在观察后及时补充，尽量避免涂改，保持页面整洁、清楚。

四、手持GPS仪

由于GPS（Global Position System）卫星定位系统可以向全球提供实时的三维位置信息，且GPS仪具有高精密性、高精度及操作简便等特征，手持GPS仪在野外地质工作中已经得到普遍的应用。

1. 手持 GPS 仪的功能

手持 GPS 仪在野外地质工作中主要用于记录观测点的空间位置（经纬度和海拔），记录考察路线，计算路线距离、速度和时间等参数，以及导航功能。

2. 手持 GPS 仪的使用方法

目前市面上的手持 GPS 仪种类繁多，但使用方法基本一致，本书以我国台湾 Garmin 生产的 GPSMAP_63sc（图 4-8）为例进行介绍。参见该设备的使用手册，具体使用方法如下：

图 4-8　手持 GPS 的构造（以 GARMIN 的 GPSMAP_63sc 为例，图片来自 GARMIN 公司官网）

Fig.4-8　Details of handheld GPS device（taking GPSMAP_63sc by GARMIN for example and figures from GARMIN.com）

（1）开机：使用前须提前打开设备（按压 "⏻" 键），并搜索卫星信号，直至搜索到的卫星数量超过 4 个。

（2）记录观测点的空间位置：该功能即航点（Waypoint）记录功能，通过点击 "存点" 键（或 "MARK" 键），并按需要编辑数据的名称等参数，然后选择 "确定"（或 "Done"），此时航点信息即以 .gpx 格式保存在设备中。

（3）记录考察路线：该功能即航迹（Track）记录功能，且航迹由航点组成，包含各个航点的时间、位置、海拔等信息。具体的操作方法是：点击 "菜单" 键（或 "MENU" 键）进入主菜单页面，选择 "设置"（或 "Setup"）→ "航迹"（或 "Track"），进入航迹记录功能的设置界面。然后，分别通过设置 "航迹记录"（或 "Track Log"）来开启或关闭航迹功能，通过设置 "记录模式"（或 "Record Method"）来选择航迹的记录方式，通过设置 "间隔设置"（或 "Record Interval"）选择按照一定的距离或时间间隔进行记录，通过设置 "自动存档"（或 "Auto Archive"）来选择自动存档的时间（每天或每周等），及通过设置 "颜色"（或 "Color"）来选择航迹记录线条的颜色等。各参数的设置应依据考察目的、个人喜好等自由选择，设置完成后，设备将自动记录航迹，航迹记录可以通过主菜单页面的 "航迹管理" 查看航迹在地图中的显示及其高度剖面图等。

（4）导航：该功能即航线（Route）功能，可以通过创建航线或者查找先存的航点或航迹来进行导航。创建航线的操作方法是：点击"菜单"键（或"MENU"键）进入主菜单页面，选择"航线管理"（或"Route Planner"）→"建立航线"（或"Create Route"）→"选择起始点"（或"Select First Point"），选择航点类型并设置第一个航点为起始点，后选择"使用"（或"Use"），再选择"添加新航点"（或"Select Next Point"）添加航点至航线，按同样步骤完成所有航点的添加，点击"退出"键（或"QUIT"键），即可完成航线的创建。导航功能的使用方法是：点击"查找"键（或"FIND"键），并选择一个新创建的航线（或"Routes"）或先存的航点（或"Waypoints"）和航迹（或"Tracks"），选择"导航"（或"Go"）即可进行导航。

（5）数据导出：手持设备中记录的航点、航迹、航线数据可以通过软件（如 BaseCamp™ 软件、MapSource 软件）导入电脑中进行管理。具体的操作方法是：将设备通过数据线连接到电脑上，并打开软件，电脑会自动识别所连接的 Garmin 设备，并读取相关的信息，筛选数据，选择"文件"（或"File"）→"导出"（或"Export"）即可导出所筛选的数据。

除了以上常用功能外，目前市面上的手持 GPS 仪大多还具备电子罗盘、行程数据记录、面积计算、照相（含位置信息）及手电筒等功能，在此不再详细赘述。

3. 使用手持 GPS 仪的注意事项

（1）必须在户外空旷处使用，建筑物内、洞穴内、密林中无法使用；

（2）GPS 仪接收的卫星信号越多，数据精确度越高，通常在接收到 3 个或 3 个以上卫星信号时，即可获取观测点的大地坐标，在接收到 4 个或 4 个以上卫星信号时，即可获取观测点的海拔高度；

（3）手持 GPS 精度一般 3~5m；

（4）准备充足的电池。

除了以上介绍的地质罗盘、手持放大镜、野外记录簿、手持 GPS 仪外，还有很多其他的野外必备用品需要野外地质工作人员不断地熟悉和改进使用方法，从而保障大家拥有一个"安全、高效、愉快"的野外地质考察旅程。

第三节 野外地质工作的基本步骤及内容

野外地质工作对系统性和综合性要求很高，且具有一定风险性，因而在开展野外地质工作时必须遵循一定的工作步骤。虽然野外地质工作的目的不同，但是大致都可以划分为以下四个基本步骤：前期准备、野外地质调查、室内整理和分析及提交成果等。

一、前期准备

野外地质工作的前期准备阶段是整个工作任务顺利完成的重要保障，此阶段需充分做好出野外的知识、思想、组织及生活等方面的准备，重点需要注意以下几个方面。

1. 资料收集

通过对考察地区相关地质、地理、气候、交通和人文等方面的资料收集，可以为野外地质工作的计划制订和开展提供可靠的依据。特别是考察地区的前人地质调查报告、区域地层表、地质

- 123 -

图、地图、卫星图、地形图等可以为野外选点、踏勘及实测等工作计划的制定提供参考。

2. 任务分配

根据野外工作的性质、目的和工作量，合理组织野外工作的分工和合作，让所有参与人员明确野外工作的整体及个人任务、目的、要求和责任，并做好充分的业务准备。

3. 思想准备

野外工作具有一定的风险性，必须牢固树立安全意识，做好突发事件的应急预案。

4. 物品配备

野外情况十分复杂，且具有不可预测性，为了确保高效、高质地开展野外工作，必须配备充足的野外用品。常用的野外地质用品已在表4-1中列出，可供考察成员参考和补充。

二、野外地质调查

由于野外工作目的和精度的差异，开展野外地质调查的基本程序和内容也不尽相同。本书将主要按精细的沉积地层调查工作为例进行详细介绍，将野外地质调查分为野外踏勘和剖面实测两个基本程序。

1. 野外踏勘

对于初次开展工作的研究区，在前期资料收集的基础上，仍需要对其地层的分布和出露情况、接触关系、构造格局、地形特征及交通状况等进行实地踏勘。野外踏勘应充分结合前期收集的资料，对初步选定的研究目标的潜在分布区域进行实地勘查，并精心寻找和选择构造简单、地层发育齐全、界线清楚、出露完好的地段，进而确定需要详细实测的剖面，并确定大致的岩性段或地层旋回，为剖面实测奠定基础。

2. 剖面实测

1）剖面实测的目的

对于沉积地层而言，通过剖面实测可以详细和全面地了解研究层的岩性特征和地层序列，是后期沉积环境及古生物、古气候、古地理等研究的基础。

2）实测剖面的选择

实测剖面选择的一般原则是：构造简单、地层发育齐全、界线清楚、出露连续和完好，且具有代表性和典型意义的剖面。然而，在实际野外工作中很难找到完全满足以上要求的剖面，故有以下几点值得注意：

（1）选择实测剖面时，应充分考虑天然断层、河流等形成的切面及人工开凿的公路、铁路、壕沟及矿坑等地方；

（2）尽可能垂直岩层走向测量，一般剖面线与岩层走向夹角不小于60°；

（3）对于构造复杂区，在实测前应对构造情况进行充分认识，以免地层的重复或缺失；

（4）在剖面出露不连续时，可以通过多条临近小剖面的拼接完成剖面实测，但应该注意拼接地层的准确性，以免地层重复或缺失。

3）剖面实测的人员分工

剖面实测的具体工作内容通常包括地形、导线及岩层产状的测量，地层的观察、描述及小层划分、记录、野外绘图以及样品和照片的采集等，任务繁杂、工作量大，需要多名成员的分工协作、互相配合。一般而言，具体的人员分配视具体情况而定，但5~8人最佳，具体的人员分工可参考表4-2。

表4-2 野外剖面实测人员分工
Table 4-2 Labor-division of staff for field section measure

分工	任务	人数	备注
测手	拉测线，测量导线的斜距、方位角和坡度角等，并承担岩层产状的测量任务	2	前、后测手各1人，需身高相近
分层员	观察、描述岩层，划分并标记小层，读取分层斜距等，协调全组工作	1~2	该任务是剖面实测的中心工作，其他工作围绕其展开，一般由经验丰富、工作细致的人员承担
填表员	填写《实测剖面记录表》	1	需要与测手、分层员、采样员、照相员、绘图员等紧密配合，避免记录错误
记录员	配合分层员，在野外记录簿中详细记录每个小层的特征	1	具体描述内容参见后面章节
绘图员	现场绘制地层剖面草图、地层柱状图草图、特殊地质现象草图等，以便室内整理参考	1	绘图细节参见后面章节
照相员	对地质现象进行照片采集，并编号	1	照相注意事项参加后面章节
采样员	采集岩石、化石等样品，并编号	1	采样工作较为危险，采样时需时刻注意自身及周边人员安全

3. 剖面实测的步骤

在前期踏勘的基础上，挑选需要实测的剖面，并大致了解剖面地层的岩性段和地层旋回，进而开展详细的剖面实测，具体步骤如下。

1）拉导线、测量导线相关数据

该工作由前、后测手拉伸皮尺或测绳来完成，通常后测手手持皮尺或测绳的0m端，并确定剖面的起始点（通常向所测层位界线下延伸一定长度），而后侧手向剖面终点方向前进，直至地形起伏变化处、剖面拐弯处或皮尺及测线的末端处时停止，两人将皮尺和测绳拉直，将导线斜距（即导线长度）报告给填表员进行记录；然后，分别使用地质罗盘测量导线的方位角（后侧手指向前侧手的方向）和坡度角（上坡为"+"，下坡为"-"），并互相比对所测数据，直至方位角数据误差小于1°后，坡度角数据相对中间值，分别取平均值，报告给填表员，记入《实测剖面记录表》（见附录1）中。其后将导线放直地面，两段用石头压实，防止挪动，直至该导线所有的分层、描述、取样等工作结束后，方可挪动导线，进行下一导线的测量。

2）分层

实测剖面时需首先对剖面进行分层，即划分小层。小层划分得越细、越薄当然越好，但工作量会相应地增加。因此，小层的最小厚度需根据需要确定，一般说来，不小于0.5m。小层划分的方法包括：

（1）根据岩石类型划分。当岩层的厚度较大时（例如超过0.5m），一种岩石即可划分为一层。岩石类型可以是根据成分划分的，如砂岩层、泥岩层、石灰岩层等；可以是根据粒度划分的，如细砂岩层、粗砂岩层；可以是根据颜色划分的，如红色砂岩层、灰色砂岩层；可以是根据沉积构造划分的，如平行层理砂岩层、交错层理砂岩层；可以是根据化石类型或含量划分的；等等。

（2）根据岩石组合划分。当岩层很薄时（例如中层、薄层），可根据两种岩石组合划分，如砂岩—泥岩薄互层、砂岩夹泥岩层。

（3）根据沉积旋回划分。当旋回的厚度不大，为几十厘米或几米时，可根据沉积旋回划分，一个沉积旋回一层。如果厚度很大，为几十米或更厚时，一个旋回可划分为一段，而不是一层。

（4）根据侵蚀面、沉积间断面划分。侵蚀面、沉积间断面通常都作为小层的界限。

在确定划分出小层后，需用阿拉伯数字对小层编号，并需在界限处用红漆标出来，即沿界限画一直线。在红漆线上、下标上相应的层号，并将分层号及其层顶面对应的分层斜距报告给填表员。测手使用地质罗盘测量各小层的层面产状，并报告填表员进行记录。

3）描述和记录

在划分小层后，分层员和记录员相互协作需要对小层进行详细描述。

描述的内容根据需要确定，例如注重沉积相分析的、注重储层的、注重石油地质特征的等等。但无论哪种目的，描述的基本内容都应包括：小层的岩性、颜色（风化色、新鲜色）、层厚（块状层，层厚＞1m；厚层，0.5～1m；中层，0.1～0.5m；薄层，0.01～0.1m；纹层，＜0.01m）、结构、构造、沉积序列、化石、接触关系、岩体形态和横向厚度变化等。

此外，还要对各小层的沉积环境进行现场初步分析，并将观点和问题一同记录下来。不能等剖面实测工作结束回去后再分析。如果那样，想起应该观察哪些现象时已经晚了，也不能立刻进行弥补了。

注重储层时，尤其要描述孔隙和裂缝的发育情况，包括孔隙的类型、大小、形态、含量；裂缝的类型、长、宽、走向、密度、有几组、裂缝发育的岩石类型等。

4）样品和照片采集

对于需要进行实验分析及进一步分析时，需要对小层进行取样，并对样品进行系统编号，并读取对应的斜距，将取样编号、取样斜距以及取样目的等报告给填表员和记录员记录。取样过程中有以下几点值得注意：（1）应采集原地未风化的样品，规格视目的而定，一般标准规格为 3cm×6cm×9cm；（2）对样品进行编号，一般编号应能够反映剖面名称、地层及样品序号等信息（如 Jz-Chc-4 代表蓟州区长城系常州沟组第4块样品），并使用防水记号笔在标签纸、样品袋上详细标注，或者直接标记在样品表面，部分样品需要标记顶底和方位；（3）对于古生物样品，应用报纸或棉花等仔细包裹，以免破碎。

数码相机在野外地质工作中的广泛应用，使得照片成为野外露头其他数据的重要补充，照片可以简单、清楚地记录相关的地质现象，方便后期的回忆和整理。在照片采集过程中应该注意以下问题：（1）拍照时应该重点突出，添加适当的比例尺；（2）及时记录照片的编号，以防发生错乱；（3）对于用于出版的照片，拍摄时还应该注意更多的细节（如光线、构图和色彩等）。

5）分段和初步沉积环境分析

对每个组要划分岩性段（如果有的话），如砂岩夹泥岩段、泥岩段等。分段可在踏勘时划分，也可在分完层后划分，但最好在踏勘时划分。

一般说来，一个段的厚度不应小于10m，一般为几十米或上百米。

6）绘制草图

野外素描、绘图结合文字描述可以更加准确、明了的表明地质现象的特征及相互接触的关系，野外绘制的图件一般包括剖面素描图、地层柱状简图及露头精细素描图等。

剖面素描图是在剖面实测的过程中，将剖面的地形轮廓、小层界线和地层的岩性、几何形态、侧向变化、岩性组合及叠覆关系等素描下来，同时标记岩层产状、取样位置等信息，并配以详细的现象文字说明（图4-9）。

地层柱状简图是在地层实测过程中，通过粗略估算地层厚度，将剖面地层以完整的垂向序列的方式绘制出来，同时标记分层界线、取样信息等，并添加详细的现象文字说明，以展示其岩性变化和沉积序列（图4-10）。

露头精细素描图主要是用于准确地记录特殊的地质现象，如沉积构造、化石和典型相标志等（图4-11）。

虽然不同类型图件在绘制技巧上具有一定的差别，但仍都需要遵循以下注意事项：（1）图件必须精确勾勒地质要素的接触关系和相对大小；（2）描绘前应估算剖面长度、地层厚度，选择适当的比例尺进行作图；（3）描绘必须简明，突出所要表现的地质现象（如构造特征等）；（4）图件应标注图名、地点、比例尺、方位、图例、地层新老关系及作图人等信息，并配以适当的文字说明。

三、室内整理和分析

室内整理和分析是野外地质工作中的十分重要、细致的工作，主要包括：整理野外原始资料（包括记录、表格、图件、照片和样品等）、核算分层厚度和编绘相关地质图件等。

1. 整理野外原始资料

在当天实测工作结束后，组员应该在回到宾馆后对野外记录、表格、草图、照片和样品等各项资料的完整性、准确性和一致性进行核实、检查、整理。如果出现差错和遗漏，应该及时设法更正和补充。不能等到回到基地后再整理，如果发现错误就晚了，难以补救。

回到基地后，对于采集的样品，应及时送出并开展预设的实验和分析，并在获取实验和分析结果后（例如薄片鉴定资料），对野外记录进行相应的校正。

2. 核算分层厚度

分层厚度的核算是绘制精细柱状图的基础，一般分层厚度计算的方法包括查表法、图解法、水平投影法和公式计算法等，最为常用的为公式计算法，且多采用利昂诺夫斯基公式：

$$D = L \cdot (\sin\alpha \cdot \cos\beta \cdot \sin\gamma \pm \cos\alpha \cdot \sin|\beta|)$$

式中　D——分层厚度，m；

　　　L——分层斜距，m；

　　　α——岩层倾角，(°)；

　　　β——坡度角，(°)；

　　　γ——导线走向与岩层走向间夹角，(°)。

在地层实测过程中，可能遇到如图4-12岩层厚度与地面坡向、坡角、地层产状的关系所示的4种情况。从图中可知，当地面坡向与地层倾向相同时，取"−"；而在地面坡向与地层倾向相反时，取"+"。

图 4-9　天津市蓟州区渔阳镇小岭子村洪水庄组剖面素描图

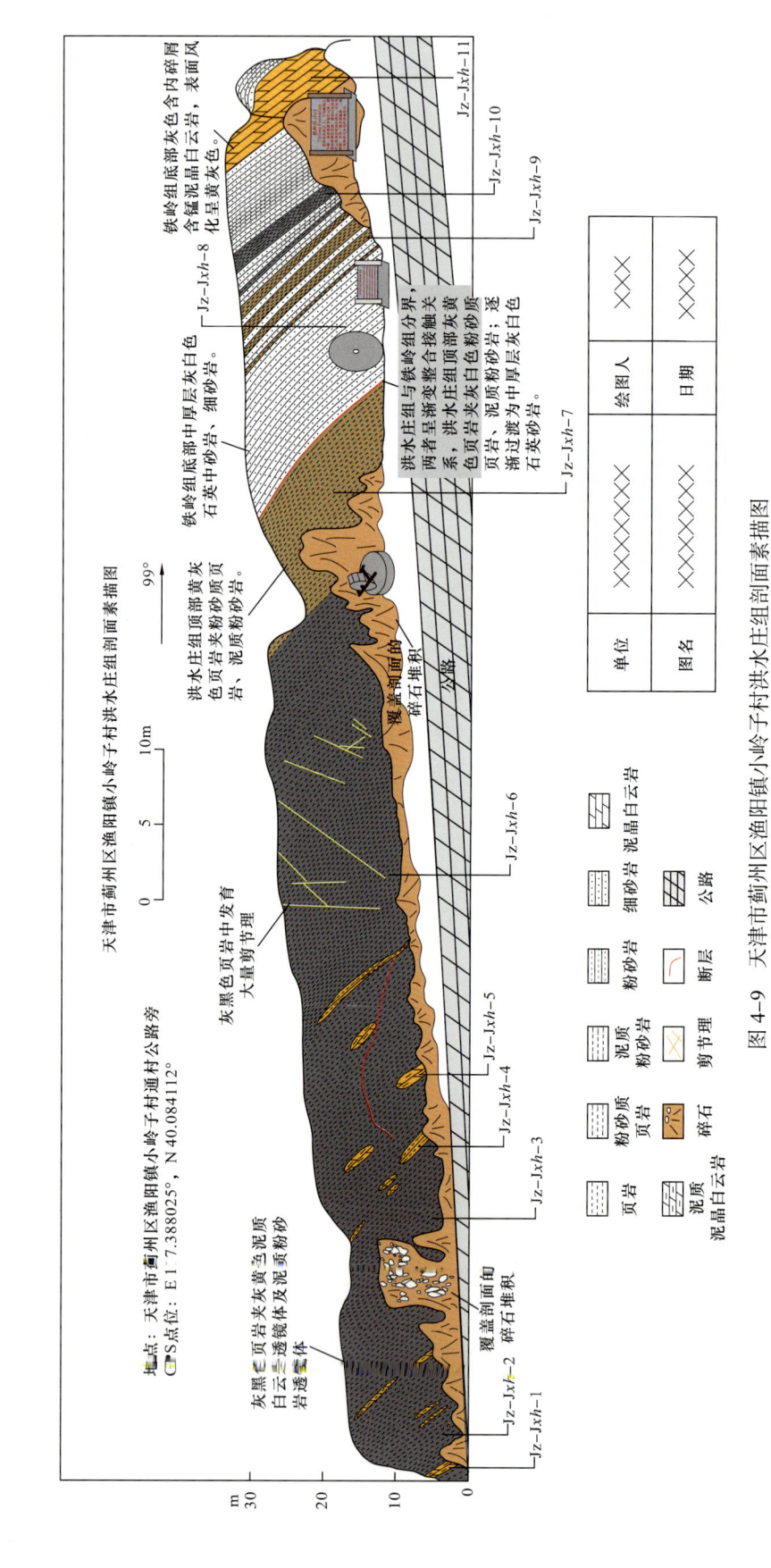

Fig. 4-9　Sketch profile of Hongshuizhuang Fm. in Xiaolingzi Village, Yuyang Town, Jizhou District, Tianjin

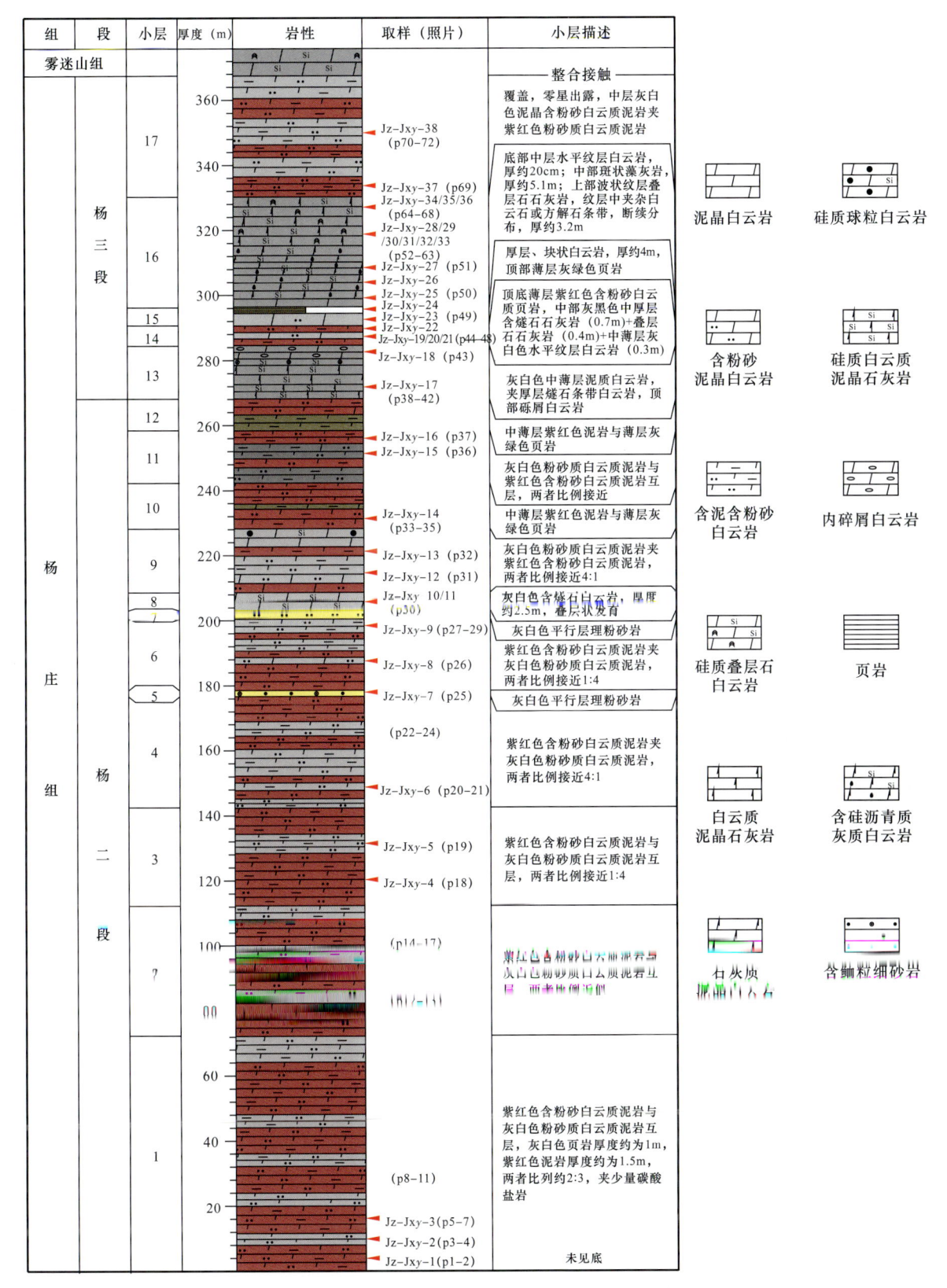

图 4-10 天津市蓟州区罗庄子镇杨庄组地层柱状简图

Fig.4-10 Stratigraphic column of the Jxy_2 and Jxy_3 in Luozhuangzi Town, Jizhou, Tianjin

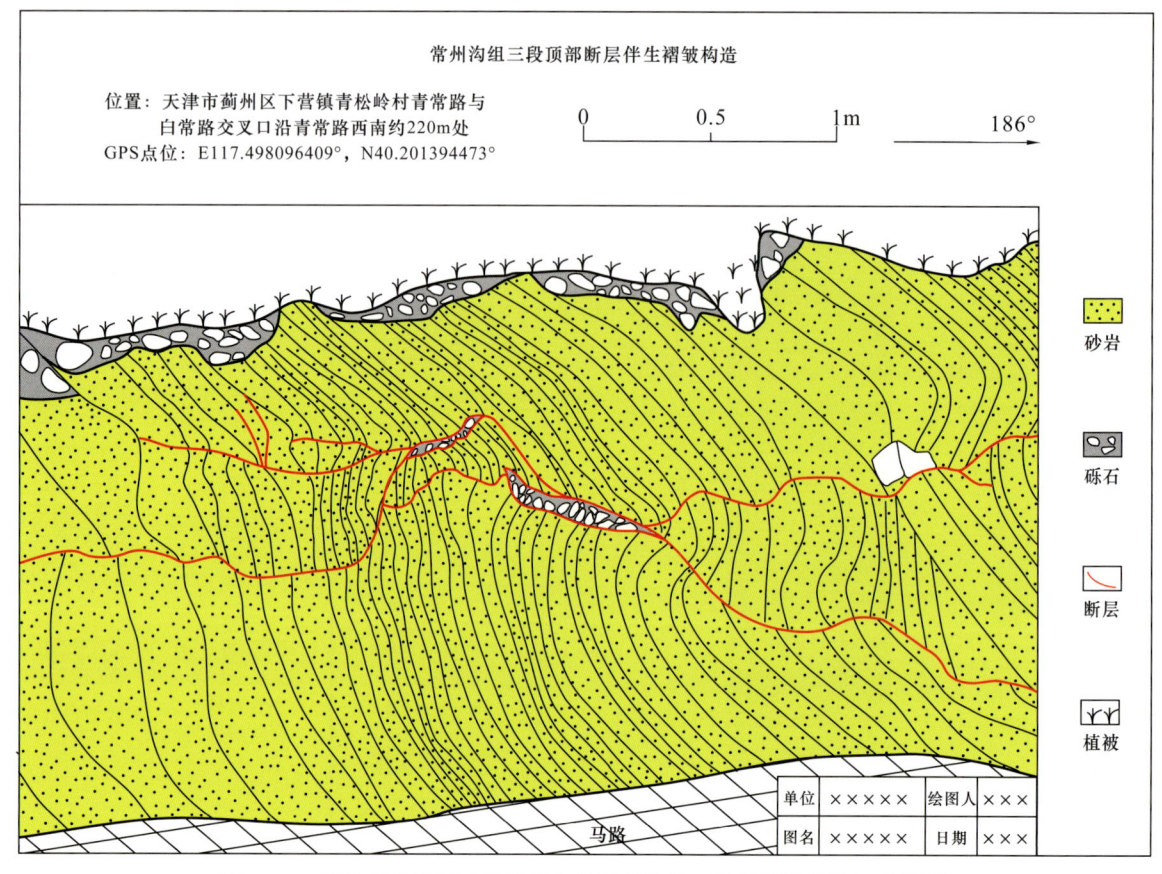

图 4-11　天津市蓟州区下营镇青山岭常州沟组三段顶部露头精细素描图

Fig.4-11　Detailed field drawing of the top of the Chc₃ at Qingshanling, Xiaying Town, Jizhou, Tianjin

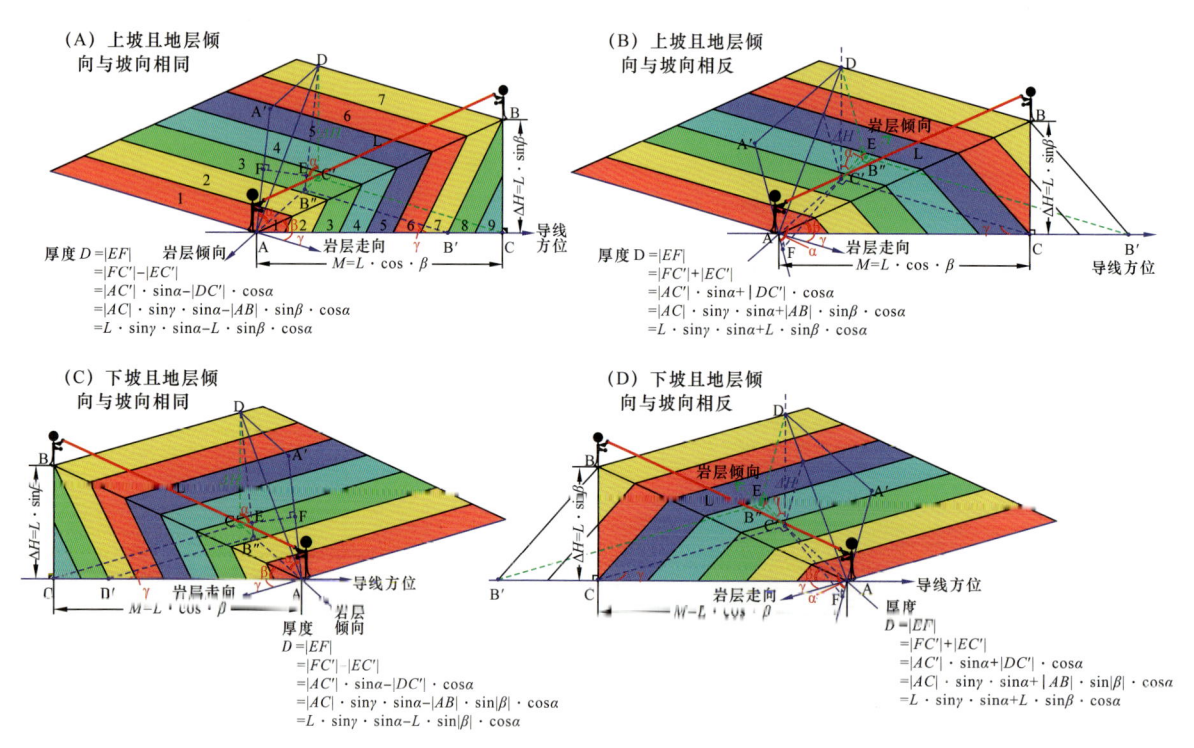

图 4-12　岩层厚度与地面坡向、坡角、地层产状的关系

Fig.4-12　The relationship among strata thickness and ground aspect, slope angle, strata occurrence

高程变化可以提供信手剖面图的地形轮廓，因此高程也是一个十分重要的参数，每段高程的计算公式为：$\Delta H = L \cdot \sin\beta$。高程的变化即为各段的累积高程。

3. 绘制相关地质图件

依据野外的剖面素描图和地层柱状简图，结合核算的数据结果，编绘更加精确、可靠的信手剖面图和地层综合柱状图。

1）编绘信手剖面图

（1）确定总导线方位。信手剖面图是地层沿某一特定方向的展布情况，这个特定的方向一般为总导线的方位。在实际操作中，总导线方位的确定方法为：选择适当的比例尺，依次绘制出各分导线，并依次首尾相接，最后连接第一分导线的起点与最终分导线的终点，连线方向即为总导线的方位，方位角可以用量角器进行测量。

（2）导线平面图的制作。在确定总导线方位后，将分导线与总导线整体旋转至总导线至水平方向，并依次标记出总导线方位、分导线号、地质界线（小层界线和组界线等）、地层单位名称等。

（3）绘制地层剖面图。在总导线下方适当位置绘制水平线作为实测剖面的高程基线，然后选择适当的垂向比例尺，参照野外绘制的剖面素描图的地形轮廓及累积高程的变化，勾绘出相应的地形轮廓线。然后，将各地质界线向下投影至地形轮廓线上，根据附录二和附录三中相关的表格和公式计算视倾角和歪曲视倾角，并绘制相关的地层界线，一般而言，小层、段、组界线的线段长度依次增长、加粗。最后，按照先前规定的岩性符号分别充填界线间的岩性，并标记取样点位置、产状数据、图例、横纵向比例尺、地点或地物名称以及责任表等。

具体的实例如图 4-13 所示。

2）编绘地层综合柱状图

地层柱状图是实测地层剖面的重要成果图件之一，也是进行地层分析和对比的基础，并可以反映地层厚度、岩性序列、接触关系等基本信息。具体的作图方法如下：

（1）根据研究目的设置综合地层柱状图的内容，一般包括地层层位（系、统、组、段、小层）、厚度、岩性柱子、取样位置、特征描述（岩石类型、颜色、结构、构造、化石、沉积序列、接触关系等）、分析解释（包括沉积相分析、沉积旋回曲线）层序及沉积相。

（2）依据分层核算的厚度，选择适当的比例尺，确定层界线的位置与类型，并填充相应的岩性符号，常见的岩性符号、地层接触类型符号等。对于部分特殊层（如矿化层、化石富集层等）厚度较小时，厚度可适度加厚，沉积构造和化石可用相应的符号进行标记。

（3）首件完成相应内容后，将图名及比例尺等放置于柱状图上方，将图例、责任表等附于柱状图下方。

具体的实例参见图 3-80。

四、剖面实测应提交的成果

剖面实测及相关分析化验全部结束后，应提交如下资料、图件和报告：

（1）野外分层描述资料。包括野外记录本、根据野外记录本整理的资料。

（2）样品及其清单。应包括日期、剖面名称和位置、样品编号、样品的岩石类型、样品的地质层位和位置、采样目的。

图 4-13 天津市蓟州区罗庄子镇青山村采摘园杨庄组二段和三段实测地层剖面图

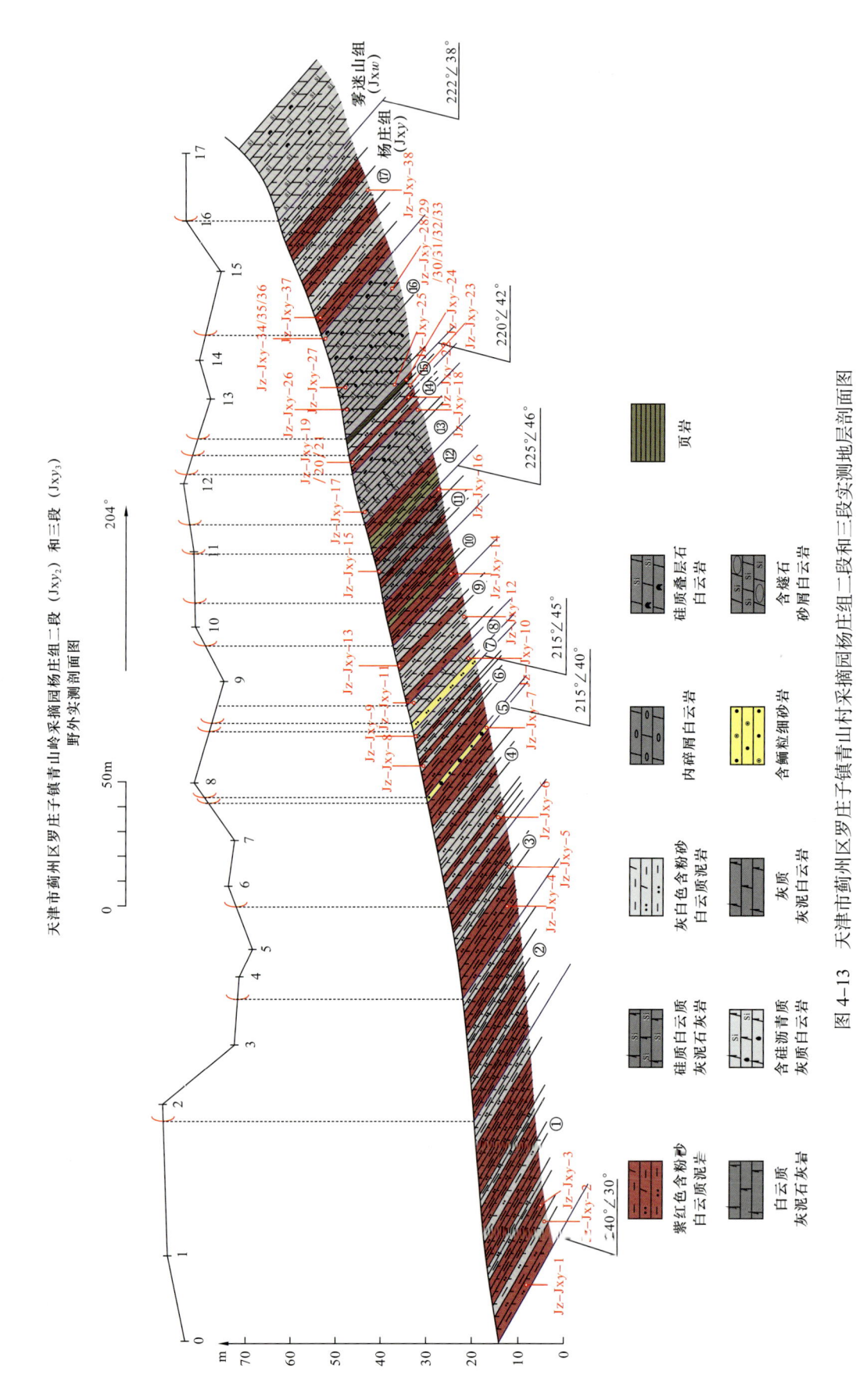

Fig 4-13 Measured stratigraphic section of the Jxy₂ and the Jxy₃ in the picking garden of Qingshan village, Luozhuangzi Town, Jizhou District, Tianjin

- 132 -

（3）照片清单。照片要挑选、汇总，整理成图版。图版中的每张照片要表明照片的位置、地质层位和小层号、展示的现象。

（4）薄片及其鉴定报告。薄片磨制是必须有的一项工作，因为在野外仅靠肉眼的鉴定是不太可靠的，必须磨片在镜下进一步鉴定、描述，补充和修正野外的观察描述。薄片磨制好后，要在薄片上标注编号。薄片的编号与手标本的完全相同。需染色的，需送样时标明。

要有薄片清单，其内容包括：日期、剖面名称和位置、薄片编号、薄片的岩石类型、样薄片的地质层位和位置。

（5）野外资料修正报告。根据薄片等鉴定报告，对野外描述资料进行修正，包括岩石定名的修正、结构特征的修正等。

（6）相关图件。实测结束后提交的图件应包括：野外地层柱状图、室内修正后的地层柱状图、信手剖面图等。

（7）研究报告。根据上述研究资料，编写研究报告。

第四节 常见岩石和矿物的野外鉴定

野外露头的观察和描述是野外地质工作的核心，其粗略程度和准确性是决定野外工作质量的关键。露头的观察和描述是每一个地质工作者必须熟练掌握的基本工作技能，通常需要地质工作者经历长期的观察、训练和总结，才能得到不断提高和完善。本节主要从野外露头描述过程中可能涉及的一些基本概念出发，简明扼要地介绍在观察野外露头时应该描述什么、如何描述，以期帮助大家提高野外工作的准确性。

由于蓟州区元古宇以沉积岩为主，本节在简单介绍不同岩石类型和常见矿物的野外鉴定方法后，重点介绍描述沉积地层的基本概念和方法。

一、常见岩石和矿物的野外鉴定方法

沉积岩、岩浆岩和变质岩是组成地壳的三大类岩石，虽然它们在成因上截然不同，但有时在野外并不易区分，现将其关键的区别特征（产状、矿物成分、结构、构造及化石等）总结于表4-3中。

表4-3 岩浆岩、变质岩和沉积岩的主要野外区分特征

Table 4-3 Main field distinguishing characteristics of igneous, metamorphic and sedimentary rocks

区分特征	岩浆岩	变质岩	沉积岩
产状	侵入岩多切割围岩地层，喷出岩多呈不规则状	多随原岩产状而定，具有一定的成层性，但层面不清晰	成层性好，层面清晰
矿物成分	多见暗色矿物（如黑云母、角闪石、辉石等）	暗色矿物常见，含特征性矿物（如红柱石、蓝晶石、堇青石、石榴子石等）	除含岩浆岩或变质岩砾石外，一般暗色矿物少见
结构	喷出岩多呈隐晶质或玻璃质，常见斑状结构；侵入岩多呈嵌晶粒状结构，矿物成分多样；矿物晶体排列多无定向性	多呈变晶结构，肉眼难与岩浆岩相区分，但矿物晶体普遍定向排列	除部分化学岩外，少见嵌晶粒状结构，且矿物成分种类较少；较少出现矿物晶体定向排列；具有碎屑结构和生物结构

续表

区分特征	岩浆岩	变质岩	沉积岩
构造	常见柱状节理、气孔和杏仁构造、流纹构造、绳状构造、枕状构造等特征性构造	常见片理、片麻理及线理等特征性构造	发育各种类型的沉积构造（如沉积层理和层面构造、生物成因构造等）
化石	无	无	仅在沉积岩中出现

常见矿物的鉴别也是区分岩性的重要手段，野外肉眼观察和鉴定矿物时，通常借助放大镜、小刀及稀盐酸等基本工具对矿物的表面光泽、颜色、硬度、解理等进行判断，从而确定矿物类型。现将常见的矿物在野外的识别特征总结归纳于表 4-4 中。

表 4-4 常见矿物野外鉴别特征

Table 4-4 The identification characteristics of common minerals in field

矿物类型	表面光泽	颜色	摩氏硬度	晶体或集合体形态	解理	其他
石英	玻璃光泽，断口呈油脂光泽	浅灰色至半透明状，含杂质后可呈紫色、烟灰色、黑色、粉红色和黄色等	7	常见六方柱、菱面体	不发育	
长石	玻璃—半玻璃光泽	斜长石常呈白色和灰白色，正长石常呈肉红色	6~6.52	斜长石多呈板状、板条状；正长石多呈板状、板柱状	两组完全解理，完全或近乎正交	
黏土	土状光泽	灰色、白色、浅红色、浅绿色、黑色等，其中黑色可能与富含有机质有关，浅绿色可能与绿泥石有关	1~2	肉眼无法观察黏土矿物颗粒，但可见其集合体，呈土状或块状		
云母	珍珠光泽	白云母多呈无色、白色，黑云母多呈褐色或浅绿色，断面呈黑色	2.5~3	多呈片状、板状或鳞片状集合体产出	一组极完全解理	
海绿石	光泽暗淡	绿色、深绿色、氧化后呈深褐色	2	砂级粒状集合体		
方解石	玻璃光泽	无色透明，含杂质后常呈灰白色、浅灰色，也可见粉色、黄色和绿色	3	晶体常呈菱面体、六方柱；集合体呈晶簇状、粒状、块状、纤维状和钟乳状等	三组完全菱面体解理，解理夹角 60° 或 120°	与稀盐酸强烈反应，气泡剧烈
白云石	玻璃光泽	白色、灰白色为主，部分含铁呈浅粉色、黄色等	3.5~4	晶体呈菱面体	同上	与稀盐酸一般不反应
石膏	玻璃光泽，纤维状为丝绢光泽	无色透明，含杂基后呈白色至灰色	2	晶体多呈板状、燕尾状，集合体多呈纤维状、块状和粒状	多个方向解理，易劈开呈薄片状	
硬石膏	玻璃光泽	无色透明，含杂基后呈白色至灰色	4	晶体呈等轴状或厚板状，集合体呈块状和粒状	三组相互垂直的完全与中等解理	
岩盐	玻璃光泽	无色透明，可见白色	2.5	晶体呈完整—不完整立方体		有咸味

续表

矿物类型	表面光泽	颜色	摩氏硬度	晶体或集合体形态	解理	其他
辉石	玻璃光泽	绿黑色或黑色	5.5~6	晶体多呈半自形柱状或他形短柱状	两组平行柱状解理	
角闪石	玻璃光泽	绿黑色或黑色	5~6	晶体呈长柱状，横断面呈假六边形，集合体呈细柱状、针状和纤维状	二组解理，解理夹角56°	
黄铁矿	金属光泽	浅黄铜色	6~6.5	晶体多呈立方体，晶面可见生长纹，集合体多呈块状、分散球状和球状结合	无解理	
磁铁矿	半金属光泽	铁黑色	5.5~6.5	晶体多呈八面体，集合体多呈块状和粒状	无解理	具强磁性
赤铁矿	金属、半金属光泽	显晶质呈铁灰色至钢灰色，隐晶质呈赤红色	5~6	单体少见，集合体多呈块状、鲕状、豆状、肾状及粉末状	无解理	无磁性
褐铁矿	半金属、土状光泽	颜色多变，黄褐色、深褐色等		集合体呈土状、葡萄状		
红柱石	玻璃光泽	灰白色、肉红色	6.5~7.5	晶体呈柱状，横切面近正方形，集合体呈放射状	平行柱状的中等解理	

注：代替硬度计的常用物品的硬度，如指甲为2~2.5，铜钥匙为3，小钢刀为5~5.5，普通玻璃为6。

二、岩石分类

在野外，碎屑岩通常按照粒度分为砾岩（72mm）、砂岩（0.1~2mm）、粉砂岩（0.005~0.1mm）、泥岩（<0.005mm）。其中，砾岩进一步划分为：细砾岩（2~10mm）、中砾岩（10~100mm）、粗砾岩（100~1000mm）、巨砾岩（>1000mm）；砂岩进一步划分为：细砂岩（0.1~0.25mm）、中砂岩（0.25~0.5mm）、粗砂岩（0.5~2mm）。

石灰岩的分类有多种方案，见表4-5—表4-7。

表4-5 方解石—黏土混合岩分类方案
Table 4-5 The classification of the calcite-clay mixed stones

岩石类型		方解石相对含量（%）	黏土相对含量（%）
石灰岩	石灰岩	>90	<10
	含黏土石灰岩	75~90	10~25
	黏土质石灰岩	50~75	25~50
黏土岩	灰质黏土岩	25~50	50~75
	含灰黏土岩	10~25	75~90
	黏土岩	<10	>90

表 4-6　邓哈姆的石灰岩分类方法（据 Dunham，1962）
Table 4-6　Classification scheme of limestone proposed by Dunham（after Dunham，1962）

沉积结构可识别				沉积结构不可识别
沉积期原始组分未粘结			沉积期原始组分互相粘结（正如共生的骨骼物质、与重力方向相反的层状物或沉积物覆盖的腔体，这些腔体被有机质或可疑的有机质覆盖且很大而不是小孔隙）	结晶碳酸盐岩
含泥（黏土或细粉砂粒级的颗粒）		无泥且颗粒支撑		
灰泥支撑	颗粒支撑			（根据与物理结构或成岩作用有关的分类方案细分）
颗粒少于10%	颗粒多于10%			
泥岩	粒泥灰岩	泥粒灰岩	颗粒灰岩	粘结灰岩

表 4-7　金振奎的石灰岩分类（据金振奎，2014）
Table 4-7　Classification scheme of limestone proposed by Jin（after Jin，2014）

划分标准	生物格架＜30%					生物格架≥30%			
	颗粒＜50%（或灰泥基质支撑）			颗粒≥50%（或颗粒支撑）		原地生物格架为主		异地生物格架为主	
	＜10%	10%～25%	25%～50%	灰泥＞亮晶	灰泥＜亮晶	灰泥＞亮晶	灰泥＜亮晶	灰泥＞亮晶	灰泥＜亮晶
类型	灰泥石灰岩	含颗粒灰泥石灰岩	颗粒质灰泥石灰岩	泥晶颗粒石灰岩	亮晶颗粒石灰岩	灰泥礁石灰岩	亮晶礁石灰岩	灰泥礁砾屑石灰岩	亮晶礁砾屑石灰岩
	灰泥石灰岩类			颗粒石灰岩类		礁石灰岩类		礁砾屑石灰岩类	
备注	可用具体的优势颗粒名称替代命名中的"颗粒"，如含砂屑灰泥石灰岩、亮晶鲕粒石灰岩等					可在"礁石灰岩"前加上具体造礁生物名称，如亮晶海绵礁石灰岩		可在"礁"前加上具体造礁生物名称，如亮晶海绵礁砾屑石灰岩	

白云岩通常按照晶粒大小进行划分（表 4-8）。在野外露头描述中，由于肉眼一般很难直接分辨出单一晶粒，通常将泥晶、细粉晶和粗粉晶白云岩统称为"微晶白云岩"，将细晶、中晶、粗晶白云岩和砾晶白云岩统称为"砂糖状白云岩"。

表 4-8　白云石晶粒大小的划分标准
Table 4-8　Classification standard of crystal size of dolomite

白云岩类型		粒级（mm）
砂糖状白云岩	砾晶白云岩	＞2
	极粗晶白云岩	1～2
	粗晶白云岩	0.5～1
	中晶白云岩	0.25～0.5
	细晶白云岩	0.1～0.25
微晶白云岩	粗粉晶白云岩	0.05～0.1
	细粉晶白云岩	0.005～0.05
	泥晶白云岩	＜0.005

至于火山碎屑岩的命名，本书提出了一个分类方案，即按照粒度进行划分，与碎屑岩相似。从细到粗划分为：火山泥岩、火山粉砂岩、火山砂岩、火山砾岩。粒度与相应的碎屑岩相同，只不过加了"火山"二字。其中，火山砂岩可进一步划分为火山细砂岩、火山中砂岩、火山粗砂岩，粒度与砂岩相同。火山砾岩可进一步划分为火山细砾岩、火山中砾岩、火山粗砾岩、火山巨砾岩，粒度与砾岩相同。

在前人的分类方案中，火山泥岩、火山粉砂岩和火山砂岩统称"凝灰岩"。火山角砾岩相当于本书的火山细砾岩和火山中砾岩，火山集块岩相当于火山粗砾岩和火山巨砾岩。

本书之所以未采用前人的分类，是因为前人的分类存在如下问题：

（1）火山碎屑也是一类碎屑物质，只不过是火山喷发形成，而非母岩风化形成。其分布遵循机械沉积分异原理，与碎屑岩相同。

（2）粒度与距离火山口的距离密切相关。但前人的方案中，粒度划分过粗，不利于精细反映火山碎屑的分带性。例如，凝灰岩包括的粒度范围就过粗。

（3）"火山角砾岩"这个术语不严谨，因为不是所有的砾级火山碎屑都是棱角状的，有些是次圆状的（塑性岩浆在空中收缩形成），如火山弹。南美很多新生代砾级火山碎屑是次圆状的。

三、常见沉积构造

按照成因，沉积构造可以分为机械成因构造、化学成因构造和生物成因构造等3大类，然后，每一大类又根据形态特征细分为若干小类，具体划分情况见表4-9。

表 4-9　沉积构造分类
Table 4-9　Classification scheme of sedimentary structures

机械成因构造	波痕	按对称度：不对称波痕、对称波痕
		按波脊形态：直脊波痕、弯脊波痕、断脊波痕（舌状、新月状、菱形波痕）
		按波高：沙纹、沙波、沙垄、沙浪、沙丘
		其他：双峰波痕、交叉叠加波痕、改造波痕、削顶波痕、冲洗波痕
	绝大部分层理	水平层理、平行层理、剥离线理、机械纹理、交错层理（板状交错层理、楔状交错层理、槽状交错层理、人字形交错层理、丘状交错层理、羽状交错层理、冲洗交错层理）、加积交错层理（侧积交错层理、前积交错层理、顶积交错层理）、沙纹层理、爬升层理、波状层理、脉状层理、透镜状层理、韵律层理（正韵律层理、反韵律层理）、粒序层理（正粒序层理、反粒序层理）、均质层理、块状层理
	其他构造	双黏土层构造、拱石卷、泥疙、层面（印模层面、波状层面、冲刷面）、再作用面、泥裂、水下泥裂、臼齿构造、马蹄纹构造、鸟眼构造、雨痕、冰雹痕、底模（槽模、沟模、锥模、刷模）、准同生变形构造（重荷模、球枕构造、火焰构造、准同生褶皱、包卷层理、准同生断层、碟状构造、柱状构造、砂脉、砂火山、泥火山）、碎裂构造（压实缝、风化缝、垮塌缝）、瘤状构造、眼球状构造、链条状构造、地花瓣构造
化学成因构造		化学纹理、结核、斑状构造（豹斑构造、晶斑构造）、铁丝鸡笼构造、散晶、晶痕、假晶、帐篷构造、叠锥构造、同心环状构造、假褶皱、假松枝化石、洞穴沉淀构造（石笋、石钟乳、石毯、石花）、缝合线（粒间缝合线、毛发丝缝合线、锯齿缝合线）
生物成因构造		遗迹构造（居住迹、爬行迹、行走迹、犁食迹、掘食迹、停息迹、印模迹、根迹）、叠层石（生物成因纹理）（层状、波状、丘状、柱状、锥状和墙状）、窗格构造、扰动斑状构造、凝块构造

下面主要对一些重要的和在蓟州区元古宙地层中常见的沉积构造进行详细描述。

1. 波痕

波痕是由风、水流等介质的运动在沉积物表面所形成的波状起伏的构造。定量描述波形的形态参数主要包括：波长、波高、波痕指数、对称度等（图4-14）。

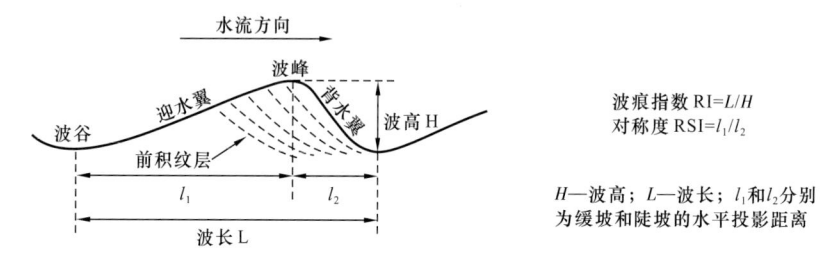

图 4-14 波痕形态参数示意图

Fig.4-14 Schematic diagram of the morphology parameters of ripple markers

按照成因的差别，波痕可以分为流水波痕、浪成波痕和风成波痕等3类。流水波痕是由单向水流形成，多见于河道和潮坪环境；其特点是波峰和波谷均比较圆滑，呈不对称状，陡坡倾向指示水流方向；浪成波痕是由波浪震荡而形成的，常见于海、湖的近岸地带，其特点是波峰尖锐，波谷圆滑，在浪基面附近呈近乎对称状，靠岸浅水带呈不对称状；风成波痕是由空气流动而形成的，常见于沙漠及海、湖的滨岸风成沙丘沉积中，其特点是波峰和波谷都比较圆滑、开阔，但常常谷宽峰窄，呈极不对称状，陡坡倾向与风向一致（图4-15）。

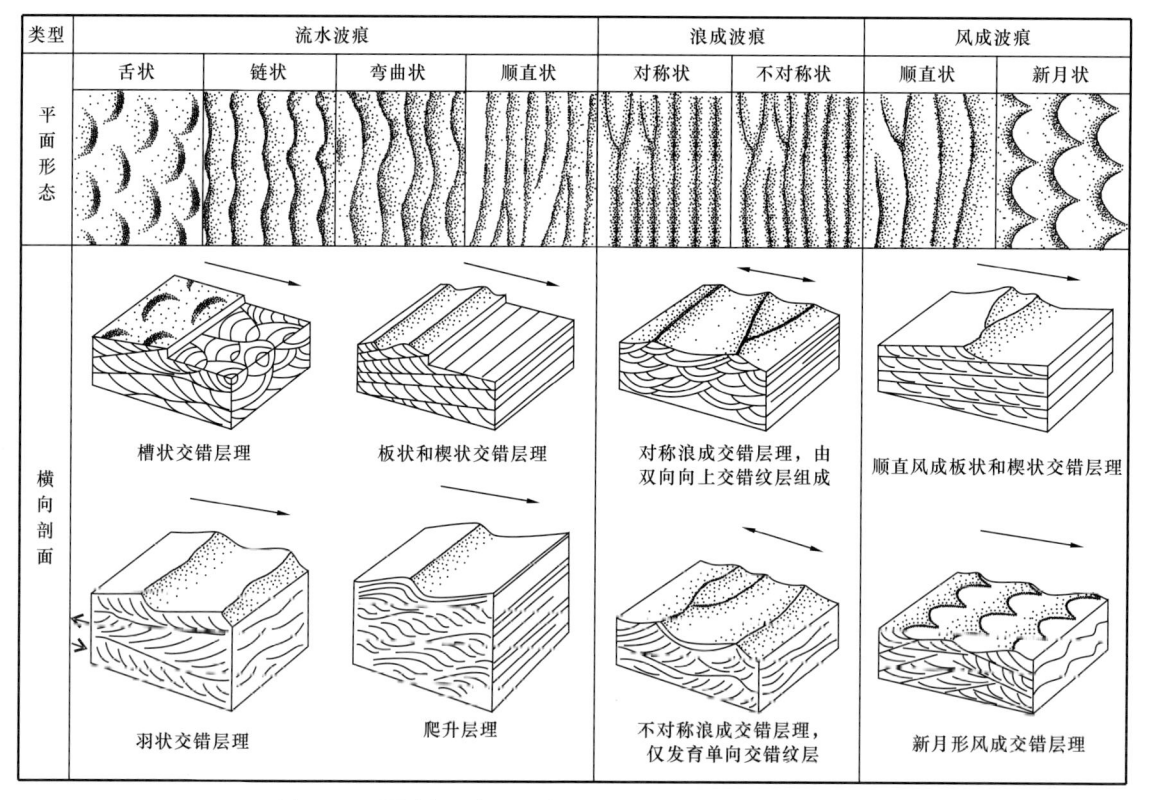

图 4-15 不同成因类型的波痕的平面形态及横向剖面形态

Fig.4-15 Plane morphology and transverse profile morphology of ripple marks formed by different origins

黑色点代表背流面，浪成波痕的两侧均有黑点

波痕在沉积环境分析中具有十分重要的作用，主要包括以下几个方面：

（1）判断水体能量大小，通常波痕形成于动荡的水体之中，且波脊的形态可以反映流体的流速，从顺直状、弯曲状、链状至舌状代表流体流速逐渐增大；

（2）判断流体运动方向，对于流水和风成波痕，波痕陡坡指示古流向，而浪成波痕的波脊方向代表海或湖的岸线延伸方向；

（3）恢复古水深，流水波痕和浪成波痕多形成于浅水环境，对于浪成波痕，水深一般小于波长的一半；

（4）判断岩层的顶底，波痕出现在层面的顶面，对于浪成波痕而言，波峰形态尖锐，且指示顶面。

在野外观测波痕时，不仅需要测量它们的形态参数，推断成因类型；还需要描述波脊的形态及走向等参数，推断流体运动方向。在天津市蓟州区元古宇中也发育大量的波痕构造，且以滨岸带的浪成波痕为主（图4-16）。

图4-16 天津市蓟州区元古宇中发育的典型浪成波痕构造

Fig 4-16 Typical wave ripples in the Proterozoic strata in Jizhou District, Tianjin

(A) 浪成波痕中砂岩，波峰和波谷十分宽缓，常州沟组一段顶部；(B) 浪成波痕泥质粉砂岩，发育泥裂，表明水深很浅，并偶尔暴露水面；(C) 浪成波痕石英中砂岩，不同层面上波脊方向不同，波脊呈弯曲状，大红峪组一段；(D) 浪滑波痕含长石石英中砂岩，波峰尖锐，波谷圆滑，高于庄组底部；(E) 圆脊波痕泥晶白云岩，波峰、波谷圆滑，波高较大，指示风急浪高的环境，高于庄组三段；(F) 浪成波痕含粉砂泥质泥晶白云岩，杨庄组二段

2. 层理

层理是沉积岩中因粒度、物质成分、颜色等有规律的垂向变化而显现的一种层状构造，是沉积岩的重要特征之一，是沉积环境解释的重要指标。

层理通常由从小至大的纹层、层系和层系组等3个单元组成。纹层是组成层理的最基本单元，纹层之内无法再细分出任何肉眼可见的层，通常为单一沉积条件的沉积产物，厚度通常小于1 cm。层系是由同类型纹层组合而成，代表一段时间内沉积条件相对稳定下的沉积产物。层理系是由多个同类型层系或者不同类型但成因上有联系的层系叠置而成，其间不存在沉积间断。

按照层理的特征及成因的差异，层理可以划分为多种类型，包括：水平层理、平行层理、交错层理、复合层理、递变层理、韵律层理和块状构造等。

1）水平层理

水平层理是由一系列平直且与岩层面平行的泥级或粉砂级沉积物（泥、粉砂、灰泥等）纹层组成的层理（图 4-17A），一般出现在泥岩、泥质粉砂岩、灰泥石灰岩等极细的沉积岩之中，是由悬浮搬运的泥级或粉砂级沉积物在较安静的沉积环境中沉积形成的，如海和湖的深水地带、潟湖、潮坪等缺氧、盐度异常的或暴露的、不适宜生物生存的环境之中。

2）平行层理

平行层理是由一系列与岩层面平行的砂级（或细砾级）沉积物纹层组成的层理（图 4-18B），一般出现在砂岩、颗粒石灰岩等沉积岩中，对应于上部流动体制下的平坦底床，层面常见剥离线理。虽然平行层理与水平层理的形态相似，但代表完全相反的水动力条件，反映一种高能环境。

在野外观察中，应该注意区分水平层理和平行层理，重点观察和描述以下几个方面的特征：

（1）观察层理的连续性和侧向变化；

（2）观察纹层的成因和特征，包括颜色、成分、粒径、分选、磨圆、颗粒的排列等；

（3）测量或估计单一纹层的厚度，观察单一纹层的内部粒序结构；

（4）观察纹层的分组特征、旋回性，考虑是否反映更长周期的沉积控制因素变化。

3）交错层理

交错层理是由一系列与岩层面斜交的纹层组成的层理，层系之间可以相互交错、切割，是最常见的层理构造。根据纹层形状、纹层与层系界面及层系的组合关系，通常将交错层理划分为多种类型：板状交错层理、楔状交错层理、槽状交错层理、爬升波纹交错层理、羽状交错层理、浪成波纹交错层理、冲洗交错层理、丘状交错层理和洼状交错层理等。

板状、楔状和槽状交错层理是最为常见的交错层理类型，多由流水波痕、风成沙丘或滩坝侧向迁移、沉积和保存而形成，纹层与层系面相斜交，多呈"底切顶截"的特征，它们的形态差异分别表现在层系顶底界面分别平整且相互平行、平整且相互斜交和弯曲且相互斜交（图 4-17C-H）。在顺水流方向上，这些层理的纹层都向下游方向倾斜，而板状和楔状交错层理在垂直水流方向可呈"假平行层理"，槽状交错层理在垂直水流方向上呈槽状。

羽状交错层理是由上下两个纹层倾向相反的层系组合而成，呈羽状或鱼骨刺状，是双向水流沉积作用的结果，多出现于潮间带的沙坪和潮汐水道之中，是潮间带的良好标志。但有些"假羽状交错层理"（图 4-17I），是上、下槽状交错层理不同部位叠置形成的，不是潮汐双向水流形成的。

浪成波痕是由浪成波痕迁移形成的，多由倾向相反、相互交错的纹层组成的交错层理，多呈束状、"人"字形，有时也可能出现前积单向倾斜纹层组成的浪成交错层理。

冲洗交错层理是由一系列低角度倾斜的平直纹层组成，倾角一般为 2°～10°，多形成与海或湖的冲洗带，是"水浅流急"的向岸涌浪和离岸回流往复冲洗作用的产物，纹层向海或湖倾斜。

丘状交错层理和洼状交错层理分别是由一系列上凸和下凹的宽缓纹层组成，顶面分别呈圆丘状和圆坑状，这两种交错层理多由海洋或湖泊风暴浪沉积而形成。

总上，在野外观察和描述交错层理时，应该注意以下几点：

图 4-17 天津市蓟州区元古宇中发育的典型水平层理、平行层理和交错层理

Fig.4-17 Typical horizontal beddings, parallel beddings and cross beddings in the Proterozoic strata in Jizhou District, Tianjin

（A）水平层理泥晶灰岩，高于庄组四段；（B）平行层理石英细砂岩，铁岭组底部；（C）板状交错层含砾粗砂岩，常州沟组一段；（D）大型板状交错层理细砂岩，常州沟组一段；（E）楔状交错层理含海绿石石英中砂岩，龙山组一段顶部；（F）槽状交错层理含砾粗砂岩，常州沟组一段；（G）槽状交错层理石英中砂岩，常州沟组一段；（H）含砾中砂岩中的"假羽状交错层理"，形态类似于羽状交错层理，但纹层倾向相反是由于下部槽状交错层理的右侧被上部槽状交错层理的左侧切割、叠置导致的，常州沟组一段；（I）石英中砂岩中的"假羽状交错层理"，常州沟组二段

（1）测量和描述交错层理的层系和层系组的厚度；

（2）尽可能从三维角度确定交错层理的形态特征，确定其类型，如板状、楔状、槽状或其他类型；

（3）测量和描述纹层的倾斜方向、最大倾角等参数，以及纹层内颗粒的粒径、排列方式及粒序等特征；

（4）综合以上，分析交错层理的成因类型，推测沉积流体的类型、能量和沉积环境。

4）复合层理

复合层理是由较大水动力波动情况下形成的混合粒径的沉积物或沉积岩频繁交互而形成的（如砂—泥、砂—灰泥或云泥、碳酸盐颗粒—灰泥或云泥）。根据粗粒和细粒组分的相对含量及连续性，复合层理可细分为压扁层理（又称脉状层理）、波状层理和透镜状层理（图4-18）。该类型层理多见于潮坪环境、正常浪基面附近、三角洲前缘与前三角洲过渡带、河流堤岸等环境，且在潮坪环境中最为常见。

图 4-18　天津市蓟州区元古宇中发育的典型复合层理

Fig.4-18　Typical complex beddings in the Proterozoic strata in Jizhou District, Tianjin

（A）压扁层理，龙山组一段顶部；（B）波状层理，由灰白色钙质石英细砂岩和黄褐色泥质泥晶白云岩频繁交互而形成，团山子组三段；（C）透镜状层理，黄褐色白云岩中夹的灰色透镜状砂岩，团山子组三段

5）递变层理

递变层理是指具有粒度递变的一种特殊层理，又称粒序层理。按照粒序变化，递变层理可以细分为正递变（或正粒序）层理、反递变（或反粒序）层理以及复合递变（或复合粒序）层理。正递变层理粒级向上变细，在自然界中最为常见；反递变层理粒级向上变粗；复合递变层理粒序有序变化或频繁交替。

正递变（或正粒序）层理通常是由流体能量逐渐衰竭而形成，根据粒径向上变细的特征，又可以细分为多种亚类（图 4-19）。形态正粒序的流体既包括水流速度和强度逐渐降低的牵引流，也包括浊流、洪水、风暴和火山碎屑流等；前者细粒基质含量低，且主要集中于上部，粗颗粒可呈定向排列，而后者细粒基质含量高，可在全层中都有分布，粗颗粒杂乱排列。

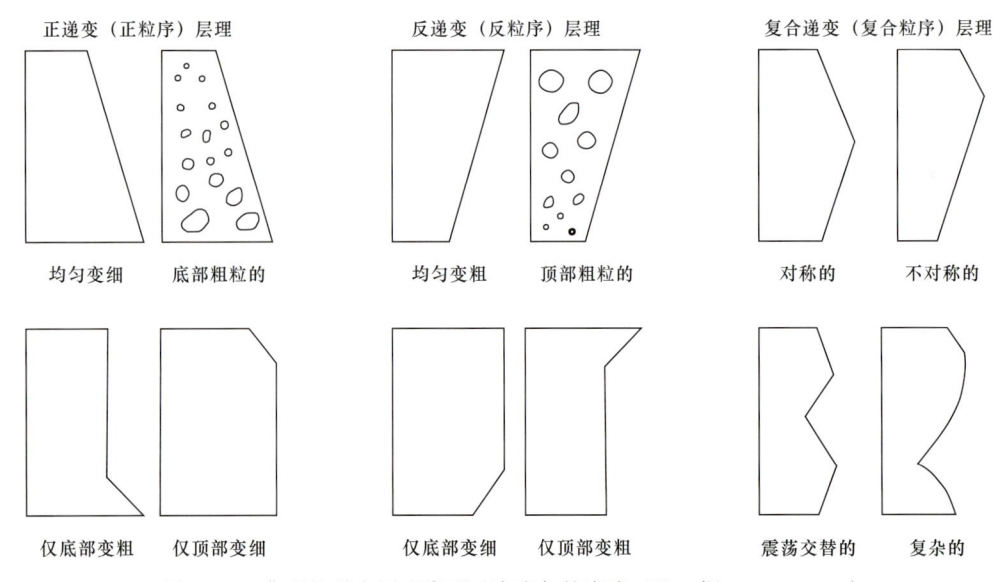

图 4-19　典型的递变层理类型（宽度与粒度成正比；据 Stow，2010）

Fig.4-19　Types of grading typical in sediments（width is proportional to grain size；after Stow, 2010）

反递变（或反粒序）层理与正递变（或正粒序）层理相反，通常是因流体能量逐渐增强而形成，或者因波浪颠选而体积大且质量轻的颗粒不断上浮等因素而形成；根据粒度递变的特征，也可以细分为多种亚类（图 4-19）。复合递变（或复合粒序）层理是由流体能量频繁交替而形成的；根据粒度递变和组合特征，也可以细分为多种亚类（图 4-19）。等深流沉积序列就是一种典型的复合

递变（或复合粒序）层理，其形成过程先经历较长时间的流速增加，再经历流速逐渐降低的过程，因而形成"细—粗—细"式的复合递变层理。

目前，较为成熟的沉积构造序列包括（Stow，2010；图4-20）：

（1）Stow序列，针对细粒浊积岩；

（2）Bouma序列，针对中粒浊积岩；

（3）Lowe序列，针对粗粒浊积岩；

（4）Dott-Bourgeois序列，针对风暴沉积岩；

（5）Sparks序列，针对熔结凝灰岩；

（6）Stow-Faugères序列，针对等深流沉积岩。

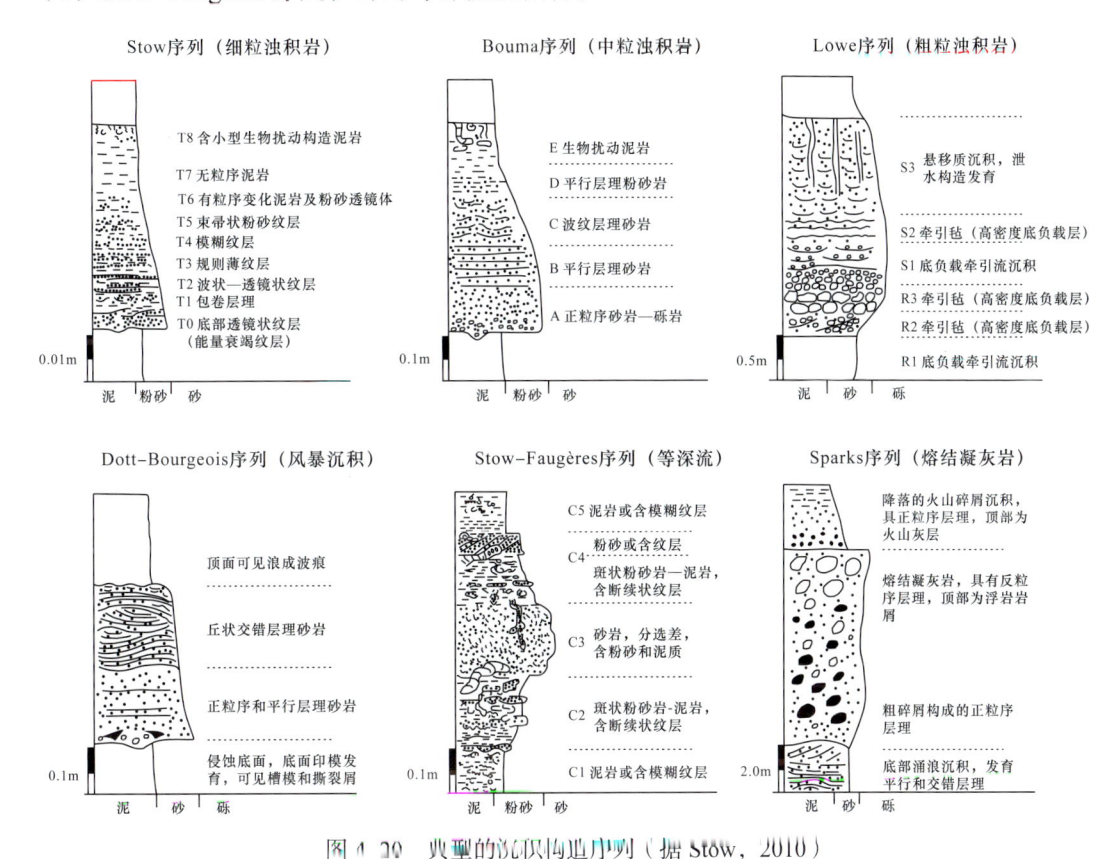

图4-20 典型的沉积构造序列（据Stow，2010）

Fig.4-20 Standard sequences of sedimentary structures that result from different depositional processes (after stow, 2010)

在野外描述递变层理时，应该注意以下几个方面：

（1）首先应该确定岩层边界，然后确定粒序是在部分层段发育，还是在整个层段发育；

（2）观察和描述粒序层理的类型及特征，包括颗粒粒径的变化范围等；

（3）观察沉积物成分、颜色及其他结构和沉积构造的变化，描述构造序列的发育情况；

（4）综合以上，推测沉积构造序列形成的环境，包括流体机制、水动力条件等。

6）韵律层理

韵律层理是指粒度、成分或颜色不同的、层面平整的纹层或薄层交替叠置而形成的层理，单层厚度很小，多为几毫米到几厘米，层面无明显波状起伏。韵律层理是流速或氧化还原条件等环境参数有规律地发生交替变化造成的，这些变化可以是短期的，也可以是长期的。例如潮坪环境中的韵律层理，涨退潮时沉积砂，平潮时沉积泥；河流堤岸或泛滥平原上季节性洪水泛滥而形成的韵律层理，洪水溢出河道时沉积粉（细）砂，洪水结束时在平静水体中沉积泥质纹层；周期性（几十年或者更长）的气候变化，使得湖泊周期性干涸，导致还原色与氧化色泥质纹层相互交替。

7）块状构造

块状构造是指一种大致均质、内部无层理的沉积构造。块状构造主要存在以下两种成因：（1）沉积物沉积时不发育任何沉积构造，一般出现在浊流、洪水、风暴、火山碎屑流、岩溶垮塌或冰川作用等快速沉积过程中形成的；（2）原始沉积构造被破坏，主要包括生物扰动、应力扰动变形和成岩重结晶等过程中形成的。该构造类型在天津市蓟州区寒武系底部的冰川角砾岩中广泛发育。

3. 其他机械成因构造

除了以上机械成因构造外，还有泥裂构造等在天津市蓟州区元古宇中也经常出现（图4-21），臼齿构造及瘤状构造虽然成因存在争议，但考虑其机械成因的可能性，因而也在这里进行讨论。

图 4-21　天津市蓟州区长城系团山子组中发育的泥裂构造

Fig.4-21　Mud cracks in the Tuanshanzi Formation in Jizhou District, Tianjin

（A-D）层面上的泥裂；（E-F）剖面上的泥裂

泥裂是由暴露于地表的泥级或泥质沉积物表面失水、收缩、破裂和充填而形成。泥裂在层面上多呈多边形状，在剖面上多呈"V"形或"U"形，但可因压实变形而呈肠状。泥裂的宽度、深度和连续度主要受暴露程度和泥质含量及厚度的影响（图4-22），一般暴露越严重，泥质含量和厚度越大，泥裂的宽度、深度越大，连续性越好。

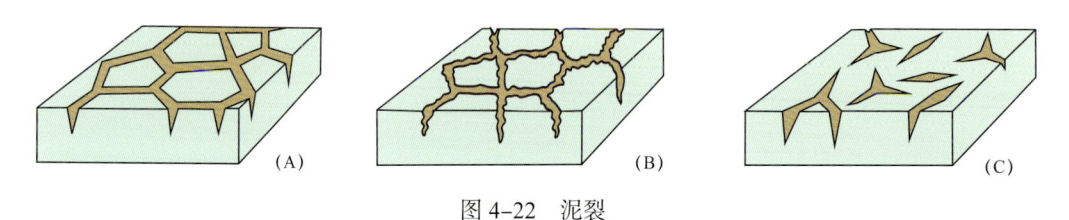

图 4-22　泥裂

Fig.4-22　Mud cracks

（A）暴露和干涸时形成的泥裂；（B）暴露和干涸时形成的泥裂，因压实变形而呈肠状；（C）浅水环境中脱水形成的泥裂

臼齿构造是指前寒武纪碳酸盐岩地层中垂直或斜交的微亮晶碳酸盐岩脉体，在平面上多呈"鸡爪"状或纺锤状。关于其成因，一直存在不同的假说，例如未固结沉积物中微生物生长的气泡膨胀、迁移成因（Smith，1968；O'Connor，1972；Frank and Lyons，1998；Furniss et al.，1998；刘燕学等，2003；旷红伟等，2004，2006，2008，2011；孟祥化等，2006；Pollock et al.，2006；梅冥相等，2007），蒸发替代作用成因（Eby，1975），水下收缩和脱水作用成因（Knoll and Swett，1990），地震液化、脱水作用成因（Song，1988；乔秀夫等，1994，2001；乔秀夫，1996；Fairchild et al.，1997；Pratt，1998；乔秀夫和李海兵，2009；武振杰等，2009）等，其中微生物活动和地震成因是目前争论的焦点。臼齿构造在天津市蓟州区蓟县系高于庄组四段顶部大量出现（图3-35A和B）。

瘤状构造是指由瘤状团块或半球形的透镜状沉积物组成的构造，主要发育于石灰岩中。关于其成因，主要存在以下观点：（1）在较深水盆地和台缘斜坡下部，由于周期性重力作用而形成海底底流作用，海底底流带来的$CaCO_3$不饱和的流体部分溶解沉积于海底的碳酸盐岩，并将剩余富含泥质物残留下来而形成的（高计元，1988）；（2）沉积成岩作用成因（张继庆，1986）；（3）压溶成岩作用成因，即由于在上覆压力或构造应力的作用下，压溶作用将弱固结的、结构成分不均一的含泥灰岩分割成大小不一、形状多变的瘤状体而形成，沿瘤状体和基质分界面（即溶解面）上产生凹凸不平的不溶残留物膜，在断面上表现为波状起伏或齿状的缝合线（Wanless，1979；郭福生，1989）；（4）周期性海底底流溶解作用和压溶作用共同作用而形成（夏丹等，2009）。瘤状构造在天津市蓟州区蓟县系高于庄组四段底部大量出现（图3-34B）。

4. 化学成因构造

化学成因构造是指在沉积期后因化学作用而形成的构造，可能扰乱、破坏或凸显原生沉积构造，或形成假的原生沉积构造，但是都会反映沉积期后的化学条件的变化。该类构造主要包括结核构造、斑状构造、晶体印痕构造、李泽岗格环（Liesegang rings）构造、树枝晶构造、铁锰侵染构造以及缝合线构造等，其中结核构造、缝合线构造、李泽岗格环以及铁锰侵染构造等在天津市蓟州区元古宇中常见（图4-23）。

1）结核构造

结核是岩石中通过沉淀或交代作用形成的结核状自生矿物集合体，其在成分、结构及颜色等方面与围岩存在差异。按照成分，结核可以划分为多种，如钙质结核、硅质结核、铁质结核等，其中硅质结核在高于庄组泥晶白云岩中极其发育（图4-23A、B和D），钙质结核和铁质结核分别在洪水庄组和下马岭组黄铁矿中发育（图4-23C）。

在野外描述结核时，应该注意判断其成分、大小、形状以及其与围岩间的关系，进而判断其成因类型。

2）缝合线

缝合线是指岩层中因压溶作用导致不溶残余物质（黏土和有机质等）在压溶缝附近聚集而形成的暗色线，缝合线的起伏高度多为几毫米至几厘米，主要发育于易溶的碳酸盐岩地层中，碎屑岩中少见（图4-23D）。

3）李泽岗格环和铁锰侵染构造

李泽岗格环是一种因氧化作用而形成的多个同心环状颜色条带的构造（图4-23E），常见于有渗透性的砂岩中,且远离节理缝、溶洞等高渗流带分布。李泽岗格环主要是在氧含量不充足的情况下,孔隙水中的 Fe^{2+}、Mn^{2+} 沿表面自由能最小的环状带氧化而形成高价 Fe、Mn 氧化物沉淀而形成。除了李泽岗格环外,在天津市蓟州区长城系常州沟组上部,还可以见到 Fe、Mn 氧化物沿裂缝壁沉淀而形成的不规则波状纹理,形态类似于波状叠层石(图 4-23F)。

图 4-23 天津市蓟州区元古宇中典型的化学成因构造

Fig.4-23　Chemogenic structures in the Proterozoic strata in Jizhou District, Tianjin

(A)硅质结核和条带构造,硅质结核切割沉积纹层,表明其为后生成岩成因,高于庄组六段;(B)硅质结核和条带构造,硅质沿着叠层石纹层交代,可能与叠层石中有机质腐烂形成的微酸性环境有关,高于庄组六段;(C)黑色页岩中碳酸盐岩结核构造,洪水庄组;(D)泥晶白云岩中的缝合线及硅质结核构造,高于庄组四段;(E)砂岩中的李泽岗格环构造,常州沟组三段;(F)砂岩断面上的铁锰质侵染构造

5. 生物成因构造

生物成因构造是指生物生长和活动形成的沉积构造,包括遗迹化石、叠层石构造、窗格构造、扰动斑状构造和凝块石构造等。由于前寒武纪缺乏大型生物,以微生物为主,因而形成大量与微生物活动相关的叠层石和凝块石构造。

1)叠层石构造

叠层石由以蓝细菌为主的微生物席与其不断捕获和粘结碳酸盐物质交替叠置而形成的一种纹层状构造。根据其形态的差异,大致可以细分为以下几类:

(1)层状叠层石,具较为平整的纹层,主要分布于潮间带上部至潮上带(图 4-24A 和 B);

(2)波状叠层石,具波状起伏的纹层,主要分布于潮间带中部(图 4-24B 和 C);

(3)丘状叠层石,具穹状起伏的纹层,主要分布于潮间带的中部至下部(图 4-24D);

(4)柱状叠层石,具丘状纹层形成的圆柱组成,柱体分散,仅底部相连,柱体可呈单柱状或分枝状,柱体之间常充填颗粒,少量充填灰泥或云泥,主要分布在潮下带上部至潮间带下部(图 4-24E 和 F);

(5)锥状叠层石,具锥状纹层形成的圆柱组成,与柱状叠层石的形成环境相似;

(6)墙状叠层石,具圆弧形纹层垂向加积而形成的墙状体组成,墙间距为数厘米,墙间充填灰泥或云泥，主要分布在潮间带（图 4-24G 和 H）。

2）凝块石构造

凝块石构造是指毫米级到厘米级大小的、暗色的、泥晶或者微晶组成的、形状和分布不规则的、被泥晶或砂级沉积物或亮晶分隔开的构造，内部结构包括叶状的、细胞状的、微球状的、凝块状的、微粒状的、球粒状的、蠕虫状的、斑点状的、块状的等，或者由肾形藻、葛万藻、努亚藻等钙化微生物组成，是不连续分布的小球状微生物群落（主要是蓝细菌）原地钙化形成的（Kennard and James，1986；吴亚生等，2018）。该类型构造在天津市蓟州区元古宇高于庄组和雾迷山组（图 4-24I）中皆有发育。

图 4-24 天津市蓟州区元古宇中的叠层石及凝块石构造
Fig.4-24 Stromatolites and colt-like structures in the Proterozoic strata in Jizhou District, Tianjin
（A）层状叠层石，注意纹层沿砾屑边缘弯曲，铁岭组；（B）层状至波状叠层石，铁岭组；（C）波状叠层石，雾迷山组；（D）丘状叠层石，雾迷山组；（E）柱状叠层石，铁岭组；（F）柱状叠层石，顶部呈分枝状，铁岭组；（G）墙状叠层石，铁岭组；（H）墙状叠层石，雾迷山组；（I）凝块石，雾迷山组

四、分选、磨圆和颗粒排列方式

1. 分选

分选是指颗粒大小的均匀程度，通常使用标准偏差和分选系数对沉积岩（或沉积物）颗粒的分选程度进行度量，在野外可以通过分选图版进行估计（图 4-25）。分选主要反映沉积水体的稳定

性，一般在水动力稳定、持续动荡的水体环境中所形成颗粒的分选相对较好，例如常州沟组二段的海相滨岸砂岩；而水动力不稳定的河流背景下的颗粒分选相对较差，例如常州沟组一段底部的辫状河相砾质粗砂岩。

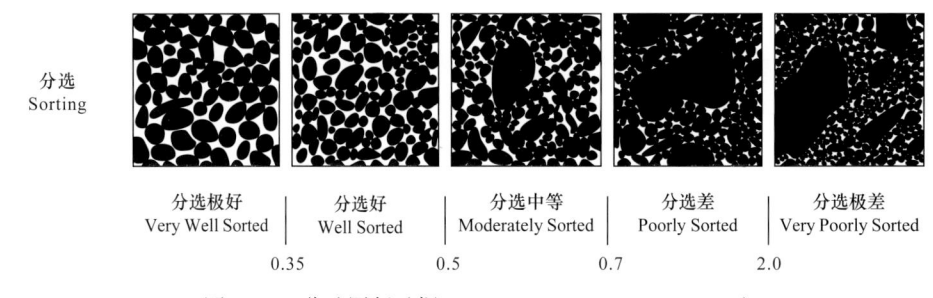

图 4-25　分选图版（据 Compton，1962；Stow，2010）

Fig.4-25　Comparator chart for estimating sorting in sediments（after Compton，1962；Stow，2010）

2. 磨圆度

磨圆度是反映颗粒的原始棱角被磨圆的程度，是由颗粒棱角的曲率决定的。在实际工作中主要通过对比图版进行估计，大致可以细分为尖棱角状、棱角状、次棱角状、次圆状、圆状和滚圆状等 6 个级别（图 4-26）。

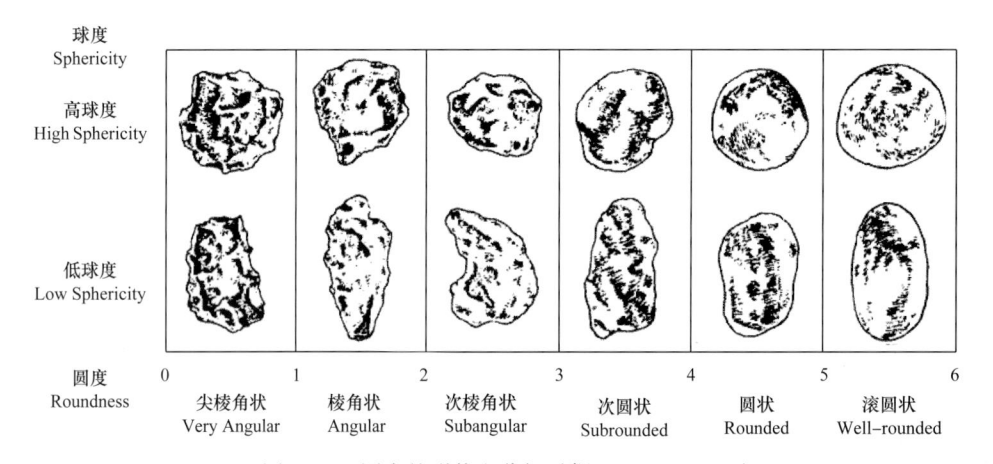

图 4-26　圆度的形状和分级（据 Powers，1953）

Fig.4-26　Comparator charts for estimating the three aspects of grain morphology–shape, roundness, and sphericity（after Powers，1953）

磨圆度主要反映颗粒搬运过程中所受的磨蚀程度，故其主要取决于搬运的时间和距离的长短、搬运方式、颗粒的物化性质、原始形状和大小等。颗粒磨圆度的总趋势是随着搬运的距离和时间的增长而增高。搬运方式也影响颗粒磨圆度的重要因素，一般滚动搬运的颗粒较悬浮搬运的颗粒的磨圆速度快，而跳跃搬运的颗粒磨圆速度居中，以块体方式搬运的颗粒磨圆速度最慢；而粗砂级及其以上颗粒（>0.5mm）多以滚动搬运为主，中—细砂级颗粒（0.5~0.1mm）以跳跃搬运为主，粉砂级颗粒（小于 0.1mm）多以悬浮搬运为主，冰川沉积物多以块体搬运为主；因此在利用磨圆度判断搬运时间和距离时，应该考虑粒度效应对搬运方式的影响，尽量使用同级粒径的颗粒进行比较，且研究粉砂级颗粒的磨圆度意义不大。例如常州沟组底部砂砾岩中砾级颗粒的磨圆度明显高于中—细砂级颗粒，而西井峪组的冰川角砾岩磨圆度多呈尖棱角状至棱角状。硬度较小的颗粒相对于硬度较大的更易磨圆，解理发育的矿物颗粒因易于沿解理面破碎而难以获得高的磨圆度。

3. 颗粒排列

沉积岩（或沉积物）颗粒的排列方式是指颗粒的定向性，其主要受沉积过程、埋藏压实过程及构造应力作用的影响。原始沉积形成的颗粒排列方式在沉积后期埋藏压实和构造应力作用过程中可能发生调整，因此在观察时应注意区分。颗粒的排列通常可以隐含以下地质信息：

（1）板状或片状颗粒通常近平行于地层层面排列，例如介壳类化石聚集的层，介壳多平行于层面分布，且介壳的凸起面多向上，可以辅助地层顶底的判断。板状或片状颗粒大角度斜交或垂直层面分布，大多是由事件性沉积作用导致的快速沉积而形成的，例如铁岭组顶部的砾屑灰岩，可能与风暴作用相关（图4-27A）。

（2）板状或碟状颗粒相互叠置而形成的叠瓦状构造，其颗粒上倾面多指示上游方向，例如常州沟组一段的砾质粗砂岩（图4-27B）。

（3）对于未经或少受改造的生物成因碳酸盐岩，保存有较为完整的生物定向生长构造，可以根据其形态分析地层顶底和水流方向，例如铁岭组的叠层石石灰岩（图4-27C和D）。

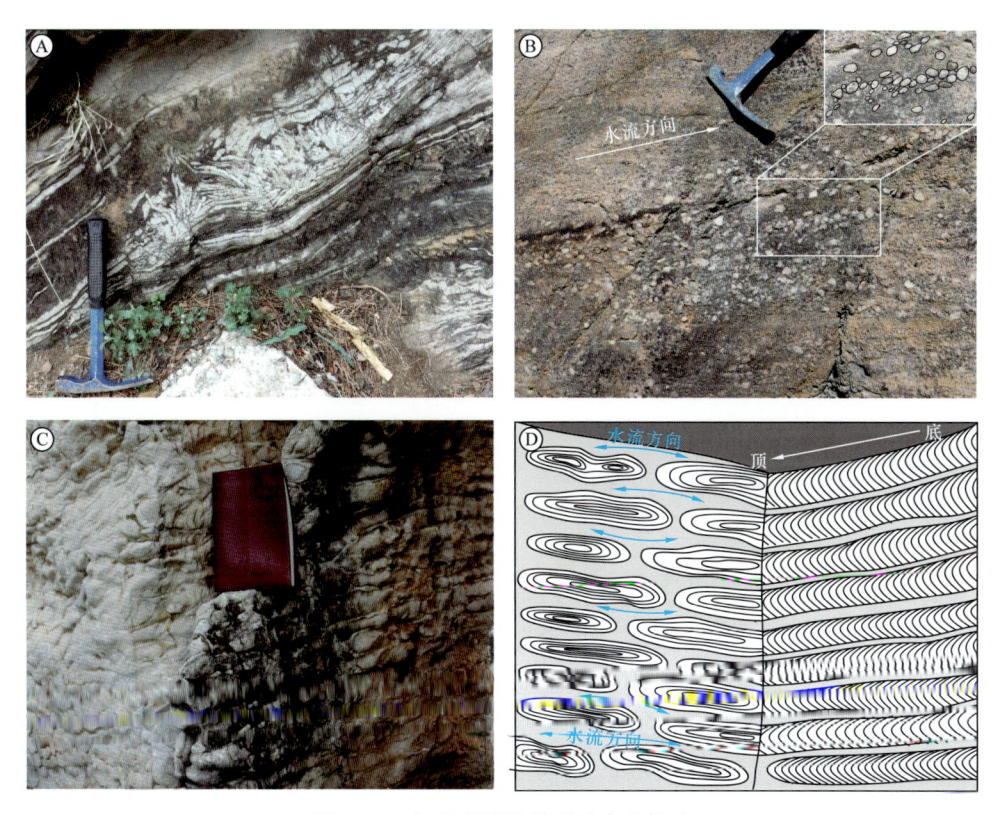

图 4-27　沉积岩颗粒排列及定向构造

Fig.4-27　Arrangement of grains and orientation structures in sedimentary rocks

（A）砾屑石灰岩，铁岭组顶部，砾屑与层面呈大角度斜交，可能与风暴有关；（B）砾质粗砂岩，常州沟组一段，砾石呈叠瓦状排列，砾石倾斜方向为上游方向；（C）柱状叠层石灰岩，铁岭组，叠层石层面向上凸起的方向为地层顶部，在层面上，叠层石沿水流方向拉长；（D）图C的示意图

五、孔隙类型

孔隙是沉积岩（或沉积物）的重要结构组分之一，对于水文地质和石油地质而言十分重要。

按照成因，孔隙可以细分为原生孔隙和次生孔隙两大类，原生孔隙和次生孔隙又可以进一步分类为多种类型（表4-10）。

表 4-10 孔隙类型的划分
Table 4-10 Classification of pore types

孔隙类型			成因及特征
原生孔隙（沉积时形成）	粒间孔		分布于颗粒之间
	泥间孔		分布于泥之间
	体腔孔		分布于颗粒内部
	格架孔		分布于生物格架之间
	遮蔽孔		分布于扁平砾级颗粒下方，由颗粒遮挡形成，可指示岩层顶底
	原生晶间孔		分布于晶体之间
次生孔隙（沉积后形成）	溶蚀孔隙	粒间溶孔	颗粒或填隙物溶蚀形成，分布于颗粒之间
		粒内溶孔	颗粒内部溶蚀形成，分布于颗粒内部
		晶间溶孔	晶体溶蚀形成，分布于晶体之间
		晶内溶孔	晶体溶蚀形成，分布于晶体内部
		铸模孔	颗粒（或晶体）全部溶解，孔隙继承了颗粒（或晶体）的形状
	次生晶间孔		分布于成岩阶段形成的晶体之间的孔隙
	干缩孔		刚沉积不久的泥级沉积物因蒸发失水收缩形成的孔，如鸟眼孔
	腐烂孔		生物有机体腐烂分解后留下的孔隙
	碎裂孔		岩石破碎形成的角砾之间的孔隙
	裂缝	构造缝 张裂缝	由张应力形成的裂缝，定向排列
		构造缝 剪裂缝	由剪切应力形成的裂缝，定向排列
		压实缝	在上覆地层重压下，岩石破裂形成的裂缝，排列无定向
		压溶缝	岩石压溶形成的缝合线，多平行层面
		风化缝	地表岩石经风化（热胀冷缩等）而形成的裂缝，排列无定向
		干缩缝	刚沉积不久的泥级沉积物因蒸发而失水收缩形成的裂缝

孔隙度是度量岩石中孔隙空间发育情况的重要参数，但在野外工作中，一般通过手标本的吸水能力进行简单的判断，精确的测量需要借助孔隙度测量仪。在野外工作中，可以通过肉眼和放大镜对孔隙成因进行观察和大致判断，通常原生孔隙与次生孔隙主要存在以下区别：

（1）原生孔隙形状较为规则，不切割颗粒（或晶粒），孔隙边缘较为平滑（或平直），分布较为均匀；

（2）次生孔隙形状通常不规则，切割颗粒（或晶粒），孔隙边缘参差不齐，分布通常不均匀，分布于颗粒（或晶粒）之间或颗粒（或晶粒）内部，部分出现直径大于颗粒的"超大孔"。

综上所述，在野外露头观察和测量过程中，对沉积岩（或沉积物）结构特征的描述应注意以下几个方面：

（1）估计沉积岩（或沉积物）的颗粒粒度（包括最大粒度、最小粒度和平均粒度）、分选情况；

（2）观察颗粒的磨圆度和球度，并寻找具有特殊意义的表面结构特征，如冰川擦痕等；

（3）观察颗粒的排列方式或原地生物生长构造的形态特征，注意挖掘其中隐含的示顶底、古流向等地质信息；

（4）观察填隙物的类型、支撑类型，区分颗粒支撑和杂基支撑；

（5）描述并估计孔隙发育情况，包括孔隙度、孔隙大小、分布及成因等信息；

（6）综合评估沉积岩（或沉积物）的结构成熟度，并尝试确定其形成环境。

参 考 文 献

Compton R R. 1962. Manual of Field Geology. New York: John Wiley, 378.

Dunham R J. 1962. Classification of Carbonate Rocks According to Depositional Texture. In: Ham W E. (Ed.), Classification of Carbonate Rocks: A Symposium, Tulsa, OK, AAPG Memoir 1, 108-121.

Eby D E. 1975. Carbonate sedimentation under elevated salinities and implications for the origin of "Molar-tooth" structure in the Middle Belt Carbonate interval (late Precambrian), northwestern Montana. Abstracts with Programs- Geological Society of America, 7:1063.

Embry A F, Klovan J E. 1971. A late Devonian reef tract on northeastern Banks Island, N.W.T. Bulletin of Canadian Petroleum Geology, 19 (4): 730-781.

Fairchild I J, Einsele G, Song T. 1997. Possible seismic origin of molar tooth structures in Neoproterozoic carbonate ramp deposits, North China. Sedimentology, 44 (4): 611-636.

Fan D L, Yang P J, Wang R. 1999. Characteristics and origin of the Middle Proterozoic Dongshuichang chambersite deposit, Jixian, Tianjin, China, Ore Geology Reviews, 15 (1-3): 15-29.

Folk R L. 1959. Practical petrographic classification of limestones. AAPG Bulletin, 43 (1), 1-38.

Folk R L. 1962. Spectral subdivision of limestone type. In: Ham W E. (Ed.), Classification of Carbonate Rocks- a symposium: Tulsa, OK, AAPG Memoir 1, 62-84.

Folk R L. 1968. Petrology of sedimentary rocks, Austin: Hemphill's, pp 170.

Frank T D, Lyons T W. 1998. "Molar-tooth" structures; a geochemical perspective on a Proterozoic enigma. Geology, 26 (8): 683-686.

Furniss C S, Rittel J F, Winston D. Gas bubble and expansion crack origin of molar-tooth calcite structures in the Middle Proterozoic Belt supergroup. Journal of Sedimentary Research, 68 (1) :104-114.

Gao L Z, Zhang C H, Shi X Y. 2007. A new SHRIMP age of the Xiamaling Formation in the North China Plate and its geological significance. Acta Geological Sinica, 81 (6) :1103-1109.

Gao L Z, Zhang C H, Shi X Y. 2008. Mesoproterozoic age for Xiamaling formation in North china Plate indicated by zircon SGRIMP dating. Chinese Science Bulletin, 53 (17) :1665-2671.

Gao L Z, Zhang C H, Liu P J. 2009. Reclassification of the Meso- and Neoproterozoic chronostratigraphy of North China by SHRIMP zircon ages. Acta Geological Sinica, 83 (6) :1074-1084.

Grabau A W. 1922. The Sinian System. Bull. Geol. Soc. China, 1, 44-88.

Hou G T, Santosh M, Qian X L, Lister G S, Li J H. 2008. Configuration of the Late Paleoproterozoic supercontinent Columbia: Insights from radiating mafic dyke swarms. Gondwana Research, 14 (3): 395-409.

James N P, Narbonne G M, Sherman A G. 1998. Molar-tooth carbonates: shallow subtidal facies of the mid- to late Proterozoic. Journal of Sedimentary Research, 68 (5): 716-722.

Kao C S, Hsiung Y H, Kao P. 1934. Preliminary notes on Sinian stratigraphy of North China. Bull.Geol.Soc.China, 13(2): 243-288.

Kennard J M, James N P. 1986. Thrombolites and stromatolites: Two distinct types of microbial structures. Palaios, 1 (5): 492-503.

Knoll A H, Sweet K.1990. Carbonate deposition during the Late Proterozoic era: An example from Spitsbergen. American Journal of Science, 290-A:104-132.

Krynine P D. 1948. The megascopic study and field classification of sedimentary rocks. Journal of Geology, 56: 130-165.

Kusky T M, Li J H. 2003. Paleoproterozoic tectonic evolution of the North China Craton. Journal of Asian Earth Sciences, 22 (4): 383-397.

参考文献

Kusky T, Li J G, Santosh M. 2007a. The Paleoproterozoic North Hebei orogen: North China craton's collisional suture with the Columbia supercontinent. Gondwana Research, 12 (1-2): 4-28.

Kusky T M, Windley B F, Zhai M G. 2007b. Tectonic Evolution of the North China Block: from Orogen to Craton to Orogen. In: Zhai M G, Windley B F, Kusky T M, Meng Q R. (Eds.), Geological Society of London Special Publication, 280: Mesozoic Sub-Continental Lithospheric Thinning Under Eastern Asia, pp. 1-34.

Kusky T M, Santosh M. 2009. The Columbia connection in North China. In: Reddy S M, Mazumder R, Evans D A D, Collins A S. (Eds.), Geological Society of London Special Publication, 323: Palaeoproterozoic Supercontinents and Global Evolution, pp. 49-71.

Lu S N, Zhao G C, Wang H C, Hao G J. 2008. Precambrian metamorphic basement and sedimentary cover of the North China Craton: a review. Precambrian Research, 160:77-93.

Meng X H, Ge M. 2002. The sedimentary features of Proterozoic microspar (molar-tooth) carbonates in China and their significance. Episodes, 25 (3): 185-196.

Meng X H, Ge M. 2003. Cyclic sequences, events and evolution of the Sino-Korean Plate, with a discussion on the evolution of molar-tooth carbonates, phosphorites and source rocks. Acta Geological Sinica (English Edition), 77 (3): 185-196.

Meng X H, Ge M. 2003. Cyclic Sequences, Events and Evolution of the Sino-Korean Plate, with a Discussion on the Evolution of Molar-tooth Carbonates, Phosphorites and Source Rocks. Acta Geological Sinica (English Edition). 77(3):382-401.

O'Connor M P. 1972. Classification and environmental interpretation of the cryptalgal organ sedimentary molar-tooth structure from the late Precambrian Belt-Purcell Supergroup. The Journal of Geology, 80:592-610.

Pettijohn F J, Potter P E, Siever R. 1972. Sand and Sandstone. New York: Springer-Verlag.

Pollock M D, Kah L C, Bartley J K. 2006. Morphology of molar-toothstructures in Precambrian carbonates:Influence of substrate geology and implacation for genesis. Journal of Sedimentary Research, 76:310-323.

Powers M C. 1953. A New Roundness Scale for Sedimentary Particles. Journal of Sedimentary Research, 23(2): 117-119.

Pratt B. R. 1998. Syneresis cracks: subaqueous shrinkage in argillaceous sediments caused by earthquake-induced dewatering. Sedimentary Geology, 117 (1): 1-10.

Richhofen F V. 1882. China. Berlin:Verlay von Dietrich Reimer, 2, PP.244.

Rogers J, Santosh M. 2003. Supercontinents in earth history. Gondwana Research, 6 (3): 357-368.

Santosh M, Maruyama S, Yamamoto S. 2009. The making and breaking of supercontinents: Some speculations based on superplumes, super downwelling and the role of tectosphere. Gondwana Research, 15 (3-4): 324-341.

Santosh M. 2010a. A synopsis of recent conceptual models on supercontinent tectonics in relation to mantle dynamics, life evolution and surface environment. Journal of Geodynamics, 50 (3-4): 116-133.

Santosh M. 2010b. Assembling North China Craton within the Columbia supercontinent: The role of double-sided subduction. Precambrian Research, 178 (1-4): 149-167.

Smith AG. 1968. The origin and deformation of some "Molar-Tooth" structures in the Precambrian Belt-Purcell Supergroup. Journal of Geology, 76:426-443.

Song T R. 1988. A probable earthquake-tsunami sequence in Precambrian carbonate strata of Ming Tombs District, Beijing. Science Bulletin, 33 (13):1121-1121.

Stow D A V. 2010. Sedimentary Rocks in the Field: A Colour Guide. London: Manson Publising, pp320.

Su W B, Zhang S H, Warren D. Huff. 2008. SHRIMP U-Pb ages of K-bentonite beds in the Xiamaling Formation: Implications for revised subdivision of the Meso- to Neoproterozoic history of the North China Craton. Gondwana Research, 14:543-553.

Wanless H R. 1979. Limestone response to stress: pressure solution and dolomitization. Journal of Sedimentary Petrology,

49（2）：437-462.

Wilde S A, Zhao G C, Sun M. 2002. Development of the North China Craton during the late Archaean and its final amalgamation at 1.8 Ga：Some speculations on its position within a global Palaeoproterozoic supercontinent. Gondwana Research, 5（1）：85-94.

Williams H, Turner F J, Gilbert C M. 1954. Petrology. San Francisco：Freeman.

Wright V P. 1992. A revised classification of limestones. Sedimentary Geology, 76（3-4）：177-185.

Yang H, Chen Z Q, Fang Y. 2016. Microbially induced sedimentary structures from the 1.64 Ga Chuanlinggou Formation, Jixian, North China. Palaeogeography, Palaeoclimatology, Palaeoecology, 474：7-25.

Zhai M G, Liu W J. 2003. Palaeoproterozoic tectonic history of the North China craton：a review. Precambrian Research, 122（1-4）：183-199.

Zhai M G, Shao J A, Hao J, Peng P. 2003. Geological signature and possible position of the North China block in the supercontinent Rodinia. Gondwana Research, 6（2）：171-183.

Zhang C. 2004. Hot tectonic events and evolution of northern margin of the North China craton in Mesoproterozoic. Acta Scientiarum Natralium, 40：232-240.

Zhao G C, Sun M, Wilde S A, Li S Z. 2003. Assembly, accretion and breakup of the paleo-mesoproterozoic Columbia supercontinent：Records in the North China Craton. Gondwana Research, 6（3）：417-434.

陈晋镳, 武铁山. 1997. 华北区区域地层. 武汉：中国地质大学出版社.

陈晋镳, 张惠民, 朱士兴, 等. 1980. 蓟县震旦亚界的研究// 中国震旦亚界. 中国地质科学院天津地质矿产研究所. 天津：天津科学技术出版社, 56-114.

陈荣辉, 陆宗斌. 1963. 河北蓟县震旦系标准地质剖面. 地质丛刊（甲种），前寒武纪地质专号（1），99-127.

陈一笠. 1992. 天津市蓟县发现沉积型海泡石矿床. 中国区域地质,（3）：204-210.

杜汝霖, 李培菊, 吴振山. 1979. 燕山西段震旦亚界. 河北地质学院学报, 4（4）：1-17.

杜汝霖, 李培菊. 1980. 燕山西段震旦亚界// 中国震旦亚界, 中国地质科学院天津地质矿产研究所. 天津：天津科学技术出版社, 341-357.

冯增昭. 1982. 碳酸盐岩分类. 石油学报, 1：11-18, 96-98.

冯增昭. 1992. 沉积岩石学（上、下册）. 北京：石油工业出版社.

冯增昭. 1994. 沉积岩石学（第二版）. 北京：石油工业出版社.

高计元. 1988. 中国南方泥盆系瘤状灰岩的成因. 沉积学报, 6（6）：77-86.

高林志, 张传恒, 史晓颖, 等. 2007. 华北青白口系下马岭组凝灰岩锆石 SHIRIMP U-Pb 定年. 地质通报, 26（3）：249-255.

高维, 张传恒, 高林志, 等. 2008. 北京密云环斑花岗岩的锆石 SHRIMP U-Pb 年龄及其构造意义. 地质通报, 27（6）：793-798.

高振家, 陈克强, 高林志. 2014. 中国岩石地层名称辞典. 成都：电子科技大学出版社.

郭福生. 1989. 下扬子地区三叠系下统瘤状灰岩成因研究. 东华理工学院学报, 12（14）：17-22.

何政军, 牛宝贵, 张新光, 等. 2011. 北京密云元古宙常州沟组之下环斑花岗岩古风化壳岩石的发现及其碎屑锆石年龄. 地质通报, 30（5）：798-802.

金振奎, 邵冠铭. 2014. 石灰岩分类新方案. 新疆石油地质, 35（2）：235-242.

金振奎, 石良, 高白水, 等. 2013. 碳酸盐岩沉积相及相模式. 沉积学报, 31（6）：965-979.

旷红伟, 刘燕学, 孟祥化, 等. 2004. 吉辽地区新元古代臼齿碳酸盐岩岩相的若干岩石学特征研究. 地球学报, 25（6）：647-652.

旷红伟, 孟祥化, 葛铭. 2006. 臼齿碳酸盐岩成因探讨——以吉林—辽宁地区新元古界为例. 古地理学报. 8（1）：63-73.

旷红伟, 金广春, 刘燕学. 2008. 吉辽地区新元古代臼齿构造形态及其研究意义. 中国科学（D辑），38（增Ⅱ）：123-130.

参考文献

旷红伟，柳永清，彭楠，等．2011．再论臼齿碳酸盐岩成因．古地理学报，13（3）:253-261．

李怀坤，李惠民，陆松年，1995．长城系团山子组火山岩颗粒锆石 U-Pb 年龄及其地质意义．地球化学，24（1）:43-48．

李怀坤，陆松年，李惠民，等．2009．侵入下马岭组的基性岩床的锆石和斜锆石 U-Pb 精确定年——对华北中元古界地层划分方案的制约．地质通报，28（10）：1396-1404．

李怀坤，朱士兴，相振群，等．2010．北京延庆高于庄组凝灰岩的锆石 U-Pb 定年研究及其对华北北部中元古界划分分新方案的进一步约束．岩石学报，26（7）：2131-2140．

李怀坤，苏文博，周红英，等．2011．华北克拉通长城系底界年龄小于 1670 Ma——来自北京密云花岗斑岩岩脉锆石 LA-MC-ICPMS U-Pb 年龄的约束．地学前缘，18（3）：108-120．

刘燕学，旷红伟，蔡国印，等．2003．辽南新元古代营城子组臼齿灰岩的沉积环境．地质通报，22（6）：419-425．

陆松年，李惠民，1991．蓟县长城系大红峪组火山岩的单颗粒锆石 U-Pb 法精确测准确定年．中国地质科学院院报，22:137-145．

梅冥相，马永生，郭庆银．2001．天津蓟县雾迷山旋回层基本模式及其马尔柯夫链分析．高校地质学报，7（3）：288-299．

梅冥相，孟庆芬，刘智荣．2007．微生物形成的原生沉积构造研究进展综述．古地理学报．9（4）:353-367．

孟祥化，葛铭，旷红伟，Nielsen J K．2006．微亮晶（臼齿）碳酸盐成因及其在元古宙地球演化中的意义．岩石学报．22（8）:2133-2143．

孟庆任，伍国利，曲永强，等．2016．华北克拉通北缘中元古代沉积盆地演化//中国东部中—新元古界地质学与油气资源．孙枢，王铁冠．北京：科学出版社．

潘家明．1992．蓟县地区新构造运动与地貌演化．天津地质学会志，10（1）:59-64．

乔秀夫．1976．青白口群地层学研究．地质科学，（3）:246-265．

乔秀夫，宋天锐，高林志，等．1994．碳酸盐岩振动液化地震序列．地质学报．68（1）:16-34．

乔秀夫．1996．中国震积岩的研究与展望．地质评论．42（4）:317-320．

乔秀夫，高林志，彭阳．2001．古郯庐带新元古界—灾变、层序、生物．北京：地质出版社，98-107．

乔秀夫，李海兵．2009．沉积物的地震及古地震效应．古地理学报．11（6）：593-610．

曲永强，孟庆任，马收先，等．2010．华北地块北缘中元古界几个重要不整合面的地质特征及构造意义．地学前缘，17（4）：112-127．

申庆荣，廖大从，1958．燕山山脉震旦纪地层及震旦纪沉积矿产．中国地质学报，38：263-278．

史书婷，王金艺，郭芪恒，等．2019．天津蓟县晚元古代冰碛岩的发现．沉积学报，37（6）：1181-1192．

苏犁，王铁冠，李献华，等．2016．燕辽裂陷带中元古界下马岭组辉长辉绿岩岩床成岩机制与侵入时间//中国东部中—新元古界地质学与油气资源．孙枢，王铁冠．北京：科学出版社．

苏文博，李怀坤，Huff W D，等．2010．铁岭组钾质斑脱岩锆石 SHRIMP U-Pb 年代学研究及其地质意义．科学通报，55（22），2197-2206．

孙善平，刘永顺，钟蓉，等．2001．火山碎屑岩分类评述及火山沉积学研究展望．岩石矿物学杂志，20（3），313-317．

孙淑芬．2000．天津蓟县洪水庄组微古植物群．前寒武纪研究进展，23（3）：165-172．

孙云铸，1957．寒武纪下界问题．地质知识（4）:1-2．

天津地质矿产研究所，南京地质古生物研究所，内蒙古自治区地质局．1979．蓟县震旦亚界叠层石研究．北京：地质出版社．

天津市地质矿产局．1992 天津市区域地质志．北京：地质出版社．

王鸿祯．1985．中国古地理图集．北京：地图出版社．

汪凯明，罗顺社．2014．河北宽城地区洪水庄组岩石特征及沉积环境．沉积与特提斯地质，34（2）：29-35．

王培君．1996．硼矿床含硼地层的二元结构模式．化工矿产地质，18（3）:58-63．

王秋舒，许虹，高燊，等．2013．稀有矿物锰方硼石的合成及其矿床地质意义．地学前缘，20（3）：123-130．

王曰伦.1960.全国震旦系对比线索.地质论评,20(5):203-205.

王曰伦.1963.中国北部震旦系和寒武系分界问题.地质学报,43(2):116-140.

闻秀明.2013.天津市大地构造相与成矿.地质调查与研究,36(2):123-130.

吴亚生,姜红霞,虞功亮,等.2018.微生物岩的概念和重庆老龙洞剖面P-T界线地层微生物岩成因.古地理学报,20(5):737-775.

武振杰,张传恒,姚建新.2009.滇中中元古界大龙口组地震灾变事件及地质意义.地球学报.30(3):375-386.

夏丹,汪凯明,罗顺社,等.2009.燕山地区高于庄组张家峪亚组瘤状灰岩成因研究.石油地质与工程,23(1):4-7.

肖荣阁,大井隆夫,侯万荣,等.2002.天津蓟县硼矿床锰方硼石矿物的硼同位素研究.现代地质,16(3):186-191.

邢裕盛,等.1989.中国的上前寒武系.中国地层3.北京:地质出版社.

阎国翰,牟保磊,许保良,等.2000.燕辽-阴山三叠纪碱性侵入岩年代学和Sr,Nd,Pb同位素特征及意义.中国科学(D辑),30(4):383-387.

杨付领,牛宝贵,任纪舜,等.2015.马兰峪背斜核部中生代侵入岩体锆石U-Pb年龄、地球化学特征及其构造意义.地球学报,36(4):455-465.

叶良辅.1920.北京西山地质志.地质专报甲种第1号.原农商部地质调查所.

于荣炳,张学祺.1984.燕山地区晚前寒武纪同位素地质年代学的研究.天津地质矿产研究所所刊,11:1-24.

俞建章,崔盛芹,仇甘霖.1964.再论辽东地区震旦地层及其与燕山地区的对比.中国地质学报,44(1):1-12.

袁鄂荣,周国华,王家生.1994.河北涿鹿地区洪水庄组底砾岩及其地质意义.河北地质学院学报,17(5):425-432.

张继庆.1986.四川盆地早二叠世碳酸盐沉积相及风暴沉积作用.重庆:重庆出版社,8-20.

张文佑,李唐泌.1935.中国北京震旦纪与前寒武纪地层之分界问题.原中央研究院院务汇报,6(2):30-50.

钟富道.1977.从燕山地区震旦地层同位素年龄论中国震旦地层年表.中国科学,(2):151-161.

朱士兴,李怀坤,孙立新,等.2016.燕山中新元古界的层序和划分//中国东部中—新元古界地质学与油气资源,孙枢,王铁冠.北京:科学出版社.

Abstract Field Guide of Stratigraphy and Sedimentary Facies of the Proterozoic in Jizhou District, Tianjin

Jin Zhenrui, Zhu Xiaoer, Wang Jinyi, Wang Xinyao, Ren Yilin, Wang Ling, Guo Qiheng,
Li Yang, Shi Shuting, Li Shuo, Yuan Kun, Li Rui, Yan Wei

China University of Petroleum-Beijing (CUPB), Changping, Beijing, China 102200

SECTION I Introduction

The site of the Proterozoic outcrop for field investigation is located in the Jixian National Geological Park, which is in the Jizhou District, Tianjin City, about 88 km east of the downtown of Beijing (Fig.1). The outcrop of the Proterozoic in this area is world renowned for its excellent exposure of continuous strata and convenient transportation.

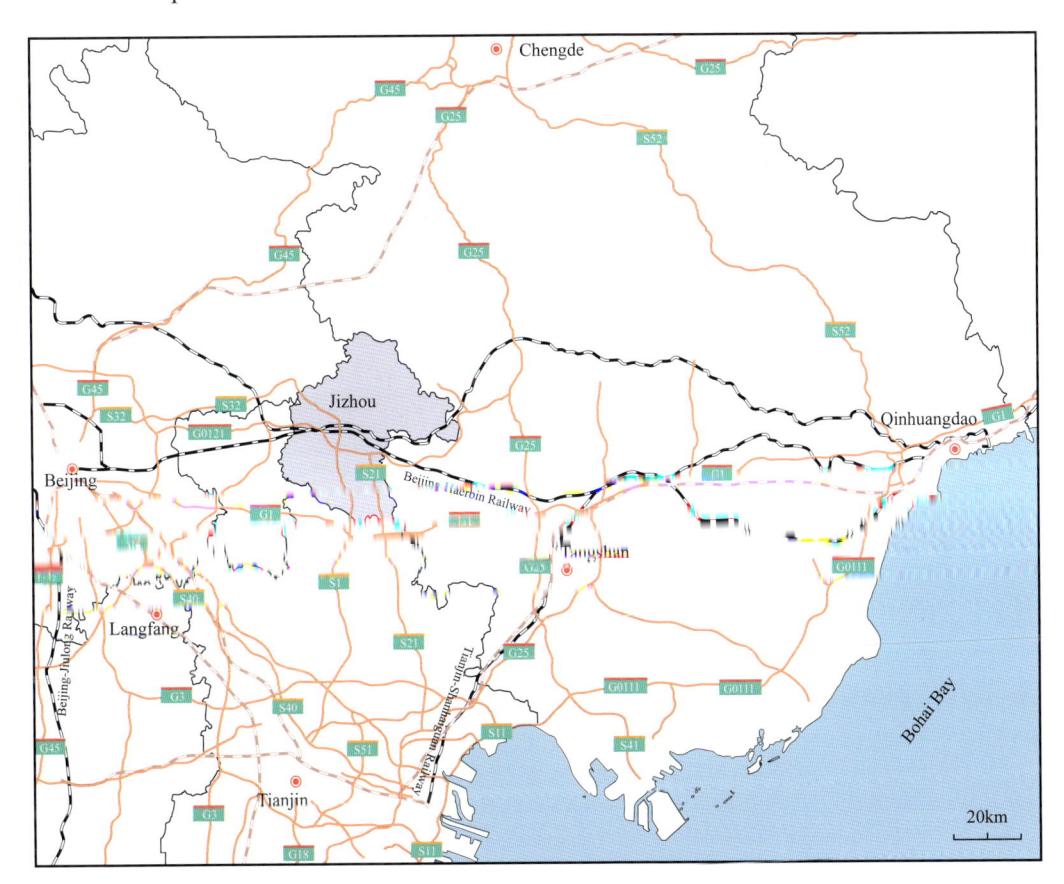

Fig.1 Location (red triangle in the map) of the site of the Proterozoic outcrop for field investigation in Jizhou District, Tianjin.

The Proterozoic profile in the Jizhou District, Tianjin City, is in the southwest wing of the Malanyu Anticline. From north to south, the Proterozoic strata are exposed from older to newer (Fig.2).

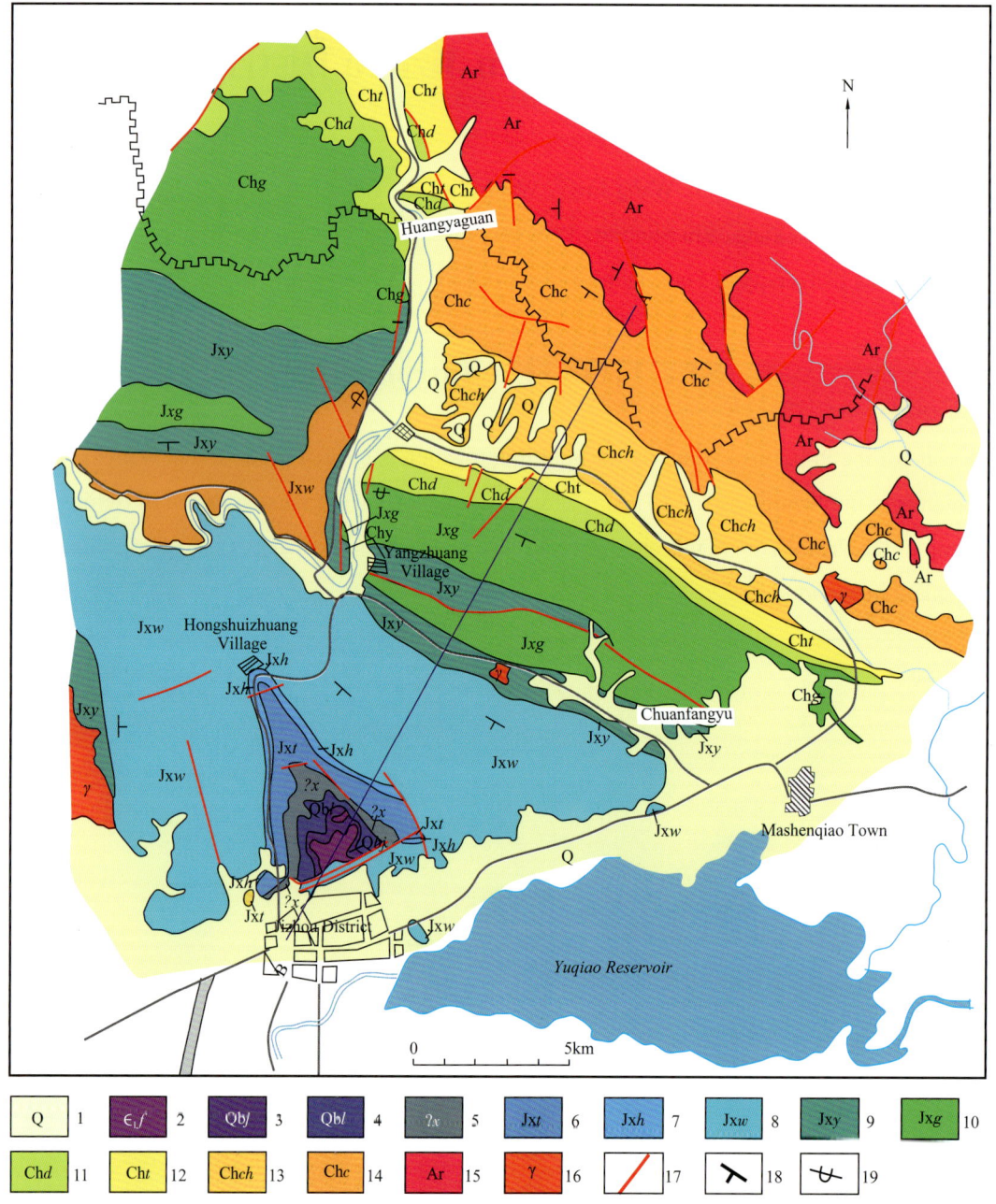

Fig.2　Geological sketch map of northern mountain area of Jizhou District, Tianjin (modified from Yang, 2010)

1—Quaternary; 2—Low Cambrian Fujunshan Fm.; 3—Neo-Proterozoic Qingbaikou System Jing'eryu Fm.; 4—Neo-Proterozoic Qingbaikou System Longshan Fm.; 5—Meso-Proterozoic Daijian System Xiamaling Fm.; 6—Meso-Proterozoic Jixian System Tieling Fm.; 7—Meso-Proterozoic Jixian System Hongshuizhuang Fm.; 8—Meso-Proterozoic Jixian System Wuhishen Fm.; 9—Meso-proterozoic Jixian Yangzhuang Fm.; 10—Meso-Proterozoic Jixian System Gaoyuzhuang Fm.; 11—Paleo-Proterozoic Changcheng System Dahongyu Fm.; 12—Paleo-proterozoic Changcheng System Tuanshanzi Fm.; 13—Paleo-proterozoic Chang System Chuanlinggou Fm.; 14—Paleo-Proterozoic Changcheng System Changzhougou Fm.; 15—Archean; 16—Granite; 17—Fault; 18—Stratigraphic atttude; 19—Stratigraphic overturning

The Proterozoic is 9545m thick, including the Paleoproterozoic, Mesoproterozoic and Neoproterozoic, and is divided into 13 formations (Fig.3). From bottom to top, the formations are: the Paleoproterozoic Changzhougou Fm. (Chc, 859 m in thickness), Chuanlinggou Fm. (Chch, 889m), Tuanshanzi Fm. (Cht, 518 m) and Dahongyu Fm. (Chd, 408m) of the Changcheng system; the Mesoproterozoic Gaoyuzhuang Fm. (Jxg, 1700m), Yangzhuang Fm. (Jxy, 773m), Wumishan Fm. (Jxw, 3416m), Hongshuizhuang Fm. (Jxh, 131m) and Tieling Fm. (Jxt, 303m) of the Jixian system, and Xiamaling Fm. (?x, 165m) of a system to be established; the Neoproterozoic Longshan Fm. (Qbl, 118m) and Jingeryu Fm. (Qbj, 110m) of the Qingbaikou system.

In this study, we established a new formation——the Xijingyu Fm., and thought it belongs to the Nanhua system. Previously the Xijingyu Fm. was thought be the lower part of the Lower Cambrian Fujunshan Fm. Based on our recent study, the Xijingyu Fm. is glacial deposits and consists of very thick carbonate breccia (155m thick, see Section 14).

Thirteen investigation routes are designed and are as following:

(1) Changzhougou Fm. from Changzhou Village to north of Qingshanling Village in Xiaying Town.

(2) Chuanlinggou Fm. from north of Qingshanling Village to Liuzhuangzi Village in Xiaying Town.

(3) Tuanshanzi Fm. west of Tuanshanzi Village, Xiaying Town.

(4) Dahongyu Fm. in Dahongyu gou valley, Xiaying Town.

(5) Gaoyuzhuang Fm. from Dahongyugou valley to Zhaizhuang Village in Luozhuangzi Town.

(6) Yangzhuang Fm. from north Zhaizhuang Village to Qinshan Village and Huaguoyu Village, Luozhuangzi Town.

(7) Wumishan Fm. along Qingshan Village—Mopanyu Village—Ershilipu Village in Luozhuangzi Town.

(8) Hongshuizhuang Fm. in Xiaolingzi Village, Yuyang Town.

(9) Tieling Fm. from Xiaolingzi Village, Yuyang Town to south Tielingzi Village in Luozhuangzi Town.

(10) Xiamaling Fm. from south Tielingzi Village to Xiazhuangzi Village in Luozhuangzi Town.

(11) Longshan Fm. in Housi Valley, Yuyang Town.

(12) Jingeryu Fm. north of Xijingyu Village, Yuyang Town.

(13) Xijingyu Fm.—Fujunshan Fm. west Fujunshan Park, Yuyang Town.

The total length of the investigation routes is about 34.4 km. It takes about three days to finish all the routes. It takes about 0.5 days from Changzhougou Fm. to Chuanlinggou Fm., about 1 day from Tuanshanzi Fm. to Gaoyuzhuang Fm., about 0.5 days from Yangzhuang Fm. to Wumishan Fm., about 0.5 days from Hongshuizhuang Fm. to Tieling Fm., and about 0.5 days from Xiamaling Fm. to Xijingyu Fm. Two days will also be ok for quick observation.

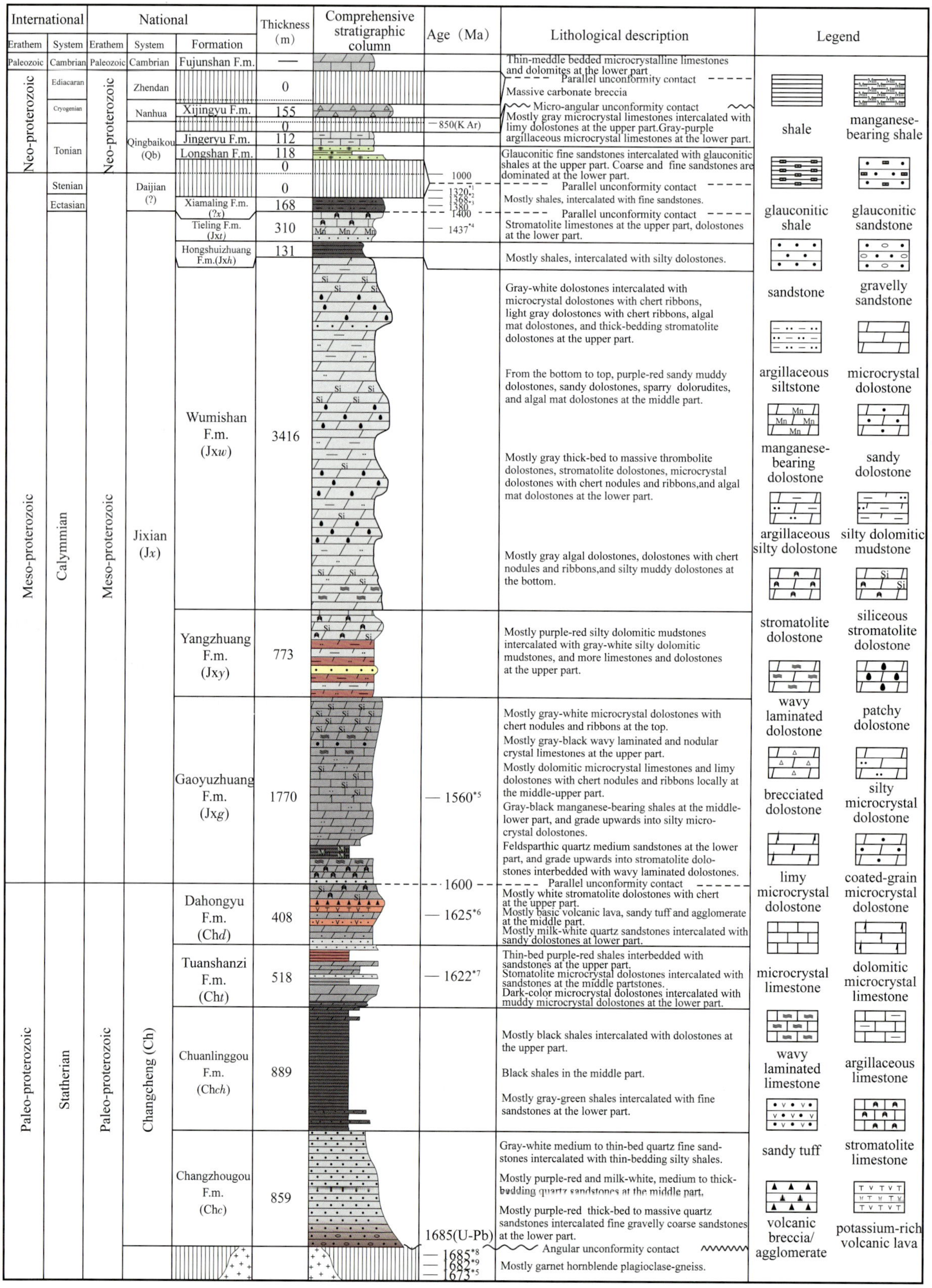

International		National		Formation	Thickness (m)	Comprehensive stratigraphic column	Age (Ma)	Lithological description
Erathem	System	Erathem	System					
Paleozoic	Cambrian	Paleozoic	Cambrian	Fujunshan F.m.	—			Thin-middle bedded microcrystalline limestones and dolomites at the lower part.
Neo-proterozoic	Ediacaran	Neo-proterozoic	Zhendan		0			Parallel unconformity contact
								Massive carbonate breccia
	Cryogenian		Nanhua	Xijingyu F.m.	155			Micro-angular unconformity contact
					0		850(K Ar)	Mostly gray microcrystal limestones intercalated with limy dolostones at the upper part. Gray-purple argillaceous microcrystal limestones at the lower part.
	Tonian		Qingbaikou (Qb)	Jingeryu F.m.	112			
				Longshan F.m.	118			Glauconitic fine sandstones intercalated with glauconitic shales at the upper part. Coarse and fine sandstones are dominated at the lower part.
					0		1000	Parallel unconformity contact
Meso-proterozoic	Stenian	Meso-proterozoic	Daijian (?)		0		1320*2 1368*2 1380*3	Mostly shales, intercalated with fine sandstones.
	Ectasian			Xiamaling F.m. (Jxx)	168		1400	Parallel unconformity contact
	Calymmian		Jixian (Jx)	Tieling F.m. (Jxt)	310		1437*4	Stromatolite limestones at the upper part, dolostones at the lower part.
				Hongshuizhuang F.m. (Jxh)	131			Mostly shales, intercalated with silty dolostones.
				Wumishan F.m. (Jxw)	3416			Gray-white dolostones intercalated with microcrystal dolostones with chert ribbons, light gray dolostones with chert ribbons, algal mat dolostones, and thick-bedding stromatolite dolostones at the upper part.
								From the bottom to top, purple-red sandy muddy dolostones, sandy dolostones, sparry dolorudites, and algal mat dolostones at the middle part.
								Mostly gray thick-bed to massive thrombolite dolostones, stromatolite dolostones, microcrystal dolostones with chert nodules and ribbons, and algal mat dolostones at the lower part.
								Mostly gray algal dolostones, dolostones with chert nodules and ribbons, and silty muddy dolostones at the bottom.
				Yangzhuang F.m. (Jxy)	773			Mostly purple-red silty dolomitic mudstones intercalated with gray-white silty dolomitic mudstones, and more limestones and dolostones at the upper part.
				Gaoyuzhuang F.m. (Jxg)	1770		1560*5	Mostly gray-white microcrystal dolostones with chert nodules and ribbons at the top. Mostly gray-black wavy laminated and nodular crystal limestones at the upper part. Mostly dolomitic microcrystal limestones and limy dolostones with chert nodules and ribbons locally at the middle-upper part. Gray-black manganese-bearing shales at the middle-lower part, and grade upwards into silty micro-crystal dolostones. Feldsparthic quartz medium sandstones at the lower part, and grade upwards into stromatolite dolostones interbedded with wavy laminated dolostones.
Paleo-proterozoic	Statherian	Paleo-proterozoic	Changcheng (Ch)				1600	Parallel unconformity contact
				Dahongyu F.m. (Chd)	408		1625*6	Mostly white stromatolite dolostones with chert at the upper part. Mostly basic volcanic lava, sandy tuff and agglomerate at the middle part. Mostly milk-white quartz sandstones intercalated with sandy dolostones at lower part.
				Tuanshanzi F.m. (Cht)	518		1622*7	Thin-bed purple-red shales interbedded with sandstones at the upper part. Stomatolite microcrystal dolostones intercalated with sandstones at the middle partstones. Dark-color microcrystal dolostones intercalated with muddy microcrystal dolostones at the lower part.
				Chuanlinggou F.m. (Chch)	889			Mostly black shales intercalated with dolostones at the upper part. Black shales in the middle part. Mostly gray-green shales intercalated with fine sandstones at the lower part.
				Changzhougou F.m. (Chc)	859		1685(U-Pb) 1685*8 1682*9 1673*5	Gray-white medium to thin-bed quartz fine sandstones intercalated with thin-bedding silty shales. Mostly purple-red and milk-white, medium to thick-bedding quartz sandstones at the middle part. Mostly purple-red thick-bed to massive quartz sandstones intercalated fine gravelly coarse sandstones at the lower part.
								Angular unconformity contact
								Mostly garnet hornblende plagioclase-gneiss.

Legend: shale; manganese-bearing shale; glauconitic shale; glauconitic sandstone; sandstone; gravelly sandstone; argillaceous siltstone; microcrystal dolostone; manganese-bearing dolostone; sandy dolostone; argillaceous silty dolostone; silty dolomitic mudstone; stromatolite dolostone; siliceous stromatolite dolostone; wavy laminated dolostone; patchy dolostone; brecciated dolostone; silty microcrystal dolostone; limy microcrystal dolostone; coated-grain microcrystal dolostone; microcrystal limestone; dolomitic microcrystal limestone; wavy laminated limestone; argillaceous limestone; sandy tuff; stromatolite limestone; volcanic breccia/agglomerate; potassium-rich volcanic lava

*Note: *1. Li et al., 2009; *2. Gao et al., 2007; *3. Su et al., 2007; *4. Su et al., 2010; *5. Li et al., 2011; *6. Lu et al., 1991; *7. Li et al., 1995; *8. Gao et al., 2008; *9. He et al., 2011.

Fig.3 Stratigraphic column of Proterozoic in Jizhou District, Tianjin

SECTION Ⅱ Investigation Route of the Changzhougou Formation (Chc) of Changcheng System from Changzhou Village to north of Qingshanling Village, Xiaying Town

The Changzhougou Formation mainly consists of sandstones and is 859m thick. According to lithological variation and cycles, this formation can be divided into three segments from bottom to top:

The Chc_1 is about 260m thick, and the lithology is mainly purplish red thick bedded sandstones intercalated with purplish red gravelly coarse sandstones and fine conglomerates. Fining-upward cycles from gravelly coarse sandstone or fine conglomerate rock to coarse sandstone are common, and the thickness of the cycles is several meters. Trough crossbeddings are developed, and the paleocurrent direction indicated by crossbeddings is to NNW. The color of sandstones is red, indicating oxidizing exposure environments. The sandstones show poor rounding and moderate sorting, and the pore-fillings are clay matrix. The Chc_1 is deposited by braided river in a rift valley.

The Chc_2 is about 220m thick. The lithology is mainly milky white colored, medium to thick bedded fine quartz sandstone. In the lower part there are some gravelly medium-coarse quartz sandstones. The grains of sandstones are mainly subrounded and subangular; and well sorted. The sandstones show massive structures, and beddings are not well developed. Cross beddings are only visible locally. The bed surface of sandstones is wavy resulting from wave action. The Chc_2 is deposited in marine shore environments. There is a fault at the contact between the Chc_1 and Chc_2, but the contact is covered by weathering products. In topography, the sudden change from the Chc_1 to the Chc_2 is a steep cliff, indicating that the rock of the Chc_2 is harder and more resistant to weathering than Chc_1.

The Chc_3 consists of grayish white thin to thick bedded quartz fine sandstones with grayish green thin silty shale layers. The thickness of these shale layers is not large, mostly from a few millimeters to several centimeters. The boundary between Chc_3 and Chc_2 is located at the north side of the intersection of Baichang Road and Qingchang Road, and the Chc_3 is marked by appearance of grayish green thin silty shale layers. The sandstones are dominated by massive structures, and the beddings are not well developed. Bed surface of sandstones is wavy. The Chc_3 is deposited in nearshore environments. From bottom to top of Chc_3, the thickness of single sandstone bed decreases from medium-thick bedded to thin bedded, and shale layers increases in abundance and thickness, indicating that water depth increases.

It can be seen that from bottom to top of the entire Changzhougou Formation, the color of the rock changes from oxidizing color to weakly reducing color, the grain size becomes finer, the beds become thinner, the shale interlayers become more abundant, and the sedimentary environments change from fluvial (Chc_1) to backshore-foreshore (Chc_2) to nearshore (Chc_3), and finally to off-shore shelf of the Chuanlinggou Fm.. The whole sequence is a well-developed transgression sequence.

The Changzhougou Formation is well exposed in the northernmost part of the Huangyaguan Great Wall and the Jiulong to Baxian Mountain in the northern part of Tianjin's Jizhou District, especially in the area from Changzhougou Village to Qingshanling. See Fig.4 and Fig.5.

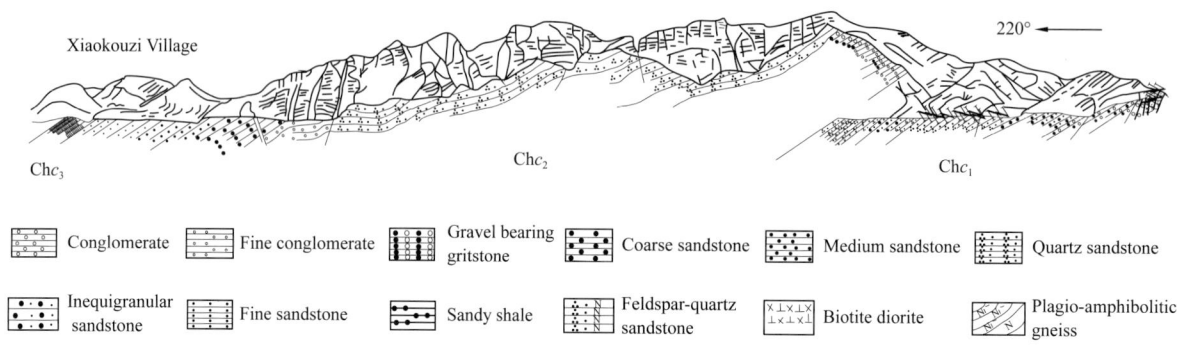

Fig.4　Sketch profile of the Changzhougou Formation (Chc) in the Jizhou District, Tianjin (modified after Tianjin Institute of Geology and Mineral Resources, 1964)

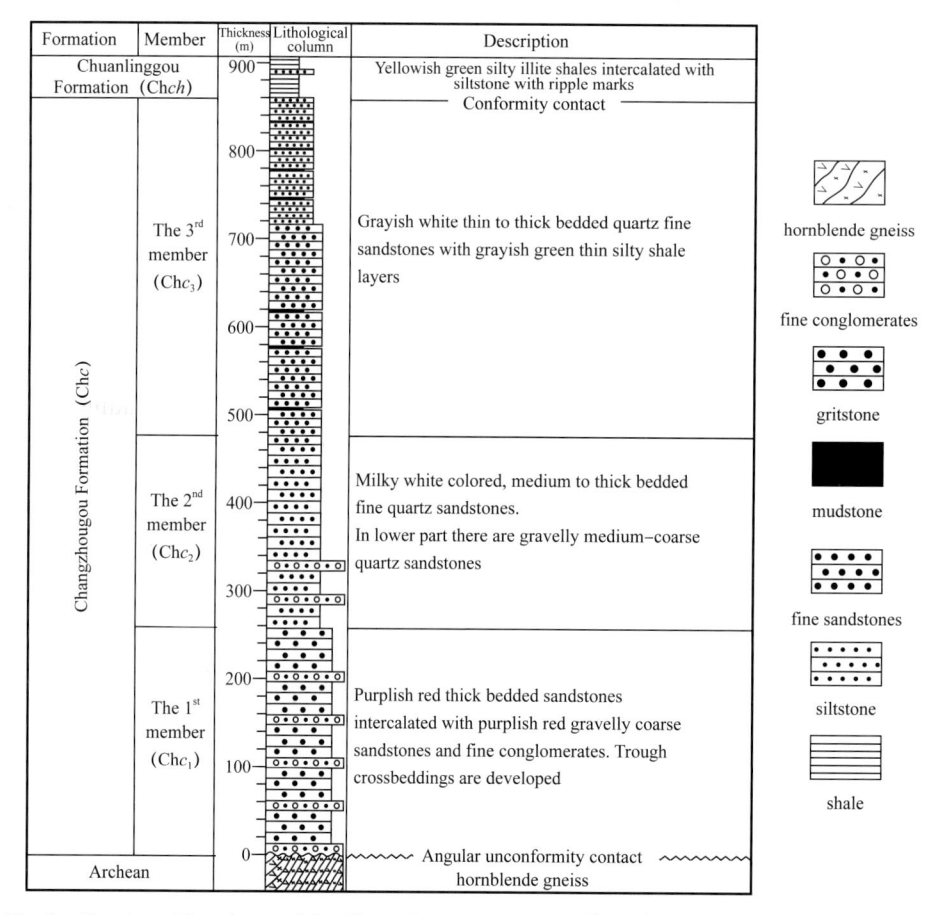

Fig.5　Stratigraphic column of the Changzhougou Formation (Chc) in the Jizhou District, Tianjin

1. Route location

The route is located in the area from Changzhou Village to Qingshanling Village, Xiaying Town, Jizhou District, Tianjin. The starting point is located on the west slope of Changzhou Village (GPS: E 117.476749°, N 40.164697°). The end point is about 650m northeast of the Qingshanling Village (GPS: E 117.498056°, N 40.200278°). The total length of the route is about 2470m, and the specific route and the points of investigation are shown in Figure 6.

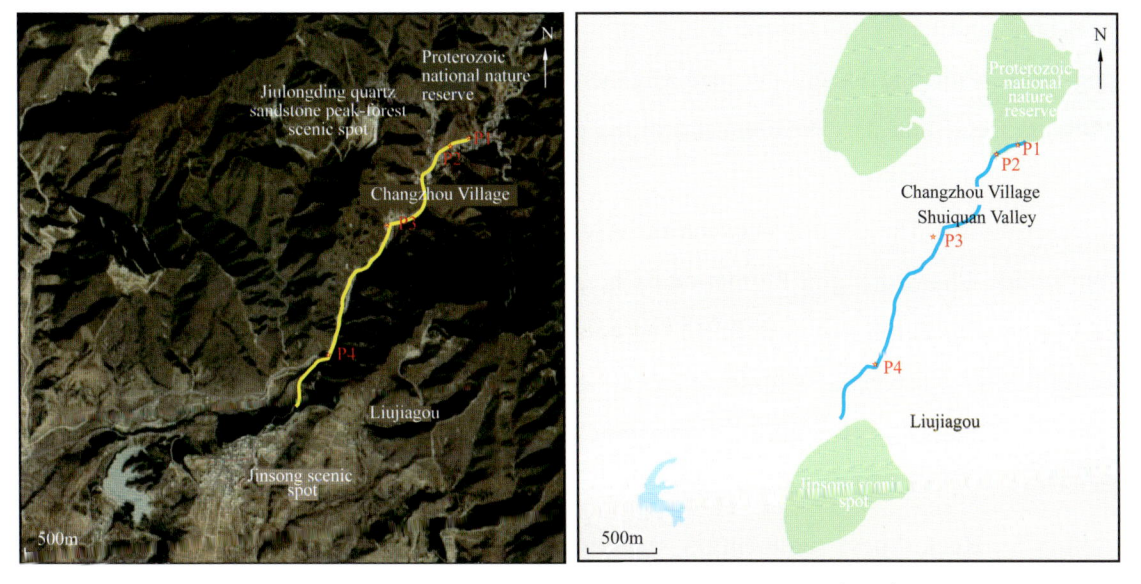

Fig.6 Distribution of investigation stops and route of the Changzhougou Formation (Chc) in the Jizhou, Tianjin

2. Objectives of investigation

(1) To observe the contact relationship between the Changzhougou Formation and the Archaeozoic and its lithologic features and sedimentary sequences;

(2) To observe the sedimentary cycles, typical sedimentary structures (including trough, tabular and wedge-shaped crossbeddings, scouring surface), quartz gravels, lag mudstone rip-clast shale, pseudofossils, faults, folds, etc. in the Changzhougou Formation;

(3) To observe the variation of color, grain size, single bed thickness and sedimentary structures of from bottom to top, and analyzing the sedimentary environments and its evolution.

3. Contents of investigation and description of stops

There are four detailed investigation stops along the route, including the stop to observe the boundary of the Archean and Paleoproterozoic (P1 in Fig.6), the stop to observe the rock types, sedimentary structures and sedimentary sequences of the Chc_1 (P2 in Fig.6), the stop to observe the rock types, sedimentary structures and sedimentary sequences of the Chc_2 (P3 in Fig.6) and the stop to observe sedimentary sequence of the Chc_3 (P4 in Fig.6). The detailed description of each stop is as follows.

Stop 1 Boundary between the Archean and Paleoproterozoic

Location: The stop is located on the west side of the Changzhougou Village. The GPS point is E 117.520037°, N 40.219031°.

Contents of investigation: (1) Angular unconformity and weathering crust development between Archean gneiss and Changzhougou Formation; (2) The crossbeddings, gravel composition and rounding in the pebbly sandstone; (3) Discuss why there is no thick weathering crust on the unconformity surface of long-term weathering and erosion, and why there is no accumulation of gravels on the unconformity; (4) Discuss the sedimentary environments of the pebbly sandstone at the bottom of the Changzhougou Formation.

Description: The contact between the Archean gneiss (Ar) and the pebbly sandstone at the bottom of the Changzhougou Formation (Chc) is angular unconformity (Fig.7A). The bottom of Changzhougou Formation is purplish red pebbly coarse sandstone, moderately sorted, and the rounding is mainly subangular. Cross bedding is developed, and the dark colored laminae results from enrichment of magnetite. The gravelly sandstone is deposit of braided river channel (see Section 15 for details).

Stop 2 Rock types, sedimentary structures and depositional sequences in the Chc_1 of the Changzhougou Formation

Location: The interval to be observed is about 20m southwest of Stop 1. GPS point: E 117.517468°, N 40.217979°.

Contents of investigation: (1) rock types; (2) types of crossbeddings and direction of paleocurrent; (3) scouring surface and rip-clasts of mudstone; (4) river sedimentary sequences and sedimentary cycles; (5) discussion of sedimentary environments.

Description: The interval to be observed is several tens of meters thick. The bottom of the interval is about 20m southwest of Stop 1. The top of this interval is the end of Stop 2, which is near a pool and a house in a valley. This interval is well exposed, and can represent rock types, sedimentary structures and sedimentary sequences of the Chc_1. Various crossbeddings are abundant and clear, and a well-developed fining-upward m-scale sedimentary sequence are developed (Fig.8 and Fig.9). The occurrence of strata is 240°∠33.5°.

The rocks types of the Chc_1 include: purplish red fine conglomerates, light red gravelly coarse sandstones and quartz sandstones. The diameter of the gravels is usually a few millimeters to about 1cm, and the composition is quartzite. Common sedimentary structures include trough, wedge and tabular crossbeddings, and the dip of laminae in crossbeddings is north-northwest, indicating that the paleocurrent flows to north-northwest (Fig.7B-D). The Chc_1 consists of stacked braided river channel deposits (see Section 15 for details).

In the lower part of Chc_1, thin shale layers are occasionally seen, which are the floodplain deposits left after erosion (Fig.7E).

Stop 3 Rock types, sedimentary structures and sedimentary sequences of Chc_2

Location: The point is located on a short path on the northwest side of Baichang Road, Changzhou Village. GPS position: E 117.512945°, N 40.213335°.

Abstract Field Guide of Stratigraphy and Sedimentary Facies of the Proterozoic in Jizhou District, Tianjin

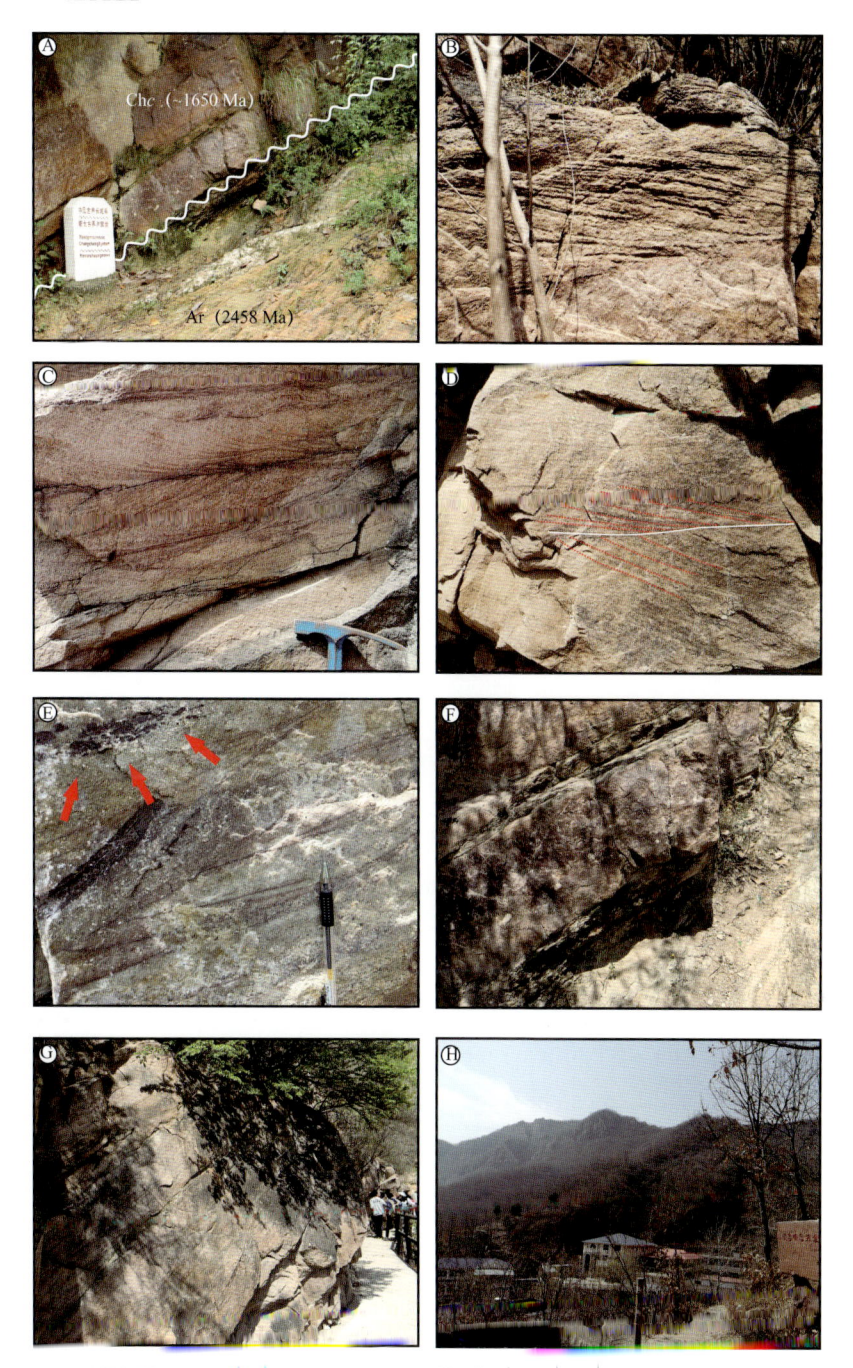

Fig.7 Geological phenomena near the boundary between the Archean (Ar) and Changzhougou Formation (Chc), Changcheng System, Paleoproterozoic, and the Chc₁ in the Jizhou District, Tianjin

(A) The unconformity between the Archaean and the Changzhougou Formation, belonging to Stop 1. (B) Large tabular crossbedding in the purplish red sandstone in the Chc₁ of the Changzhougou Formation. This point is 20m southwest of the proterozoic/archaean boundary, belonging to Stop 2. The small tree is the scale. (C) "fake herribone crossbedding" in the sandstone of the Chc₁. The opposite dip direction of laminae is not due to bidirectional tidal current, but due to superposition of the left side of a trough crossbedding over the right side of the underlying crossbedding. This point belongs to Stop 2. (D) Tabular crossbedding in the purplish red sandstone of the lower part of the Chc₁. In the photo, the laminae in the lower part intersect the bottom at a high angle, while that in the upper part is tangential to the bottom, reflecting higher current velocity for crossbedding in the upper part than that in the lower part. This point belongs to Stop 2. (E) Purplish red mudstone rip-clasts (upper left of the photo) in the purplish red sandstone of the Chc₁. This point is belongs to Stop 2. (F) Thin red shale layer (in tree shadows) deposited in the floodplain. This kind of interlayer is rare in the Changzhougou Formation. This point belongs to Stop 2. (G) The purplish red sandstone in the lower part of the Chc₁. This point belongs to Stop 2. (H) The end view of the interval of Stop 2. The steep cliff in the far is the sandstone of the Chc₂. Its bottom is the boundary between the Chc₂ and Chc₁

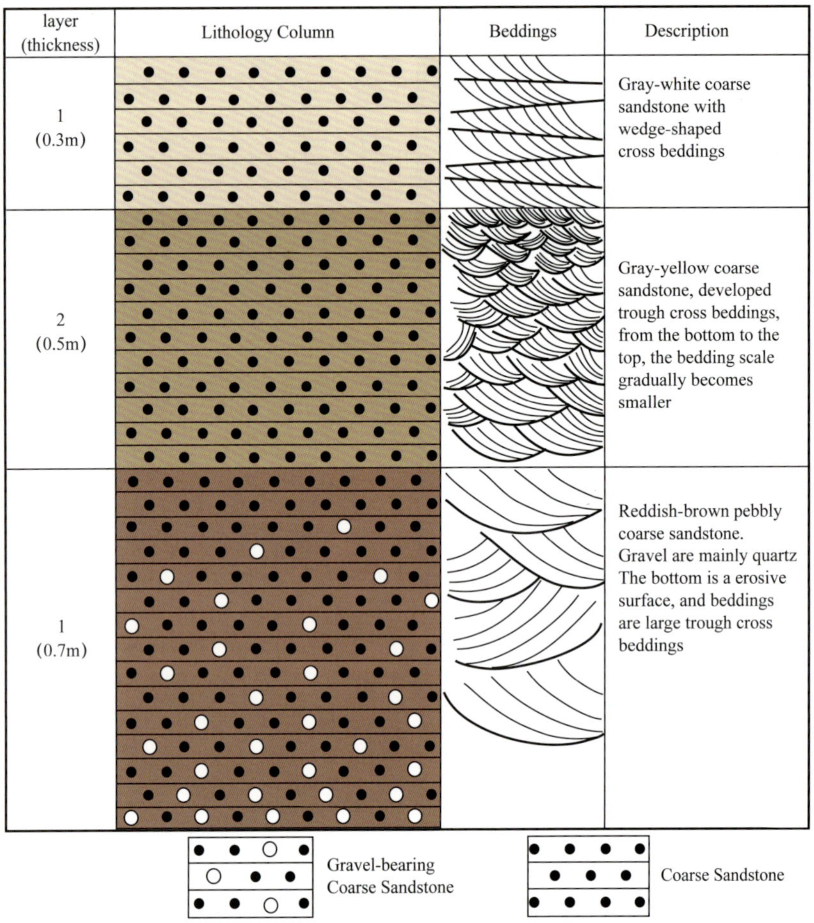

Fig.8 Schematic diagram of the lithofacies association in the Chc_1

Contents of investigation : (1) rock types ; (2) types of crossbeddings and palaeocurcurrent direction ; (3) the lithological difference between the Chc_2 and the Chc_1 ; (4) discussing the difference of sedimentary environments between the Chc_2 and the Chc_1.

Description : The Chc_2 consists of milky white quartz fine sandstones with some purplish red and milky white gravelly medium-fine sandstone and fine conglomerate (Fig.10、Fig.11). The sandstones show good sorting and rounding, and the pore-fillings are quartz cement. The sandstones are deposited in nearshore environments.

Stop 4 Sedimentary sequence of the Chc_3

Location : This point is located near the intersection of Baichang Road and Qingchang Road (Fig.12A). GPS position : E 117.506340°, N 40.204052°.

Contents of investigation : (1) rock types ; (2) changes in bed thickness of sandstone and abundance of shale interlayers ; (3) types of crossbeddings and palaeocurcurrent direction ; (4) origin of wavy bed surfaces of sandstones ; (5) upward coarsening sedimentary cycles ; (6) fake fossils and "Lizegang Ring" ; (7) small folds ; (8) discussing the difference and evolution of sedimentary environments between the Chc_2 and the Chc_3 ; (9) discussing the conditions for very thick successive deposition of sandstones in the Changzhougou Formation.

Abstract　Field Guide of Stratigraphy and Sedimentary Facies of the Proterozoic in Jizhou District, Tianjin

Fig. 9　Major lithofacies sequence and microscopic features of the Chc_1 of the Changzhougou Formation

(A) Fining-upward sequence in the lower part of Chc_1, from gravelly coarse sandstone to coarse sandstone. This point belongs to Stop 2. (B) Photomicrograph of thin section of purplish red coarse sandstone in the Chc_1; dark minerals are magnetite. Plane polarized. Scale bar is 1 mm. (C) Crossbedding in the purplish red coarse sandstone in the lower part of the Chc_1. Dark colored laminae rich in magnetite. This point belongs to Stop 2. (D) Photomicrograph of light red coarse sandstone in the Chc_1. Pore-filling is mud. Plane polarized. Scale bar is 1 mm. (E) Ripple marks on bed surface of grayish white coarse sandstone in the Chc_1. This point belongs to Stop 2. (F) Photomicrograph of thin section of grayish white coarse sandstone in the Chc_1. Plane polarized. Scale bar is 1 mm. (G) Stacked channel deposits. The upper channel deposits consist of purplish red gravelly coarse sandstones, with less gravels in the lower part and more gravels in the upper part, reflecting that the current becomes faster. The deposits of the underlying channel are purplish red sandstone. This point belongs to Stop 2. (H) Purplish red gravelly coarse sandstone in the lower part of the Chc_1 (magnified part of Fig.9 F). The gravels are quartzite, with a diameter of about 1 cm, and the rounding is subangular to subrounded. This point belongs to Stop 2

Fig.10 Grayish white fine sandstone with crossbeddings, Chc_2

Fig.11 Microscopic photo of quartz sandstone in middle Chc_2. (A) Plane polarized; (B) Cross polarized, same view to A. Well sorted, well rounded, quartz cemented. Stop P3

Description: The Chc$_3$ is grayish white medium-thick bedded quartz fine sandstone with some thin shale layers. The bed surfaces of sandstones are wavy, resulting from storm waves (Fig.12B and C). The upper part is thin bedded quartz fine sandstones intercalated with thin silty shales (Fig.11B-D). Coarsening-upward sequences are common (Fig.12B and C). Some sandstone beds are lenticular in shape. Large crossbeddings can be seen inside the sandstone, dipping to the south. "Pseudofossils", like stromatolites (Fig.13A), can be seen on the joint surfaces of sandstones in the middle part of Chc$_3$. They are formed by the deposition of iron and manganese oxide on the joint surfaces when groundwater flows along the joints. There are small folds in the thin bedded fine sandstones in the upper part Chc$_3$, and locally strata are almost upright (Fig.13B). The grayish green shale interlayers become more and more abundant toward top of the Chc$_3$. Sedimentary environments of the Chc$_3$ is lower nearshore. From the Chc$_2$ to Chc$_3$, the environments gradually gets deeper and finally changes to shale shelf of the overlying Chuanlingou Formation (see section 15 for details).

Fig.12 Sedimentary sequence of the Chc$_3$ of the Changzhougou Formation

(A) The boundary between the Chc$_2$ and the Chc$_3$ is located near the signpost and belongs to Stop 4 (B) Coarsening-upward sedimentary sequence in the Chc$_3$. The hammer is a scale, belonging to Stop 4. (C) Coarsening-upward sedimentary sequence in the Chc$_3$. The hammer is a scale, belonging to Stop 4

Fig.13　Typical rock types and microscopic characteristics in the Chc_3 of the Changzhougou Formation

(A) "Fake fossils" on the sandstone joint surface, looking like a hemispherical stromatolite, the middle part of the Chc_3, belonging to Stop 4. (B) Folded thin bedded fine sandstone, the upper part of the Chc_3, belonging to Stop 4. (C) Thinly interbedded fine sandstone and silty shale, the top of the Chc_3, belonging to Stop 4. (D) Quartz fine sandstone with mud matrix, plane polarized light, scale bar 1 mm, top of the Chc_3

SECTION Ⅲ　Investigation Route of the Chuanlinggou Formation (Ch*ch*) of Changcheng System from north of Qingshanling Village to Liuzhuangzi Village, Xiaying Town

The Chuanlinggou Formation of the Changcheng System of the Paleoproterozoic is a set of sedimentary strata dominated by silty illite shale, with a small amount of siltstone and dolomite, and volcanic rock dikes and rock sills. The total thickness of the formation is 889 m. According to the lithological combination characteristics, it can be subdivided into three segments: the lower segment (Chch_1) is mainly composed of grayish green and yellowish green silty illite shales; the middle segment (Chch_2) is dominated by black shales; the upper segment (Chch_3) is mainly composed of black illite shales sandwiched thin iron-rich dolostones. The Chuanlingou Formation is widely exposed in the Qingshanling Village, Guojiagou and Liuzhuangzi, Xiaying Town, Jizhou District, Tianjin. However, due to the soft shale lithology, it forms valleys after weathering.

The Chuanlingou Formation is well exposed along the Qingchang Road and the Guoma Road. The sketch profile and stratigraphic columns of the Chuanlinggou Formation are shown in Fig.14 and Fig.15.

Abstract Field Guide of Stratigraphy and Sedimentary Facies of the Proterozoic in Jizhou District, Tianjin

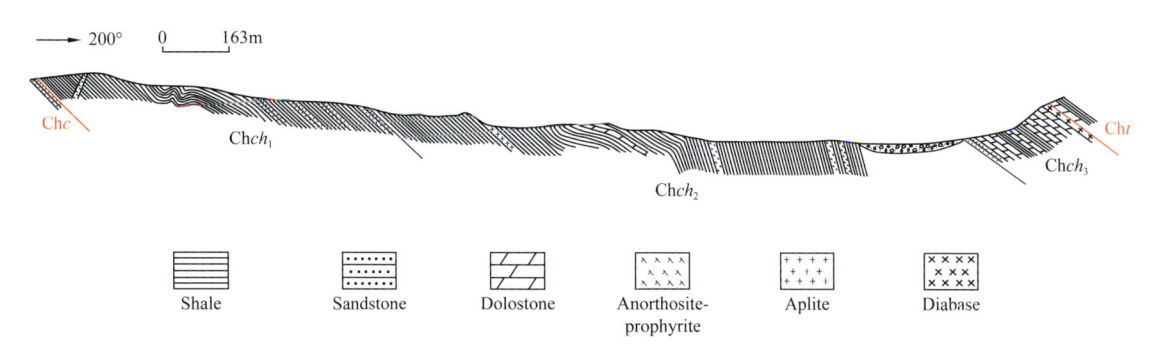

Fig.14 Sketch profile of the Chuanlinggou Formation from Qingshanling to Liuzhuangzi, Xiaying Town, Jizhou District, Tianjin (modified after Tianjin Institute of Geology and Mineral Resources, 1964)

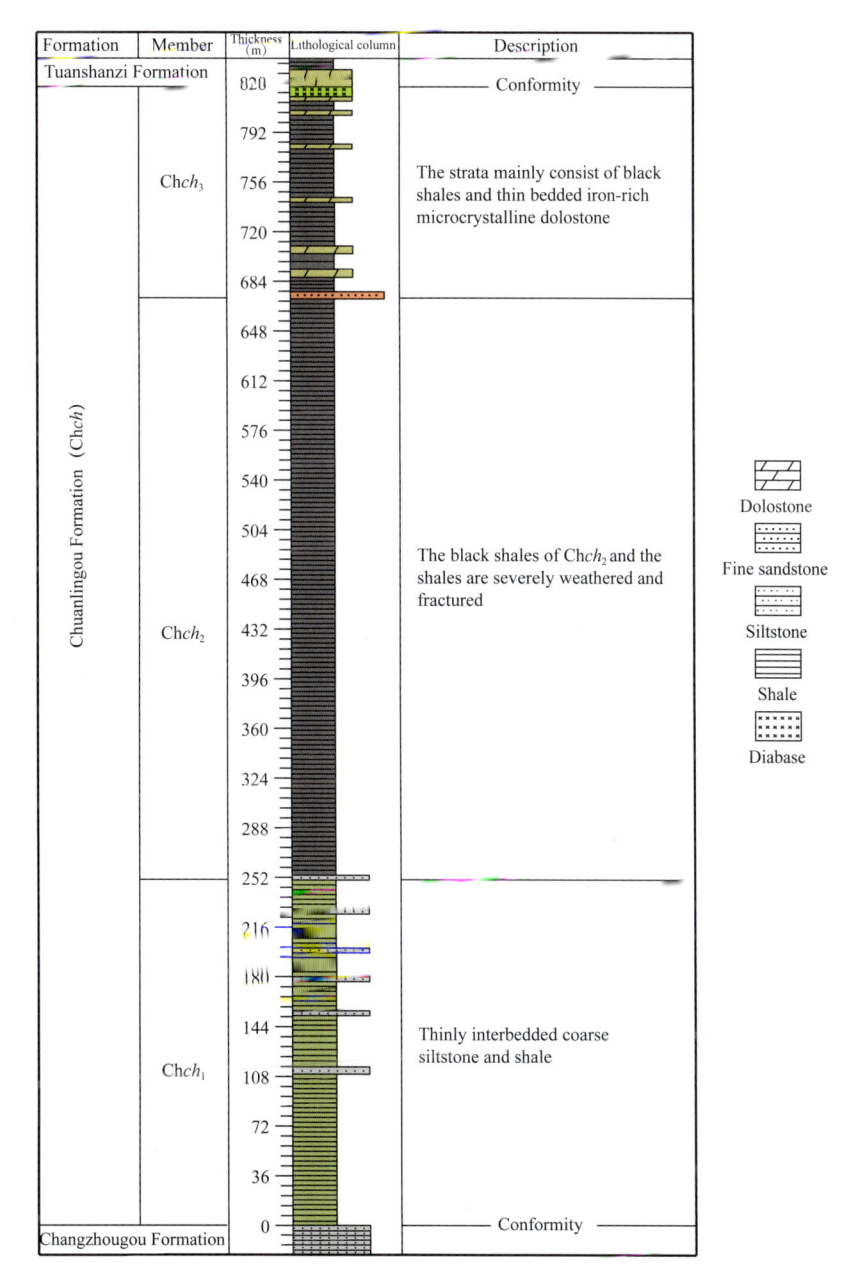

Fig.15 Stratigraphic columns of the Changcheng System Chuanlinggou Formation (Chch) in Goujiagou, Jizhou District, Tianjin (modified after Yang et al., 2016)

1. Route location

The route is located in the area from Qingshanling Village to Tuanshanzi Village, Xiaying Town, Jizhou District, Tianjin. The starting point is located in the northeast corner of Qingshanling Village (GPS position: E 117.498056°, N 40.200278°). The end point is located in the west of Tuanshanzi Village, Maying Road (GPS position: E 117.474444°, N 40.176389°). The total length of the route is about 5600m, and the specific route and stops are shown in Figure 16. The route roads are generally wide, large and small vehicles are accessible, and there are fewer daily vehicles. Parking is convenient.

2. Objectives of investigation

(1) To observe the lithologic characteristics and stratigraphic sequence of the Chuanlinggou Formation;

(2) To analyze the sedimentary environments and evolution of the Chuanlingou Formation, especially the difference between the sedimentary environments of the Chch_1, Chch_2, and Chch_3.

3. Contents of investigation and description of stops

This route is from the Qingshanling village to Tuanshanzi village, and 5 stops are selected along the route. They include the stop to investigate the boundary between Changzhougou Formation and Chuanlinggou Formation (P1 in Fig.16); the stop to investigate the sedimentary characteristics of the lower part of the Chch_1 (P2 in Fig.16); the stop to investigate "mud crack" (?) and ripple marks of the upper part of the Chch_1 (P3 in Fig.16); the stop to investigate black shales of the Chch_2 (P4 in Fig.16) and the stop to investigate microcrystalline dolostones and black shales (P5 in Fig.16). The specific description of each stop is as follows.

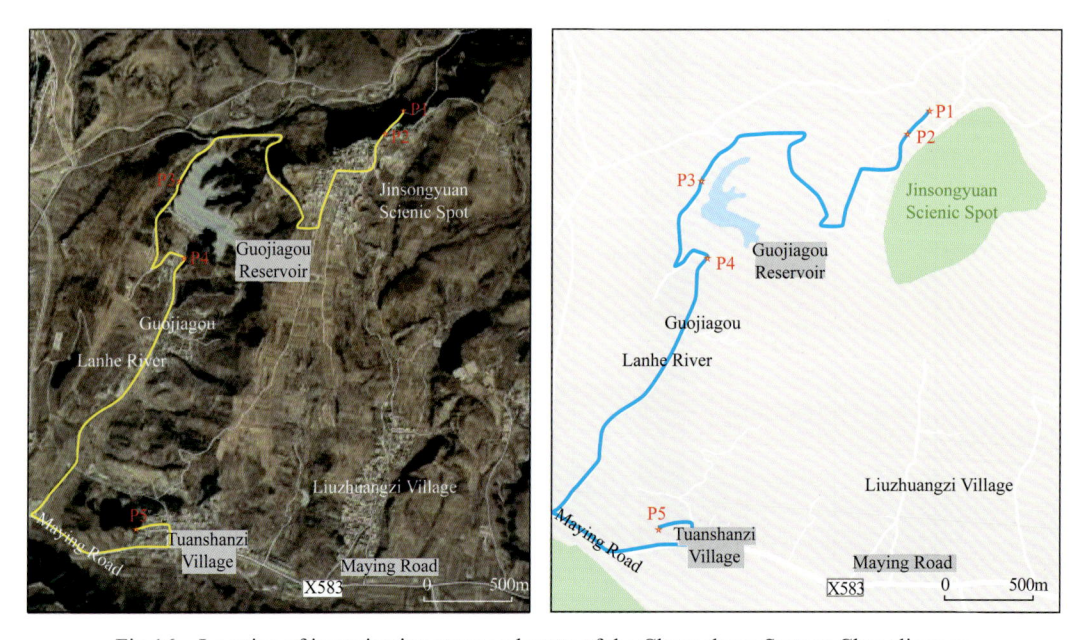

Fig.16 Location of investigation stops and route of the Changcheng System Chuanlinggou Formation (Chch) in Jizhou District, Tianjin

Stop 1　Boundary between Changzhougou Formation (Chc) and Chuanlinggou Formation (Chch)

Location: The boundary between Changzhougou Formation (Chc) and Chuanlinggou Formation (Chch) is located in the northeast of Qingshanling village. GPS positions: E 117.498056°, N 40.200278°.

Contents of investigation: (1) The boundary between Changzhougou Formation and Chuanlinggou formation; (2) Observe how the sandstone at the top of Changzhougou Formation change to shales at the bottom of Chuanlinggou Formation; (3) The type and origin of the ripple marks on bed surfaces of thin fine sandstone at the bottom of the $Chch_1$, and palaeocurrent direction indicated by the ripple marks.

Description: The boundary between the Changzhougou Formation (Chc) and the Chuanlinggou Formation (Chch) is conformity (Fig.17A). But due to weathering, the contact is covered by debris (Fig.17B). The $Chch_1$ is dominated by grayish green shale (Fig.17C and E). The bottom several meters of the formation is thinly interbedded coarse siltstones and shales. Ripple marks are common (Fig.17D). The Chuanlinggou Formation is generally deposited on shale shelf. The color is grayish green, representing weak reducing environments (see Section 15 for details).

Stop 2　Characteristics of shales of $Chch_1$

Location: The northeast corner of Qingshanling Village, GPS position: E 117.496111°, N 40.198611°.

Contents of investigation: (1) sedimentary characteristics of the shales of the $Chch_1$; (2) analyze the sedimentary environments of grayish green shale; (3) observe the joints in shales and discuss its role in oil and gas migration.

Description: This stop is a typical shale investigation point of $Chch_1$, and the color of the shales is mainly grayish green. At the bottom of $Chch_1$, there are some thin beds of siltstones which should be deposited by storm currents. In the shales of the lower part of $Chch_1$, there are "fake mud cracks" (Fig.17F), in which silts seem to fill cracks. In fact, the "silt fillings" are actually exposed crests of wave ripple marks of the underlying thin siltstone beds. Mud laminae over crests of wave ripple marks are eroded during weathering, making crests exposed while valleys are still covered by mud laminae. The shales of $Chch_1$ are shale shelf deposits (see Section 15 for details).

Stop 3　Characteristics of mud cracks and wave ripple marks in shales of $Chch_2$

Location: on the west side of the road just west of the Guojiagou Reservoir. GPS position: E 117.480000°, N 40.195278°.

Contents of investigation: (1) "mud cracks", wave ripple marks and algae fossils in the shales of $Chch_2$; (2) discussing whether "mud cracks" are true mud cracks and their origins; (3) distinguishing between the primary and secondary colors of shales.

Description: The lithology at this stop is grayish green shales, which develops sedimentary structures such as fake mud cracks and wave ripple marks on the bed surfaces of shales (Fig.18A and B). The surface of the shale is brown after weathering. The shales at this location are shale shelf deposits.

Fig.17 Geological phenomena of the lower strata of the Chuanlinggou Formation, Changcheng System

(A) The boundary between the Changzhougou Formation and the Chuanlinggou Formation, Stop 1; (B) The boundary stela near the boundary between the Changzhougou Formation and the Chuanlinggou. The bottom of the Chuanlinggou Formation is severely weathered and covered. Stop 1; (C) The grayish green shale in the lower part of the Chuanlinggou Formation dip to the south. Stop 1; (D) Thinly interbedded coarse siltstone and shale in the lower part of the Chuanlinggou Formation (part of E). Stop 1; (E) Thinly interbedded coarse siltstone and shale change upward to thick grayish green shale in the lower part of the Chuanlinggou Formation. Stop 1; (F) "Fake mud crack" (shown by the red arrow) in the shale of upper Chch_1

Stop 4　Characteristics of black shales in the Chch_2

Location: on the east right side of the road just south of the Guojiagou Reservoir. GPS positions: E 117.480556°, N 40.191111°.

Contents of investigation: (1) Characteristics of the black shales; (2) discussing the origin of folds, whether are they penecomtemporaneous slumping deformation folds on slopes or structural folds formed during structural movement; (3) the difference between the shales and the sedimentary environments between the Chch_2 and Chch_1.

Description: At this stop, black shales typical of the Chch_2 are observed. The shales are severely weathered and fractured (Fig.18C). The occurrence of shales changes and there are folds. The shale color of the Chch_2 is darker than the Chch_1, indicating more reducing environments, and the shales are more pure. Thus the sedimentary environments of the Chch_2 should be deeper than that of the Chch_1, and should be deep water gentle slope-basin environments (see Section 15 for details). The folds seen may be penecomtemporaneous slumping deformation folds.

Stop 5 Characteristics of the Chch_3

Location: Hillside just west of the Tuanshanzi Village on the north side of the Maying Highway. GPS position: E 117.474444°, N 40.176389°.

Contents of investigation: (1) The change of lithology occurred in the transitional zone between the Chuanlinggou Formation and the Tuanshanzi Formation; (2) the characteristics of the thin bedded muddy dolostones within the shales; (3) analyzing the origin of the dolostones.

Description: Strata at this stop is typical of the Chch_3. The strata mainly consist of black shales and thin bedded iron-rich microcrystalline dolostone (Fig.18D). The dolostone is brown after weathering. The Chch_3 is gentle slope deposits. Upwards water depth of the depositional environments becomes shallower, because the dolostone interlayers become more abundant (see Section 15 for details).

Fig.18 Geological phenomena of the middle and upper Chuanlinggou Formation

(A) Fake mud cracks on the surface of the shales (yellow-brown surface), the Chch_2, Stop 3; (B) Wave ripple marks on the surface of shales, the Chch_2, Stop 3; (C) Black shales, the Chch_2, Stop 4; (D) The black shales and thin bedded microcrystalline dolostones (the weathered surface is brown), the Chch_3, Stop 5

SECTION Ⅳ Investigation Route of the Tuanshanzi Formation (Cht) of Changcheng System west of Tuanshanzi Village, Xiaying Town

The Tuanshanzi Formation is mixed deposits of carbonate rocks and clastic rocks. The total thickness of this formation is 518m. According to the difference in lithology, it can be subdivided into four members from bottom to top : Cht_1, Cht_2, Cht_3, and Cht_4.

The Cht_1 of the Tuanshanzi Formation is composed of dark grey (yellowish brown after weathering) thin-medium bedded microcrystalline dolostones intercalated with muddy microcrystalline dolostones and dolomitic shales, and horizontal beddings are developed.

The Cht_2 of the Tuanshanzi Formation is composed of dark grey medium-thick bedded iron-rich microcrystalline dolostones interbedded with muddy microcrystalline dolostones, and horizontal beddings are not well developed.

The Cht_3 of the Tuanshanzi Formation is dominated by the interbedded of thin-medium bedded microcrystalline dolostones, muddy microcrystalline dolostones and dolomitic sandstones. Stromatolites and mud cracks are common in microcrystalline dolostones. Ripple marks and small crossbeddings are common in dolomitic sandstones.

The Cht_4 of the Tuanshanzi Formation consists of thinly interbedded purplish red silty dolomitic shales and grayish white sandstones.

The Tuanshanzi Formation is widely exposed in Tuanshanzi Village, Dahongyugou Valley, Chuancangyu, and Daoguyu, Xiaying Town, Jizhou District, Tianjin, especially in the area along the west side of Tuanshanzi Village. The sketch profile and the stratigraphic column of the Tuanshanzi Formation are shown in Fig.19 and Fig.20.

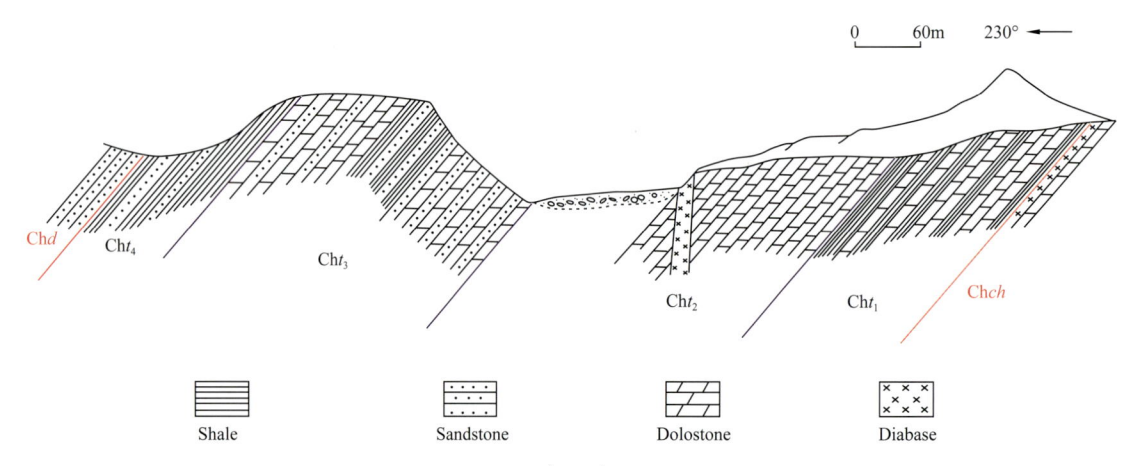

Fig.19 Sketch profile of the Tuanshanzi Formation (Ch*t*) in the west of Tuanshanzi village, Xiaying Town, Jizhou District, Tianjin (modified after Tianjin Institute of Geology and Mineral Resources, 1964)

1. Route location

The route is located near Tuanshanzi Village, Xiaying Town, Jizhou District, Tianjin. The starting stop of the route is on the west side of Tuanshanzi Village. GPS position : E 117.474444°, N 40.176389°. The end is on the west side of Yingshu Road, Dahongyugou Valley. GPS position : E 117.466389°, N 40.174167°. The total length of the route is about 1670m.

The road of this route is relatively wide and can be reached by cars and buses.

Form-ation	Member	Thickness (m)	Lithological column	Description
Dahongyu Formation		486		Conformity
Tuanshanzi Formation (Cht)	Cht₄	468, 450, 432		The thinly interbedded sandstones and purplish red shales. From the bottom to the top, purplish red shales become less and thinner
	Cht₃	414, 396, 378, 360, 342, 324, 306, 288, 270, 252, 234, 216, 198, 180		The medium-thick bedded yellowish brown microcrystalline dolostones and medium-thick bedded of grayish white fine sandstones. stromatolites and mud-cracks are abundant in dolostones

The thinly interbedded black shales and grayish white fine sandstones

The thinly interbedded yellowish brown microcrystalline dolostones and grayish white fine sandstones |
	Cht₂	162, 144, 126, 108, 90, 72, 54		The medium-thick iron-rich and silty dolostones
	Cht₁	36, 18, 0		The microcrystalline dolostones intercalated with silty shale and horizontal beddings are well developed
Chuanlinggou Formation				Conformity

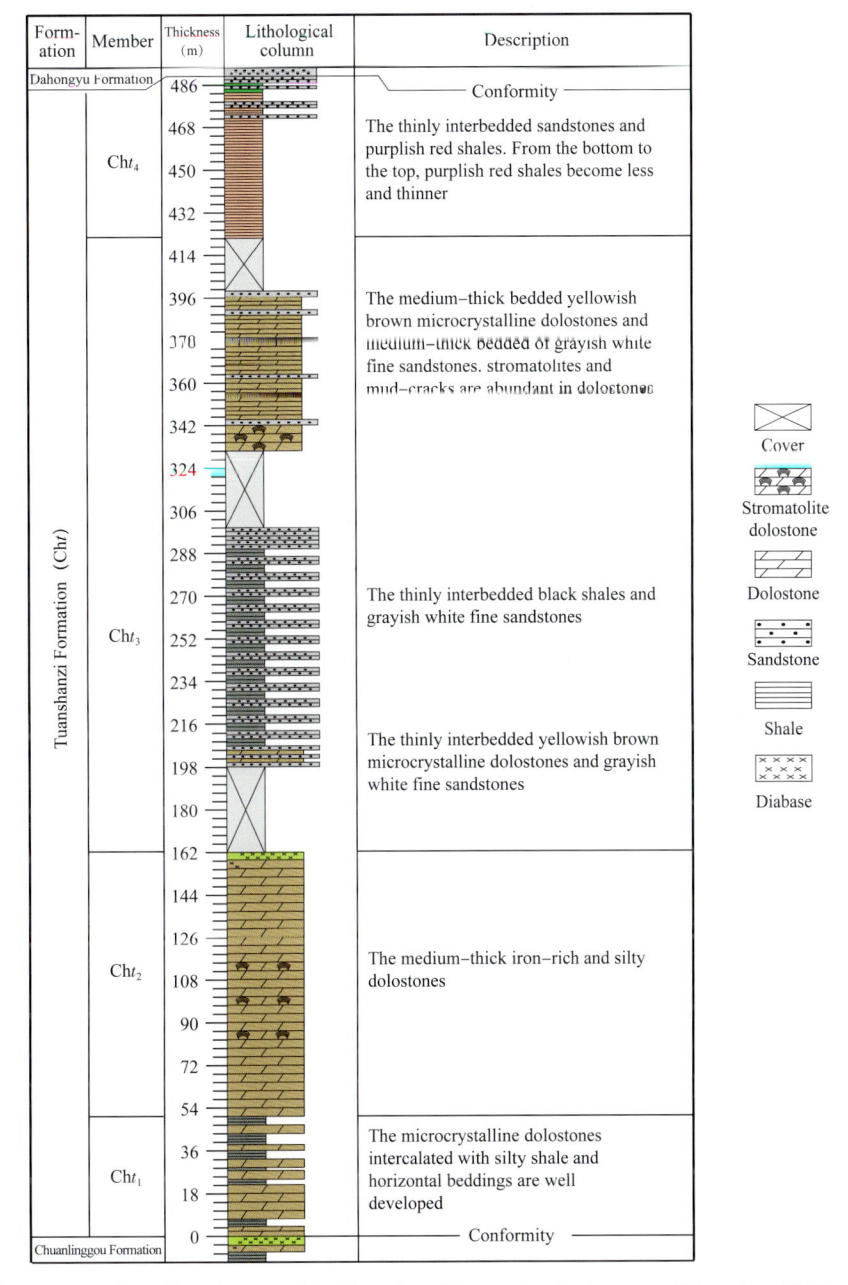

Legend: Cover, Stromatolite dolostone, Dolostone, Sandstone, Shale, Diabase

Fig.20 Stratigraphic column of the Tuanshanzi Formation in the west of Tuanshanzi Village, Xiaying Town, Jizhou District, Tianjin

2. Objectives of investigation

(1) To observe the lithologic characteristics and stratigraphic sequences of the Tuanshanzi Formation;

(2) To observe typical depositional structures such as ripple marks, cross beddings, stromatolites, and mud cracks;

(3) To analyze the depositional environments and its evolution in each member of the Tuanshanzi Formation.

3. Contents of investigation and description of stops

The route is along the west side of the Tuanshanzi Village, Xiaying Town, and along the Yingshu Road. 7 stops are selected in the route, including: the boundary between the Chuanlinggou Formation and Tuanshanzi Formation and characteristics of Cht_1 (P1 in Fig.21), the iron-bearing dolostones and characteristics of Cht_2 (P2 in Fig.21), interbedded dolostones and sandstones in the lower part of Cht_3 (P3 in Fig.21), interbedded black shales and sandstoness in the middle part of Cht_3 (P4 in Fig.21), thick sandstones in the middle to upper part of Cht_3 (P5 in Fig.21), the dolostones intercalated with sandstoness in the upper part of Cht_3 (P6 in Fig.21) and the thinly interbedded shales and sandstones of Cht_4 (P7 in Fig.21). The specific descriptions of each stop is as follows.

Fig.21 Location of description stops and route of the Tuanshanzi Formation in the west of Tuanshanzi Village, Xiaying Town, Jizhou District, Tianjin

Stop 1 Boundary between the Chuanlinggou Formation and Tuanshanzi Formation

Location: West side of the Tuanshanzi Village. GPS position: E 117.474444°, N 40.176389°.

Contents of investigation: (1) The boundary between the Chuanlinggou Formation and Tuanshanzi Formation; (2) Intrusive igneous dyke at the boundary; (3) Rock types, depositional structures and depositional sequences of Cht_1; (4) Analyze the depositional environments.

Description: This stop is to see the boundary between the Chuanlinggou Formation (Ch*ch*) and the Tuanshanzi Formation (Ch*t*). The contact between the two formations is a conformity, and a diabase dyke occurred at the boundary between them (Fig.22A and B). Above the boundary is Cht_1, which is mainly composed of dark gray (yellowish brown after weathering) microcrystalline dolostones intercalated with silty shales. Horizontal beddings are well developed (Fig.22C and D). The silty shales (Fig.22E and F) gradually changes from grayish black at the bottom to grayish green. The Cht_1 of the Tuanshanzi Formation is deposited on deep water gently-tilted slope (see Section 15 for details).

Stop 2 Characteristics of medium-thick bedded iron-rich dolostones in Cht_2

Location: Next to the stone monument of the iron-bearing dolostones on the north side of the Maying Highway, GPS: E 117.475556°, N 40.176667°.

Contents of investigation: (1) The differences between the color of fresh surface of dolostones and the color of weathered surface and its causes; (2) Why do some dolostones have a massive structure, and the others have horizontal beddings? (3) Discuss the lithological differences and depositional environments differences between Cht_1 and Cht_2; (4) Discuss the influences of diabase intrusion on the upper part of Cht_2 on dolostones.

Fig.22 Geological phenomena at the bottom of the Tuanshanzi Formation and photomicrographs of thin rocks (A) The diabase dyke just below the Tuanshanzi Formation, stop 1; (B) Diabase, cross-polarized light, scale bar is 1 mm, stop 1; (C) Interbedded muddy microcrystalline dolostones, muddy dolostones and dolomitic mudstones in the lower part of Cht_1, stop 1; (D) Thin bedded microcrystalline limestones, plane polarized light, scale bar is 1 mm, stop 1; (E) Black silty shales, in the middle part of Cht_1, stop 1; (F) Photomicrograph of the black silty shales, containing lots of quartz silt, plane polarized light, scale bar is 1 mm, in the middle part of Cht_1, stop 1

Description: This stop is to see the medium-thick bedded iron-rich and silty dolostones in Cht_2 (Fig.23A and B). Due to iron content in the dolostones, the weathered surface appears yellowish brown. Observed under the microscope, the dolostones contains lots of scattered quartz silts (Fig.23C), and locally siltstones interlaminated with microcrystalline dolostones (Fig.23D). There is a diabase dyke in the upper part of Cht_2, and the baking edge is not obvious. The Cht_2 is deposited alternatively on open platform and the gentle slope (see Section 15 for details).

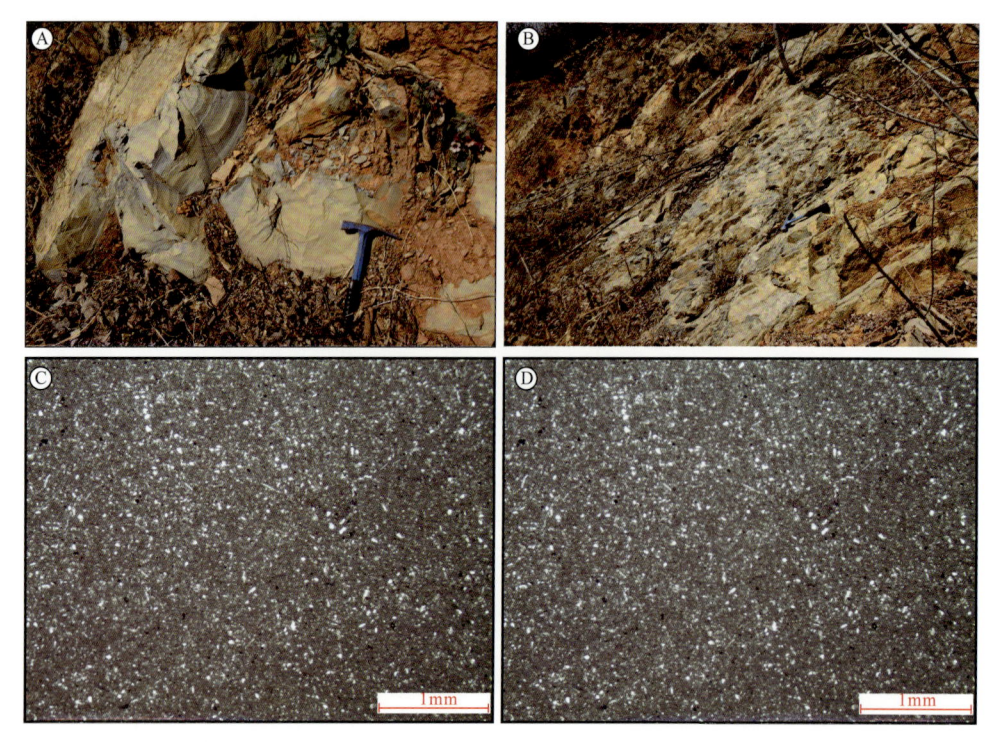

Fig.23　Geological phenomena of upper part of Cht_2 and photomicrographs of their rocks

(A and B) Thick iron-rich silty microcrystalline dolostones, Cht_2 of the Tuanshanzi Formation, stop 2; (C and D) Photomicrographs of the iron-rich silty microcrystalline dolostones, plane polarized light, scale bar is 1 mm, Cht_2, stop 2

Stop 3　Rock characteristics in the lower part of Cht_3 of Tuanshanzi Formation

Location: South side of the Maying Highway. GPS position: E 117.474444°, N 40.176389°.

Contents of investigation: (1) Describing the types of rocks; (2) Observing the ripple marks and cross beddings in thin fine sandstones, and analyze the direction of paleocurrent; (3) Analyze the depositional environments.

Description: This stop is to see typical rocks in the lower part of Cht_3. The Cht_3 is mainly composed of thinly interbedded yellowish brown microcrystalline dolostones and grayish white fine sandstones. The fine sandstones are lenticular, ribbonned or thin-bedded, and small asymmetrical or symmetric ripple marks with cross-beddings are developed (Fig.24A-C). The lower part of Cht_3 is mainly composed of carbonate-clastic mixed tidal flat deposits (see Section 15 for details).

Stop 4　Rock characteristics in the middle part of Cht_3 of Tuanshanzi Formation

Location: The south side of the valley by the Maying Highway, Tuanshanzi Village. GPS position: E 117.469167°, N 40.177222°.

Contents of investigation: (1) Observe the types and characteristics of the rocks; (2) Depositional structures in thinly interbedded grayish white fine sandstones and black shales; (3) Analyze the direction of paleocurrent; (4) Analyze the depositional environments; (5) Observe the stoma structure in the basalt and observe whether there is a baking edge.

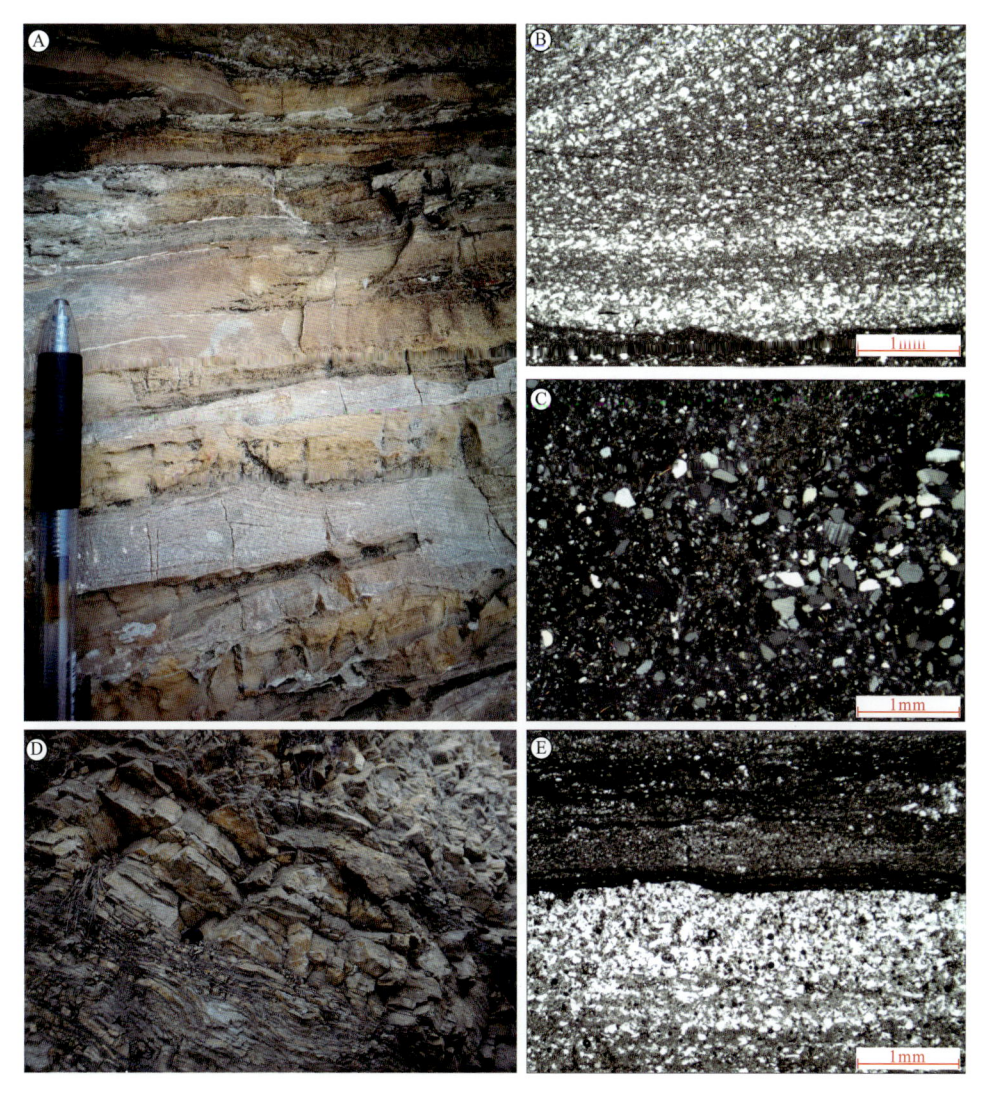

Abstract Field Guide of Stratigraphy and Sedimentary Facies of the Proterozoic in Jizhou District, Tianjin

Fig.24 Geological phenomena of the Cht_3 of Tuanshanzi Formation and photomicrographs of their rocks

(A) Interbedded dolomitic sandstones and microcrystalline dolostones, and crossbeddings in the sandstones, the lower part of the Cht_3, stop 3; (B) The dolomitic sandstones, with obvious crossbeddings, plane polarized light, scale bar is 1 mm, the lower part of the Cht_3, stop 3; (C) Photomicrograph of calcareous sandstones, cross-polarized light, scale bar is 1 mm, the lower part of the Cht_3, stop 3; (D) Interbedded sandstones and the shales, the Cht_3, stop 4; (E) Photomicrograph of interbedded sandstones and the dolostones, plane polarized light, scale bar is 1 mm, the Cht_3, stop 4

Description: This stop is to see typical rocks in the middle part of Cht_3. The main lithology of this part is thinly interbedded black shales and grayish white fine sandstones (Fig.24D), and locally interbedded microcrystalline dolostones and sandstones (Fig.24E). Here you can see several beds of basalt and stoma structures. The middle part of Cht_3 is mainly composed of sand-mud mixed subtidal deposits (see Section 15 for details).

Stop 5 Rock characteristics in the upper part of Cht_3 of Tuanshanzi Formation

Location: About 50m northeast of the stone monument at the entrance of the scenic spot, on the east side of Yingshu Road. GPS position: E 117.465833°, N 40.176667°.

Contents of investigation: (1) Observe the types and characteristics of the stromatolites;

(2) Observe the characteristics of sandstones interlayers; (3) Analyze the depositional environments.

Description: This stop is to see typical rocks in the middle-upper part of Cht_3. This part is dominated by medium-thick bedded microcrystalline dolostones intercalated with yellowish gray fine sandstones. The dolostones show laminar, wavy or hemispherical stromatolites, and are deposited on tidal flat environments. Quartz sandstones are beach deposits. Carbonate platform deposits alternate with clastic beach deposits, and should be a mixed platform.

Stop 6　Rock characteristics in the top part of the Cht_3

Location: The stop is on the east side of the Yingshu Road, about 30m south of Stop 5. GPS position: E 117.466111°, N 40.175833°.

Contents of investigation: (1) Observe the types and characteristics of the stromatolites; (2) Observe the morphology of mud cracks in the dolostones from plane view and vertical view to see the differences; (3) Analyze the origin of the dolostones; (4) Observe the characteristics of sandstones interlayers; (5) Analyze the depositional environments and the mechanism for carbonate to alternate with sandstones, and establish a mixed model of carbonate and clastic rocks.

Description: The top part of Cht_3 is mainly composed of medium-thick bedded yellowish brown microcrystalline dolostones and medium-thick bedded of grayish white fine sandstones. Stromatolites and mud-cracks are abundant in microcrystalline dolostones which should be deposited on intertidal and supratidal flats and probably result from evaporative pumping dolomitization (Fig.25A) (see Section 15 for details). The sandstones are marine beach deposits.

Stop 7　Rock types and characteristics of the Cht_4

Location: On the west side of Yingshu Road, the boundary between the Tuanshanzi Formation and the Dahongyu Formation. GPS position: E 117.466389°, N 40.174167°.

Contents of investigation: (1) Observe the types and characteristics of the rocks of Cht_4. (2) Observe the beddings and ripple marks in thin sandstones, and analyze the direction of paleocurrent. (3) Observe the changes of the sand/mud ratio from bottom to up, and analyze the transitional relationship with the overlying Dahongyu Formation. (4) Analyze the depositional environments and its evolution.

Description: the Cht_4 consists of thinly interbedded sandstones and purplish red shales. From the bottom to the top, purplish red shales become less and thinner, while the grayish white fine sandstones become more and thicker (Fig.25B). Close to the top of Cht_4, the color of shales change from purplish red to grayish green. This part is a typical transgression sequence, from supratidal flat (dominated by purplish red shales) in the lower part of Cht_4 to sand-mud mixed intertidal flat (thin bedded sandstones intercalated with thin bedded purplish red and grayish green shales) in the upper part of Cht_4, finally to thick bedded fine sandstones of shallow subtidal shoals at the bottom of the Dahongyu Fm.

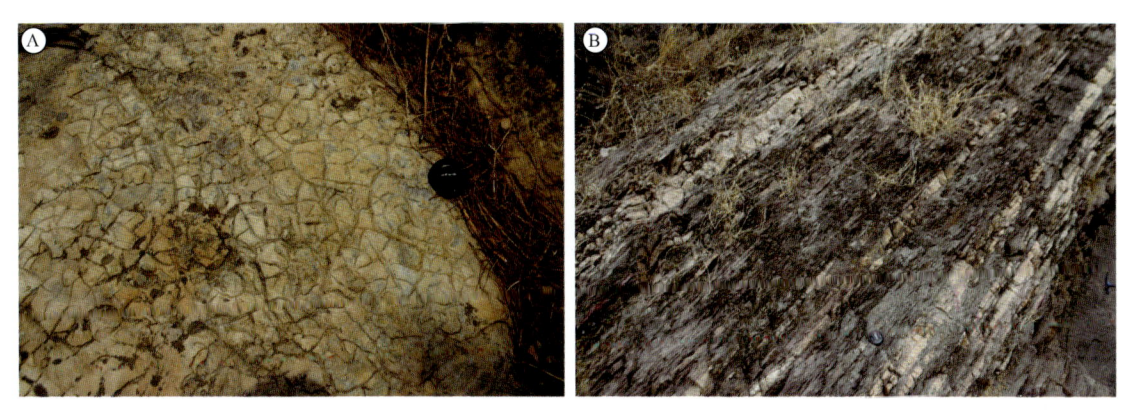

Fig.25 Mud cracks in the upper part of Cht_3 and characteristics of Cht_4

(A) Mud cracks in the microcrystalline dolostones, top part of Cht_3, stop 6;

(B) The grayish white sandstones interbedded with purplish red shales, Cht_4, stop 7

SECTION V Investigate Route of the Dahongyu Formation (Ch*d*) of Changcheng System in Dahongyugou Valley, Xiaying Town

The Dahongyu Formation consists of sandstones, dolostones, pyroclastic rocks and volcanic rocks. This formation can be subdivided into three members from bottom to top: Chd_1, Chd_2 and Chd_3.

The Chd_1 mainly consists of grayish white fine quartz sandstones. The Chd_2 mainly consists of pyroclastic rocks (volcanic breccia) and volcanic rocks. The Chd_3 mainly consists of grayish white stromatolite dolostones with abundant chert nodules and ribbons.

The sketch profile and the stratigraphic column of the Dahongyu Formation are shown in Fig.26 and Fig.27.

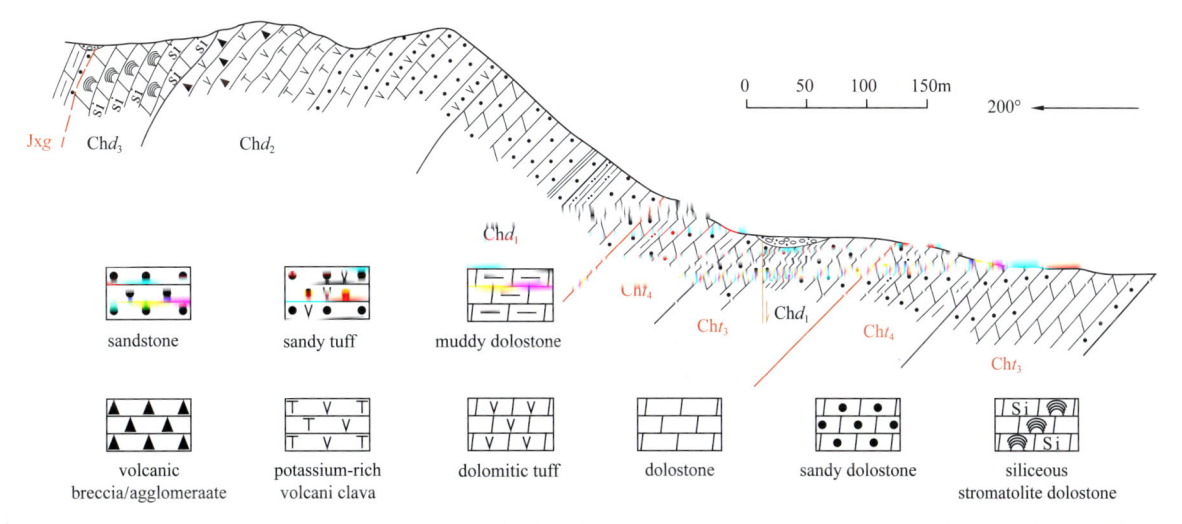

Fig.26 Sketch profile of the Dahongyu Formation in Dahongyugou Valley, Xiaying Town, Jizhou District, Tianjin

(Modified from Tianjin Institute of Geology and Mineral Resources, 1964)

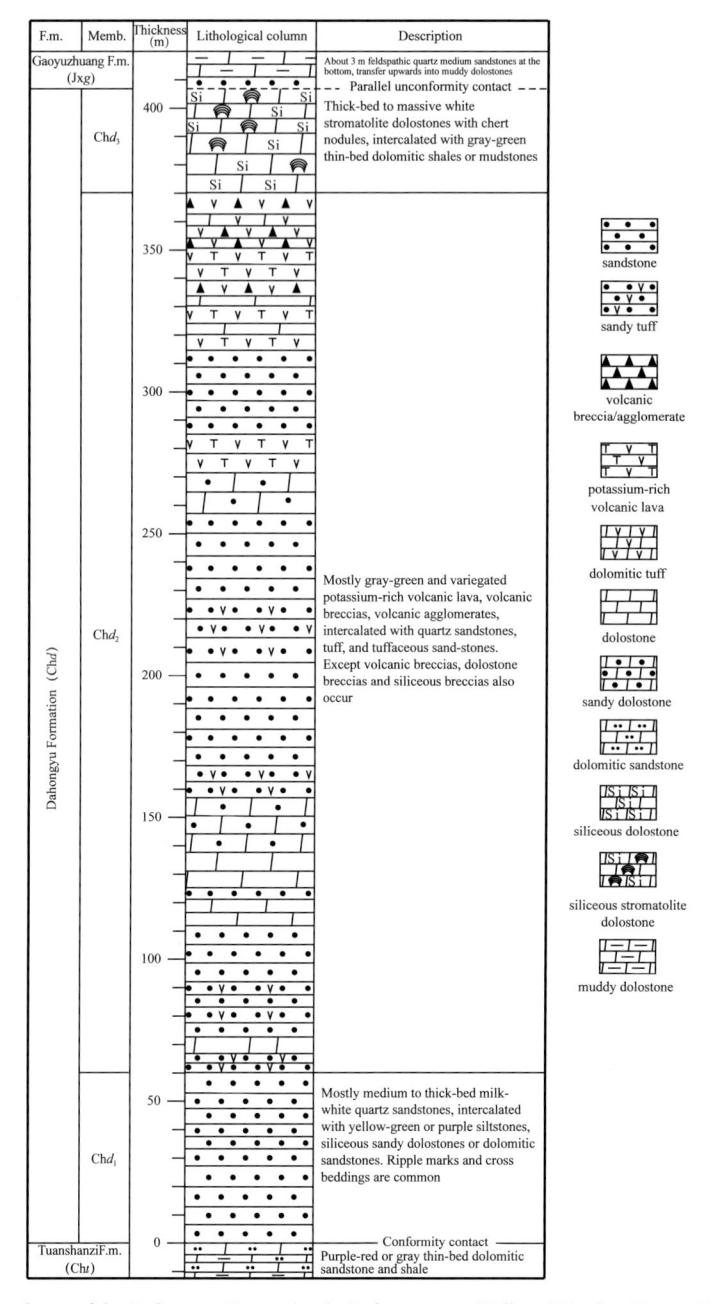

Fig.27　Stratigraphic column of the Dahongyu Formation in Dahongyugou Valley, Xiaying Town, Jizhou District, Tianjin

1. Route location

The route is located in Dahongyugou valley, Xiaying Town, Jinzhou District, and the starting point is located at the intersection of Maying Road and Yingshu Road. GPS position：E 117.466482°, N 40.174527°). The end point is at the south mouth of the Dahongyugou valley. GPS position：E 117.476749°, N 40.164697°.

The total length of the route is about 1613m. The specific stops are shown in Fig.28. The road along this route is wide, and both large and small vehicles are accessible. There are few daily vehicles and parking is convenient.

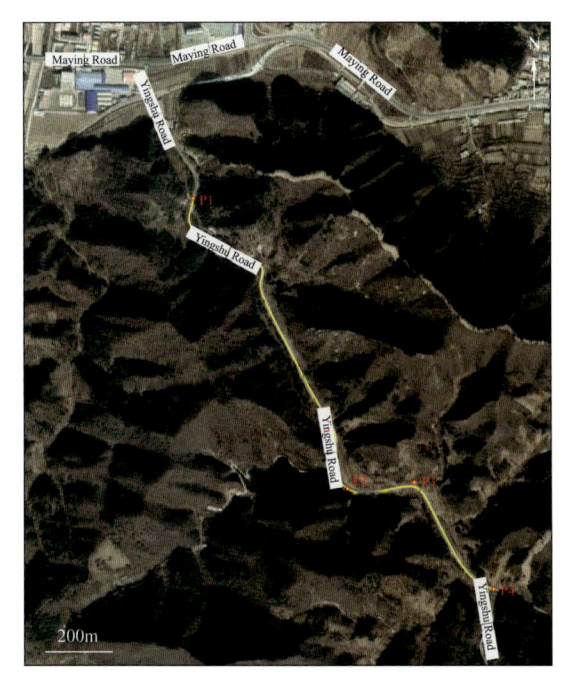

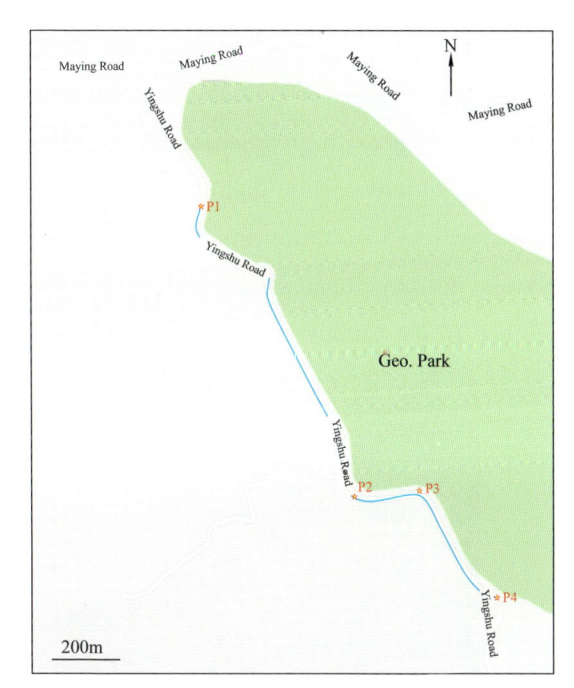

Fig.28 Distribution of description stops and route of the Dahongyu Formation in Jixian District, Tianjin

2. Objectives of investigation

(1) To observe the boundary between the Tuanshanzi Formation and Dahongyu Formation.

(2) To observe the lithologic characteristics and stratigraphic sequences of the Dahongyu Formation;

(3) To observe the rock characteristics of the potassium-rich volcanic rocks and the pyroclastic rocks in the Chd_2, such as stoma and almond structures, etc.

(3) To analyze the depositional environments and establish the depositional model.

3. Contents of investigation and description of stops

The starting point of the route is the north end of the Dahongyugou valley, and the end point is its south end. Three stops in this route are selected to investigate the lithology and stratigraphic characteristics of the Dahongyu Formation; including the boundary between the Dahongyu Formation and the Tuanshanzi Formation (P1 in Fig.28), the characteristics of the pyroclastic rocks (P2 in Fig.28) and the potassium-rich volcanic lava and the volcanic breccias (P3 in Fig.28).

Stop 1　Boundary between the Tuanshanzi Formation (Cht) and Dahongyu Formation (Chd)

Location: This stop is located near the intersection of Maying Road and Yingshu Road. GPS position: E 117.466482°, N 40.174527°.

Contents of investigation: (1) The boundary between the Dahongyu Formation and the Tuanshanzi Formation and the variation in lithology under and above it. (2) Sedimentary structures in the sandstones. (3) Analyze the depositional environments of the sandstones.

Description: This stop is to see the boundary between the Tuanshanzi Formation (Ch*t*) and the Dahongyu Formation (Ch*d*). Below the interface is thinly interbedded grayish white quartz sandstones and grayish green shales at the top of the Tuanshanzi Formation. Above the interface is thickly bedded milky white quartz sandstones at the bottom of the Dahongyu Formation. The contact is transitional and is conformity (Fig.29A and B). A number of different ripple marks in different directions are developed in the quartz sandstones (Fig.29C), in which the parallel beddings are developed in the thickly bedded quartz sandstones (Fig.29D). The sandstones are subtidal shoal deposits (see Section 15 for details).

Fig.29 Geological phenomena near the boundary between Tuanshanzi Formation (Ch*t*) and Dahongyu Formation (Ch*d*) of Changcheng System

(A) Conformity contact between the Tuanshanzi Formation (Ch*t*) and the Dahongyu Formation (Ch*d*). Stop 1; (B) Red dolomitic shales gradually change into gray-white fine quartz sandstones in the top of the Tuanshanzi Formation, stop 1; (C) A number of different ripple marks in different directions are developed on the bottom surfaces of thickly bedded quartz sandstones at the bottom of the Dahongyu Formation. Stop 1; (D) The parallel beddings are developed in the grayish white thickly bedded quartz sandstones, Stop 1

Stop 2 Typical volcanic breccias of the Chd_2

Location: This stop is located at the volcanic breccias about 1km south of stop 1. GPS position: E 117.471845°, N 40.167216°.

Description: This place is the typical volcanic breccias of the Dahongyu Formation (Chd_2) (Fig.30A). The breccias are of different sizes and complex compositions, including breccias, volcanic rock and tuff.

Stop 3   The volcanic lava and breccia of Chd_2

Location: This stop is about 200m southeast of Stop 2.

Description: This stop is to see the typical volcanic rocks and breccia in the top of Chd_2. Stoma and almond structures are developed in the volcanic lavas (Fig.30B and C). The volcanic breccias are of different sizes. The breccia composition is mainly basalts (Fig.30D).

Fig.30   Characteristics of volcanic breccias and lavas in the Chd_2

(A) Volcanic breccias, the breccias are of different sizes and complex composition, including cherty dolostones, volcanic rock and tuff, Chd_2, Stop 2. (B) Stoma structures in the basalts. The red arrow indicates the almond structures, Chd_2, Stop 3. (C) Almond structures in the basalts, Chd_2, Stop 3. (D) The volcanic breccias, and the breccias are of different sizes and complex composition. Chd_2, Stop 3

SECTION VI   Investigation Route of the Gaoyuzhuang Formation (Jxg) of Jixian System from Dahongyugou Valley to Zhaizhuang Village, Xiaying Town

The Gaoyuzhuang Formation is 1770 m thick which is mainly composed of carbonate rocks. And it disconformably overlies the Dahongyu Formation. The entire Gaoyuzhuang Formation is divided into six members from bottom to top: Jxg_1, Jxg_2, Jxg_3, Jxg_4, Jxg_5, and Jxg_6.

The Jxg_1 mainly consists of cherty microcrystalline dolostones with stromatolites intercalated with grayish green shales, and the bottom is 3m thick quartz sandstones. The Jxg_2 consists of grayish brown manganese silty shales and manganese silty dolostone and manganese siltstone. The Jxg_3

consists of medium-thickly bedded microcrystalline dolostones. The Jxg_4 consists of medium-thinly bedded limy dolostones and dolomitic limestones with a layer of nodular limestone at the bottom. At the top, there are molar-tooth limestones. The lower part of the Jxg_5 is mainly composed of asphalt laminar or massive fine crystalline limestones. The upper part is composed of coarse crystalline oncoid dolostones, medium to fine crystallized dolostones with ripple marks, and the fine crystalline dolomite with laminar stromatolites. The Jxg_6 consists of medium-thickly bedded microcrystalline dolostones with chert nodules and ribbons.

The Gaoyuzhuang Formation is widely exposed in Dahongyu valley north of the Zhaizhuang village. Sketch profile and stratigraphic column of the Gaoyuzhuang Formation are shown in Fig.31 and Fig.32.

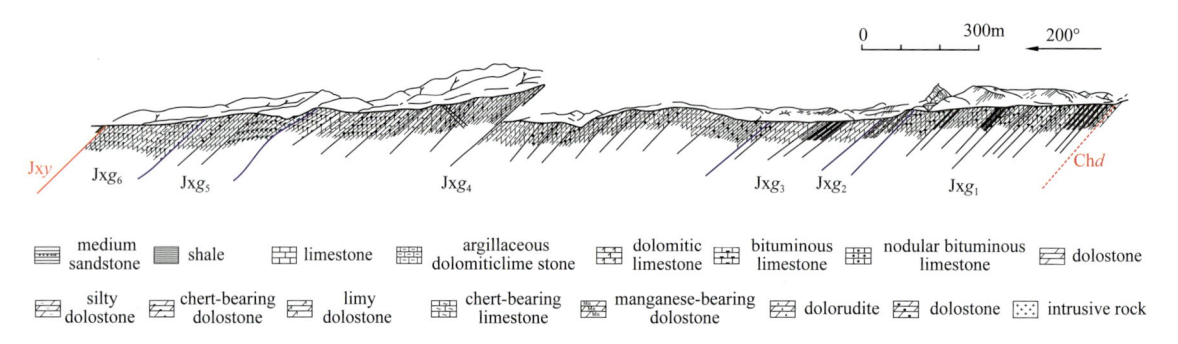

Fig.31 Sketch profile of the Gaoyuzhuang Formation, Jixian system, Mesoproterozoic from Dahongyugou Valley to north of Zhaizhuang Village in Xiaying Town, Jizhou District, Tianjin (modified from Tianjin Institute of Geology and Mineral Resources, 1964)

1. Route location

The first segment of the route is in the south part of the Dahongyugou valley, and the starting point is at the fork of Yingshu Road. GPS position: E 117.476749°, N 40.164697°. The end point is north of Zhaizhuang village in Luozhuang Town. GPS position: E 117.467395°, N 40.141950°. The total length of the route is about 4107.5 m.

The stops selected are shown in Fig.33. The first half of the route (Yingshu Road) is wide, both large and small vehicles can pass, and the observation stops are along the road; the second half (Mazhai Road) is along a mountain road through which only SUV cars can pass. But there are pavement roads in the geological park suitable for walking. Drivers can wait and park cars at the intersection of Chenglong Road and the Mazhai Road.

2. Objectives of investigation

(1) To observe the lithologic characteristics and depositional sequences of Gaoyuzhuang Formation.

(2) To observe the disconformity between the Gaoyuzhuang Formation and Dahongyu Formation.

(3) To observe the typical structures such as stromatolites, round ridge ripple marks, nodular structure, molar tooth structure, etc.

Abstract Field Guide of Stratigraphy and Sedimentary Facies of the Proterozoic in Jizhou District, Tianjin

F.m.	Memb.	Thickness (m)	Lithological column	Description
Yangzhuang F.m. (Jxy)		1800		Purple argillaceous microcrystal dolostones interbedded with microcrystalline dolostones.
				——— Conformity contact ———
Gaoyuzhuang F.m. (Jxg)	Jxg₆	1700–1500		Gray-white laminated microcrystalline dolostones and dolarenites with chert nodules and ribbons.
	Jxg₅	1400–1300		Gray-black wavy and nodular bituminous limestones at the bottom, coated-grain (like oncolite) coarse crystalline dolostones, interbedded with wavy dolostones and horizontal laminated dolostones.
	Jxg₄	1200–600		Nodular limestones, medium to thin-bedding silty dolomitic microcrystalline limestones, and microcrystal limestones at the bottom; medium to thick-bedding horizontal laminated gray silty limy microcrystalline dolostones, interbedded with microcrytalline dolostones, limy dolostones, and locally occured with siliceous nodules or ribbons and intercalated with dolomitic calcirudites.
	Jxg₃	500–400		Mostly medium to thick-bedding gray-white slity microcrystalline dolostones, occurred with medium to thin-bedding gray-white dolomitic limestones and limy dolostones at the top.
	Jxg₂			Thin bed gray-black manganese-bearing silty shales interbedded with …
	Jxg₁	300–0		About 3 m feldspathic quartz medium sandstones at the bottom, grade upwards into muddy dolostones interbedded with dolomitic mudstone, and cyclic sequences consisting of stromatolite dolostones and laminated dolostone, cone stromatolite dolostones with chert nodules and ribbbons in the top.
Dahongyu F.m.				——— Parallel unconformity contact ——— Medium to thick-bed gray-white cone stromatolite dolostones with chert nodules

Legend:
- shale
- manganese-bearing shale
- manganese-bearing silty shale
- manganese-bearing siltstone
- sandstone
- microcrystalline limestone
- wavy laminated limestone
- Wavy laminated bituminous limestone
- nodular bituminous limestone
- dolomitic microcrystalline limestone
- microcrystalline dolostone
- limy microcrystalline dolostone
- stromatolite dolostone
- siliceous stromatolite dolostone
- argillaceous silty dolostone
- wavy laminated dolostone
- dolostone with chert nodules or ribbons
- intraclast dolostone
- manganese-bearing dolostone
- silty microcrystalline dolostone
- coated-grain (like oncolite) microcrystalline dolostone
- covered

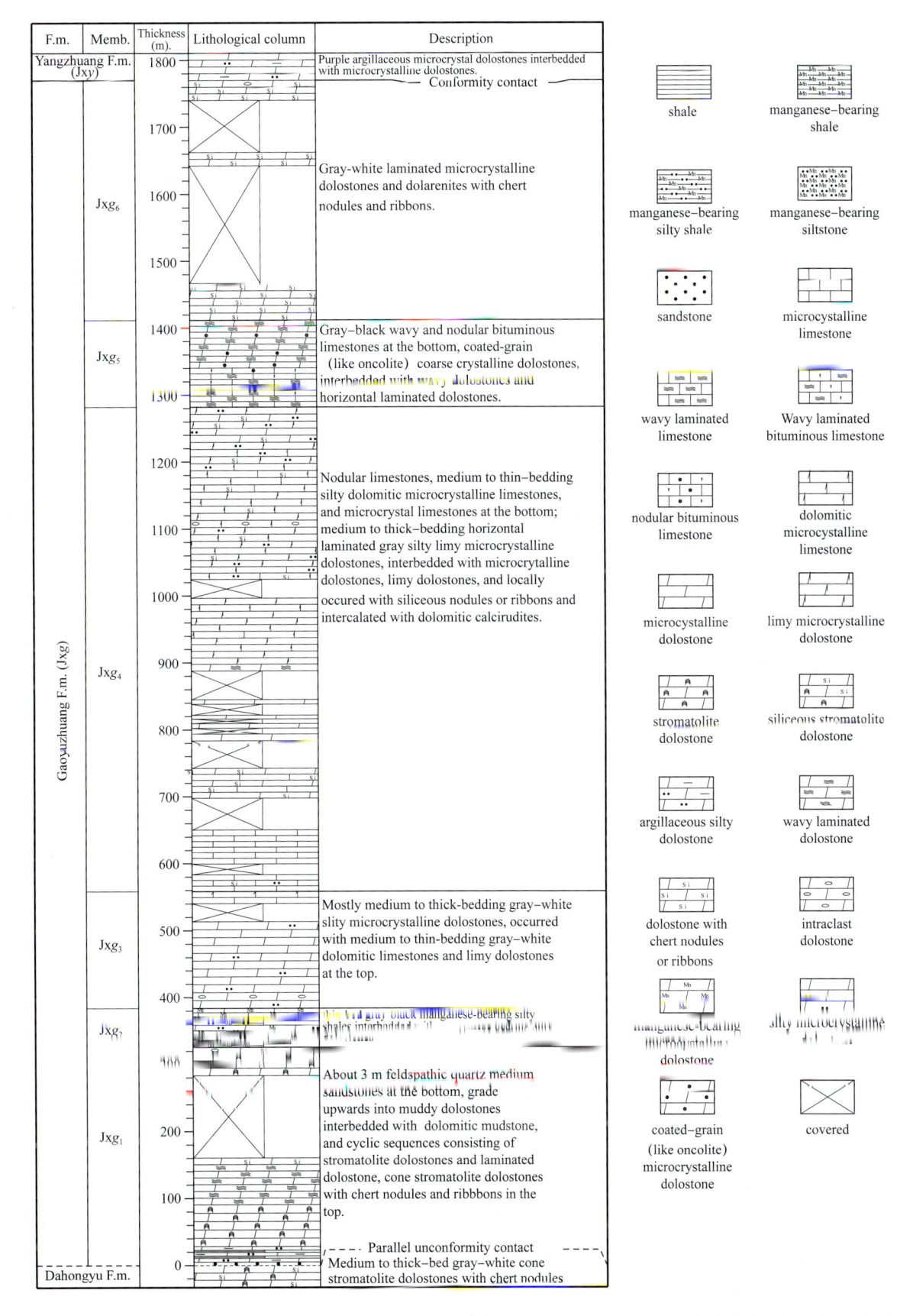

Fig.32 Stratigraphic column of the Gaoyuzhuang Formation, Jixian System, Mesoproterozoic in Jizhou District, Tianjin

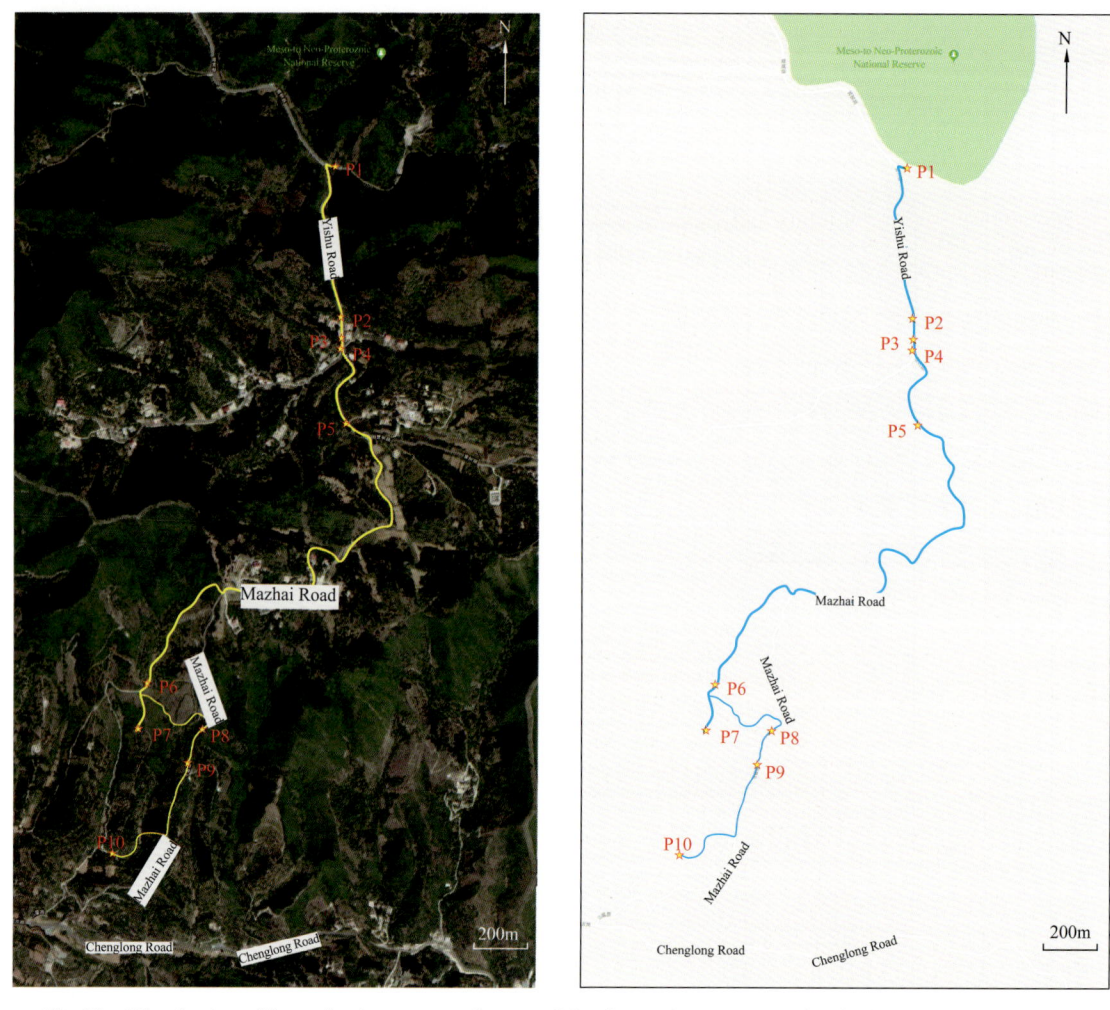

Fig.33   Distribution of investigation stops and route of the Gaoyuzhuang Formation in Jixian District, Tianjin

(4) To analyze sedimentary environments and its evolution.

3. Contents of investigation and description of stops

The lithology and stratigraphy characteristics of Gaoyuzhuang Formation are investigated along Dahongyugou Valley to Zhai Zhuang village, and nine stops are selected in the route, including the boundary between Dahongyu Formation and Gaoyuzhuang Formation (P1 in Fig.33), the boundary between the Jxg_1 and the Jxg_2 (P2 in Fig.33), manganese boron deposits (P3 in Fig.33), ripple marks (P4 in Fig.33), nodular limestone (P5 in Fig.33), the boundary between the Jxg_4 and the Jxg_5 (P6 in Fig.33), typical lithology in the Jxg_5 (P7 and P8 in Fig.33) and typical lithology in the Jxg_6 (P9 in Fig.33), etc. The descriptions of each stop are as follows.

Stop 1   The boundary between the Dahongyu Formation and the Gaoyuzhuang Formation

Location: The stop 1 is located at the fork in Yingshu Road where there is a monument, marking the boundary between the Dahongyu Formation and the Gaoyuzhuang Formation. GPS position: E 117.476749°, N 40.164697°.

Contents of investigation : (1) Observe the parallel disconformity between the Gaoyuzhuang Formation and Dahongyu Formation. (2) Cherty dolostone with cone stromatolites in the Jxg$_4$. (3) The origin of ripple marks in basal sandstone and infer the direction of the coastline. (4) Analyze sedimentary environments above and below the disconformity.

Description : The boundary between the Dahongyu Formation and the Gaoyuzhuang Formation is a parallel disconformity (Fig.34 A and B) . Above the boundary are 3m thick fine sandstones at the bottom of the Gaoyuzhuang Formation. Symmetrical ripple marks are formed on the surface of the sandstones. The wave ripples are continuous and very small, the wave height is about 1cm, and the trend of wave crests is approximately 290° (Fig.34 C and D) . The sandstones are thought to be shallow subtidal sand shoal. Under the boundary, there is a cherty dolostone with cone stromatolites in the Dahongyu Formation. The stromatolite is mostly silicified, and represents shallow subtidal to lower intertidal flat. A thin weathering crust can be seen on the interface, which represents a depositional hiatus resulting from structural movement "Qinglong rise".

Fig.34　Geological phenomena near the boundary between the Changcheng System Dahongyu Formation (Chd) and the Jixian System Gaoyuzhuang Formation (Jxg), and photomicrographs of their rocks
(A) Contact between Jxg and Chd. (B) Contact between Jxg and Chd at another site near by. (C) Ripple marks in sandstone at bottom of Jxg. (D) Microphotograph of Sandstone in C.

Stop 2　The boundary between the Jxg$_1$ and the Jxg$_2$ and the lithologic features of the Jxg$_2$

Location : The Stop 2 is located at the monument marking the boundary between the Jxg$_1$ and the Jxg$_2$ near the village committee of Sangshu'an Village. GPS position : E 117.477235°, N 40.159696°.

Contents of investigation : (1) Rock types and features ; (2) Sedimentary structures in shales ; (3) The lithology of the upper part of the Jxg$_1$ and the origin of wavy bed surfaces ; (4) Abrupt contact between Jxg_1 and Jxg_2 and between Jxg_2 and Jxg_3; (5) Characteristics and origin of chert nodules and ribbons at the top of the Jxg_1; (6) Rock types of the Jxg_2; (7) Analyze the sedimentary environments of Jxg_1 and the Jxg_2 and what caused the abrupt change in environments.

Description: The Jxg_1 seems to be conformable with the Jxg_2 (Fig.35A), but the contact is very abrupt. Below the boundary are dark gray cherty dolostones (Fig.35B), which is thought to be deposited on restricted platform and tidal flat. Above the boundary is dark grayish brown manganese silty shale and manganese muddy siltstone (Fig.35C) interbedded with microcrystalline dolostones (Fig.35D). The Jxg_2 is a deep-water shelf deposit (see Section 15 for details).

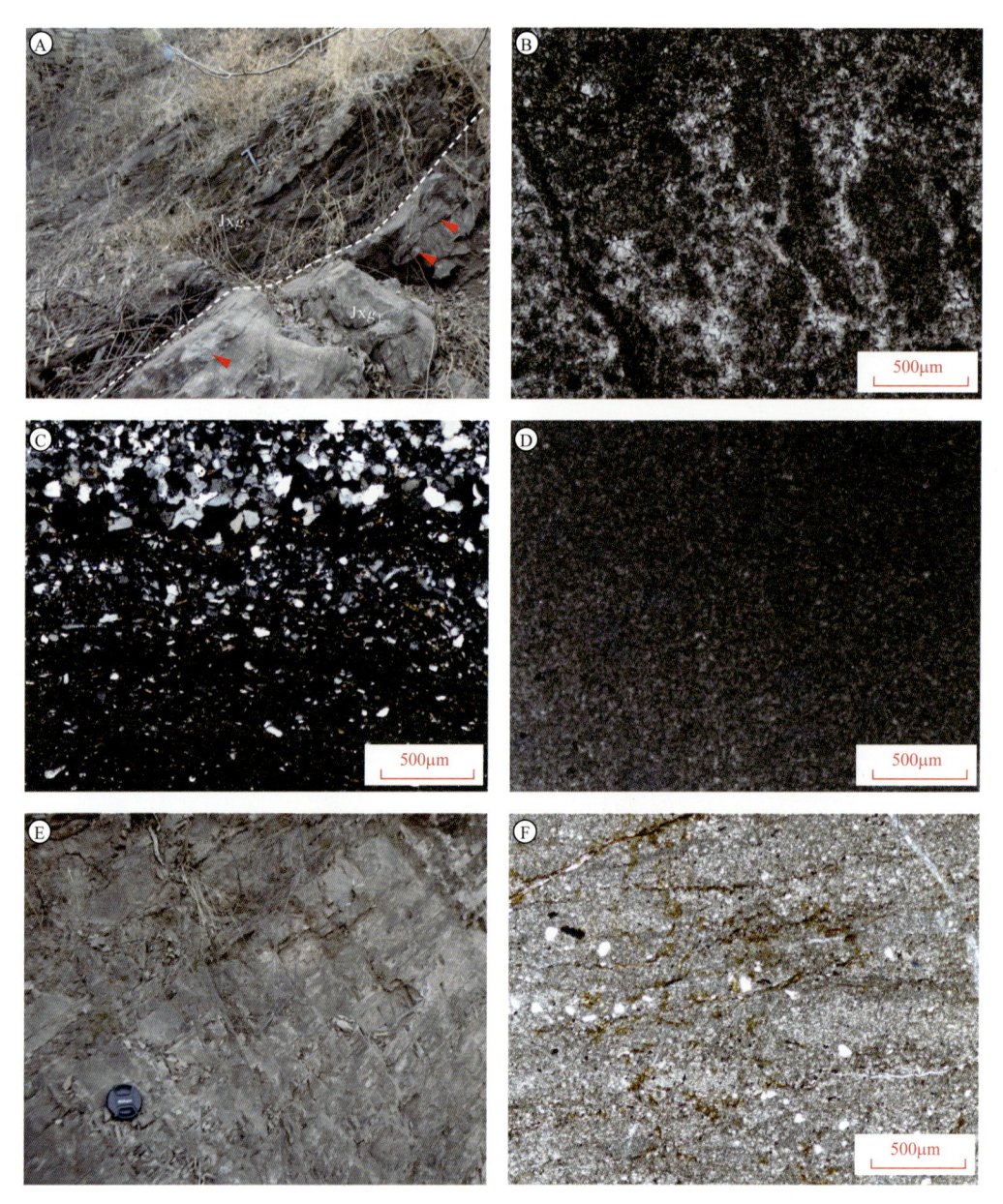

Fig.35 Geological phenomena near the boundary between the Jxg_1 and the Jxg_2 and photomicrographs of their rocks
(A) Sharp contact between Jxg_2 and Jxg_1.stop 2. (B) Microphotograph of algal dolostone at top of Jxg_1, Plane polarized.
(C) Microphotograph of shale and siltstone in Jxg_2, Crossed polarized. (D) Microphotograph of dolostone in middle Jxg_2. (E) Shale intercalated with muddy dolostone. (F) Microphotograph of silty dolostone at top of Jxg_2.

Stop 3 Manganese (boron) deposits

Location: The stop 3 is located at the monument marking manganese (boron) deposits near the Sangshu'an village committee. GPS position: E 117.477340°, N 40.158943°.

Contents of investigation: The features of the Manganese (boron) deposit.

Description: It is about 50 m south of the stop 2, and is the upper part of the Jxg_2. The lithology is grayish brown silty microcrystalline dolostone (Fig.35E and F) with chambersite which constitutes the famous "Jixian" manganese (boron) deposits. Regarding the origin of the manganese boron deposit in this area, the predecessors mainly have the following views: a) Some scholars believe that the formation of the manganese (boron) deposit is in lagoon facies during the largest marine regression. The organic matter and pyrite in the microcrystalline dolostone are spatially and temporally related to chambersite. b) Wang (1996) believes that manganese (boron) deposit the in Jizhou District conforms to the binary deposit type of depositional ore deposit, and the source of boron is related to volcanic eruption. c) Fan et al. (1999) believes that the ore-forming materials of manganese (boron) deposit are derived from submarine volcanic eruptions and the weathering of a small amount of early Proterozoic submarine hot water depositions. d) Wang (2013) studied the mineralogical characteristics of chambersite, and determined that boron and manganese originated from submarine volcanic eruption. The ore-forming temperature was lower than 400°C, and environments is slightly acid, highly saline and anaerobic. It is speculated that it is formed by seafloor hydrothermal deposition.

Stop 4 Lithology changes of the Jxg_3 and ripple marks

Location: The stop 4 is just south of the Sangshu'an Village. GPS position: E 117.47725°, N 40.158758°.

Contents of investigation: (1) Round crest ripple marks on top surface of dolostone; (2) Rock types and depositional sequences of the Jxg_3; (3) Sedimentary environments of the Jxg_3.

Description: This stop is about 28m south of stop 3, and it belongs to the Jxg_3. The lithology is medium-thickly bedded dolostones on whose bed surface the round crest ripple marks develops (Fig.36A). The characteristics of the round crest ripple marks is that the shape of the wave mark cross section is symmetrical sinusoidal shape, which cannot indicate the direction of water flow and is formed by storm waves. The whole Jxg_3 consists of medium-thickly bedded massive iron-bearing microcrystalline dolostones and laminated microcrystalline dolostones, interbedded with grayish green shales. Laminations are due to laminar stromatolites. A small amount of chert nodules and ribbons can be seen in the laminated microcrystalline dolostones. The sedimentary environments of Jxg_3 are tidal flats and restricted (or open) platform. Shales represent a transgression event and are deposited in shale shelf environments.

Fig.36 Round-ridge ripple marks (A) at stop 5 and nodular limestone (B) at stop 6

Stop 5 Nodular limestone

Location: The stop 5 is located about 340 m south of the village committee of Sangshu, an Village. GPS position: E 117.477429°, N 40.156295°.

Contents of investigation: (1) The feature of nodular limestone; (2) Origin of nodular limestone.

Description: The nodular limestone is developed at the bottom of Jxg_4 (Fig.36B). Under the nodular limestone, the rocks are mainly the thin-medium bedded microcrystalline dolostones. Limestone nodules are mainly composed of calcite (90%±), dolomite (5%~10%) and a small amount of organic matter and terrigenous mud. The matrix between nodules is light gray, and its main composition is calcite micrite. The crystal size is finer than that of the nodules, and the clay content is much higher than that of the nodules. Predecessors believe that the main causes of the nodular structure are as follows: a) In the deep-water basin and the lower part of platform margin, the $CaCO_3$-unsaturated bottom current partially dissolves the carbonate rocks deposited on the sea floor to form nodules (Gao, 1988). b) Sedimentary diagenesis. c) Pressure solution. That is, due to the overlying pressure or tectonic stress, pressure solution cuts the weakly consolidated limestones with clay into different sizes and shapes of nodular-body (Wanless, 1979; Guo, 1989). d) Both periodic sea bottom currents dissolution and pressure solution work together (Xia *et al.*, 2009).

In this book, we believes that limestone nodules are formed by microbes.

Stop 6 The boundary between the Jxg_4 and the Jxg_5

Location: The stop 6 is about 2.4km south of Stop 5 along the Mazhai Road. GPS position: E 117.468656°, N 40.147629°.

Contents of investigation: (1) The rock types of the Jxg_4; (2) The depositional sequences; (3) Molar-tooth structures and its origin; (4) The depositional environments of the Jxg_4.

Description: The Jxg_4 is characterized by limestones, rare chert nodules and lack of stromatolites. Most bed surfaces are flat. There are three rock types: medium-thick bedded limemudstones wihout stratification, limemudstones with horizontal bedding, and shaly argillaceous limemudstones. These three types of rocks constitute two types of meter-scale cycles. One is shaly argillaceous limemudstones-

limemudstones with horizontal bedding-medium-thick bedded limemudstones wihtout stratification from bottom to top; the other is limemudstones with horizontal bedding-medium-thick bedded limemudstones without stratification. Chert nodules mainly appear in the medium-thick bedded limemudstones and are mostly pipe-like, and the interior is not silicified. This type of chert nodules may be associated with burrows or a large worm organism (such organisms were probably present in the Proterozoic), and the flesh of the organism partially rots to produce an acidic microenvironment, resulting in selective silicification.

The stromatolites are generally absent in the Jxg_4, which may be due to the deep water. Limestones with flatbed surfaces or horizontal bedding are deposited in deeper environments and the waves cannot affect the sea floor. Some shaly argillaceous limesmudtones and some medium-thick limesmudtones have wavy bed surfaces (actually wave marks), indicating that waves can sometimes affect the sea floor. Therefore, the sedimentary environments of Jxg_4 is generally deep, and is near the storm wave base, which are thought to be relatively deep open platform to gentle slope. The water becomes shallower near the top of Jxg_4.

The boundary between the Jxg_4 and the Jxg_5 of the Gaoyuzhuang Formation is marked by the appearance of the molar-tooth limestone, and the molar-tooth limestone belongs to the Jxg_4. Below the interface is dark gray medium-thick bedded dolomitic limestones with chert nodules, and above the interface are laminated or oncoid bituminous crystalline limestones (Fig.37D-F).

The limemudstone in the top part of the Jxg_4 contains a large number of "Molar-tooth structures". The Molar-tooth structure is characterized by different shapes of veins in the limemudstones, and the veins are about several centimeters to ten centimeters long. The middle part of the veins is thicker, and the ends are gradually sharpened. Smaller vein branches and isolated spots can be seen along the margin of the vein (Fig.37A and B). Generally, the microcrystalline limestones of veins are coarser than the host rocks which are composed of cryptocrystalline calcite (Fig.37C).

The origins of the "Molar-tooth structures" mainly include the following : a) Some people believe that the formation of molar-tooth structure is related to the life activities of algae (James et al., 1998; Frank and Lyons, 1998; Liu et al., 2003). b) Some people believe that the liquefaction induced by earthquakes causes the underwater sedimentation of the unconsolidated limemud sediment to develop the Molar-tooth structures (Pratt, 1998; Fairchild et al., 1997).

In this book, we believe that molar-tooth structue is mud cracks which are not fully developed. The limemud sediment deposited underwater is exposed later, forming mud cracks due to evaporation, and is submerged again under water. Then the mud crack is filled by limemud sediment. For example, the limemud deposited in the extremely shallow lagoon is often exposed during the unusually low tide, forming mud cracks; when the mud cracks have not yet developed fully, tide rises again, the mud cracks are submerged under water, and the mud crack is filled with limemud. Due to the restricted and reducing microenvironments in the mud cracks, the sediment in the mud cracks is darker in color.

The laminar structure in the recrystallized limestone at the bottom of the Jxg_5 is laminar stromatolites

(Fig.37D). The laminae are mainly composed of fine crystalline calcites, which are formed by recrystallization. The inter-lamina space is filled with pores or dolomite (Fig.37E). Coated grains in limestones are oncoids (Fig.37F).

Fig.37 Geological phenomena near the boundary between the Jxg_4 and the Jxg_5, and photomicrographs of their rocks at stop 6 (A) and (B) Molar limestones at top of Jxg_4. (C) Microphotograph of molar limestone. (D) oncolite limestone at bottom of Jxg_5. (E) Microphotograph of laminated limestone of Jxg_5. (F) Oncolite limestone at bottom of Jxg_5.

SECTION Ⅶ Investigation Route of the Yangzhuang Formation (Jx*y*) of Jixian System from Zhaizhuang Village to North Qingshan Village and Huaguoyu Village, Luozhuangzi Town

The Yangzhuang Formation is 773 m thick, and is mainly composed of interbedded purplish red dolomitic mudstone and grayish white dolomitic mudstone, which look like the color of "marbled meat".

In the upper part of this formation, there are some limestones and dolostones.

The purplish red color is a strongly oxidizing color, indicating subaerial exposure. The purplish red dolomitic mudstones are deposited on supratidal flats.

The grayish white color is weakly reducing, indicating shallow water. The grayish white dolomitic mudstones are deposited on intertidal flats which are frequently under water.

The limestones and dolostones which contain laminar stromatolites are deposited on carbonate tidal flats. Those with thrombolites are deposited in shallow subtidal environments.

According to the lithological association, Yangzhuang Formation can be divided into three sections: the 1st member (Jxy_1) in the bottom, consists of the "marble meat mudstone", limestone with chert nodules and dolomite; the 2nd member (Jxy_2) in the middle, consists of "marble meat mudstone"; the 3rd member (Jxy_3) in the top, consists of "marble meat mudstone" interbedded with limestone and dolomite with chert nodules.

The Yangzhuang Formation widely exposes in Yangzhuang Village, Zhaizhuang Village, Qingshan Village and Huaguoyu Village of Luozhuangzi Town. However, considering the impact of the faults in the center of the distribution area and the occurrence, the inspection stops are dispersed and mainly divided into three investigating routes: The 1st route is located in the north of Zhaizhuang Village, mainly to investigate the lithological characteristics of the Jxy_1; the 2nd route is located in the picking garden of Qingshan Village, mainly to observe the lithological characteristics of the Jxy_2 and the Jxy_3; the 3rd route is located in the southeast of Huaguoyu village, mainly to observe the lithological characteristics of the Jxy_2 and the Jxy_3. The 2nd route runs along the hillside, and the outcrops are more continuous and complete, and we can see the boundary of the Wumishan Formation. Stratigraphic column of Yangzhuang Formation is shown in Fig.38.

1. Route location

Affected by faults and the occurrence, the inspection stops are dispersed and mainly divided into three routes to investigate (Fig.39). The starting point of the 1st route is located in the north of Zhaizhuang Village, Luozhuangzi Town (GPS: E 117.467395°, N 40.141950°), and the end point is about 125 m to the east of the intersection of Mazhai Road and Chenglong Road (GPS: E 117.465003°, N 40.139757°); The 2nd route is located in Qingshan Village, Luozhuangzi Town. The starting point is at the gate of Qingshan Picking Garden (GPS: E 117.445361°, N 40.124436°). The end point is at the half hillside of Wuming Mountain (GPS: E 117.443030°, N 40.120781°); The 3rd route is located in Huaguoyu Village, Luozhuangzi Town, about 920 m southeast along the Maping Highway (GPS: E 117.464169°, N 40.118511°). The length of the route is about 11.2 km, and all three routes are beside of the main road. It is convenient for transportation. Only the 2nd route requires walking

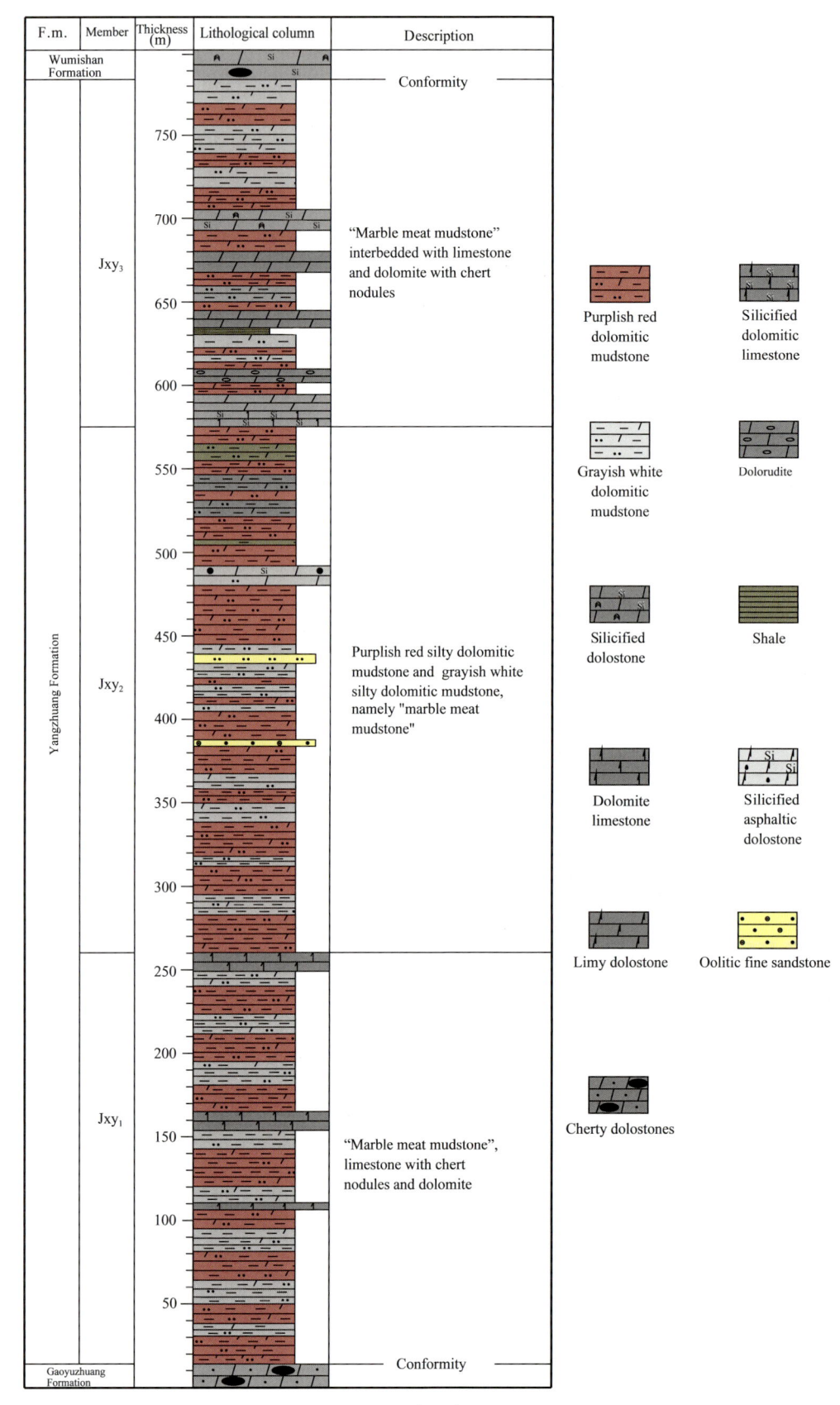

Fig.38　Stratigraphic column of the Yangzhuang Formation (Jxy), Jixian system, Mesoproterozoic in the picking garden of Qingshan Village, Luozhuangzi Town, Jizhou District, Tianjin

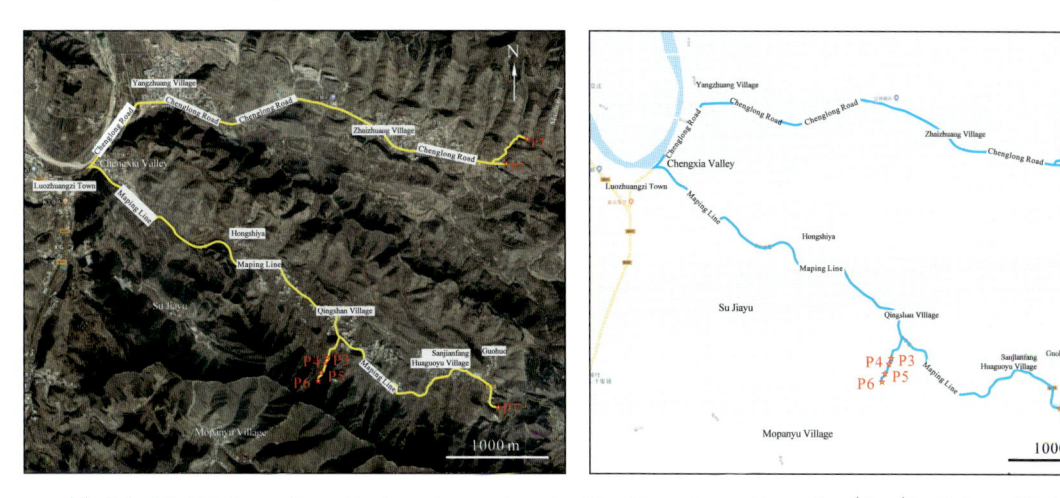

Fig.39 Distribution of investigation stops and route of the Yangzhuang Formation (Jxy) in Jizhou, Tianjin

2. Objectives

(1) To observe the lithologic features and stratigraphic sequence of the Yangzhuang Formation;

(2) To observe rocks with typical sedimentary structures such as graphophyric dolomites and stromatolite dolomites;

(3) To analyze the sedimentary environment and its evolution.

3. Contents of routes and investigation stops

The route runs along Dahongyu Valley to the north of Zhaizhuang Village for investigating the lithology and stratigraphic characteristics of the Yangzhuang Formation. We select seven inspection stops, including the boundary of Gaoyuzhuang Formation and Yangzhuang Formation in the north of Zhaizhuang Village (P1 in Fig.39), the typical lithology of the Jxy_1 (P2 in Fig.39), the typical lithology of the Jxy_2 in Qingshan Village (P3 in Fig.39), the boundary between the Jxy_2 and the Jxy_3 (P4 in Fig.39), the typical lithology in the Jxy_3 (P5 in Fig.39), the boundary between the Yangzhuang Formation and the Wumishan Formation (P6 in Fig.39) and the boundary between the Jxy_2 and Jxy_3 in Huaguoyu Village (P7 in Fig.39). The description of each inspection point is as follows:

Stop 1 Boundry between the Gaoyuzhuang Formation and the Yangzhuang Formation.

Location: The stop 1 is located in the north of Zhaizhuang Village, and about 400m to the northeast of the intersection of Chenglong Road and Mazhai Road. GPS: E 117.467395°, N 40.141950°.

Contents of investigation: The boundry between the Gaoyuzhuang Formation and the Yangzhuang Formation.

Description: Here is the boundary between the Gaoyuzhuang Formation (Jxg) and the Yangzhuang Formation (Jxy) (Fig.40A). Below the boundary is medium bedded cherk nodules, ribbons dolostones at the top of the Jxg_6 (Fig.40B). Above the boundary is the purple-red dolomitic mudstone and gray dolomitic limestone with siliceous tuberculosis at the bottom of the Yangzhuang Formation (Fig.40A and C).

Two formations' boundary is continuous deposition.

Stop 2　The typical lithology combination of Jxy₁

Location：The Stop 2　is located about 125 m east of the intersection of Mazhai Road and Chenglong Road. GPS：E 117.465003°，N 40.139757°.

Contents of investigation：The characteristics of "marble meat mudstone" and its sedimentary environment.

Description：The typical rock of the Jxy₁ is well exposed, and the lithology is purple-red silt-laden dolomitic mudstone interbedded with gray lumpy dolomitic mudstone, and the thickness ratio of them is about 3∶1（Fig.40D）.

Fig.40　Geological phenomena near the boundary between the Gaoyuzhuang Formation（Jxg）and the Yangzhuang Formation（Jxy）of Jixian System, and micrographs of the rocks

（A）The boundry between the Gaoyuzhuang Formation and the Yangzhuang Formation, Stop 1；（B）Microcrystalline dolostones, plane polarized light, a scale of 0.5 mm, Stop 1；（C）Gray dolomitic limestone with chert nodules, Stop 1；（D）Typical lithology combination of Jxy₁, Stop 2

Stop 3　The typical lithology combination of Jxy₂

Location：The Stop 2　is located about 250 m above the slope of Qingshan Picking Garden. GPS：E 117.444195°，N 40.122783°.

Contents of investigation：（1）The characteristics of "marble meat mudstone". （2）Characteristics of sandstone. （3）Sedimentary environment analysis.

Description：The typical lithology of the Jxy₂ consists of a mixture of purple-red silt-laden dolomitic mudstone and gray-white silt-laden dolomitic mudstone, namely "marble meat mudstone" (Fig.41A). A layer of medium quartz sandstone can be seen in the middle part (Fig.41D and E). The striking purple color is mostly caused by Fe^{3+}, which generally indicates the exposure in the hot, dry and strong oxidizing environment. Gray-green spots and bands are also common in purple-red silt-laden dolomitic mudstone (Fig.41B), and the spots and bands are relatively bright under the microscope (Fig.41C). This phenomenon is formed by the partial reduction of purple-red mudstone. The reducing groundwater seeped along the high permeability layer or the crack, which reduced Fe^{3+} to Fe^{2+}, and the color turns red to gray green. Ripple marks develop on rock layer surface (Fig.41F).

Fig.41 Typical rocks of the Jxy_2 and their micrographs

(A) The typical lithology of the Jxy_2, purple-red silty dolomitic mudstone and gray-white silty dolomitic mudstone, Stop 3; (B) Gray-green spots and bands are in purple-red silty dolomitic mudstone, Stop 3; (C) Purple-red silty dolomitic mudstone, plane polarized light, with a scale plot of 2 mm, Stop 3; (D) Medium quartz sandstone, Stop 3; (E) Medium quartz sandstone, plane polarized light, with a scale plot of 0.5 mm; (F) Purple-red silty dolomitic mudstone and gray-white silty dolomitic mudstone, ripple marks developing on rock surface, Stop 3

Stop 4   The boundry between the Jxy_2 and Jxy_3

Location: The stop 4 is located about 300 m above the slope of Qingshan Picking Garden. GPS: E 117.443833°, N 40.122186°.

Contents of investigation: The boundry between the Jxy_2 and Jxy_3. The lithological changes above and below the boundry.

Description: Here is the boundary between the Jxy_2 and the Jxy_3, marked by the presence of thick carbonate rocks. Below the boundary, there are purple-red silt-laden dolomitic mudstone interbedded with gray silt-laden dolomitic mudstone. Above the boundary there are thick gray, gray-black dolomite, gray-scale dolomite and stromatolite dolomite with a small amount of gray-green or purple-red silt-laden dolomitic mudstone (Fig.42). The common carbonate sequence consists of thick clot-like limestone

Fig.42   Typical rocks of the Jxy_3 and their photomicrographs

(A) Carbonate cyclic sequence in the Jxy_3, B-siliceous clot-like limestone, C-laminar stromatolite limestone with chert ribbons, D-chert ribbons with residual ooids, E- dolorudite; (B) Siliceous clot-like limestone, plane polarized light; (C) Stromatolite limestone, cross polarized light, note that the different crystal arrangement in different laminae; (D) Dolomitic oolitic chert, cross polarized; (E) Dolorudite; (F) Dolorudite, plane polarized light (subtidal) and medium chert ribbon stromatolite limestone (tidal flat). The stromatolite limestone intercalated in silicified oolitic dolomite and dolorudite (Fig.42), which are transported by the storm stream to the granular carbonate deposits on the tidal flat.

Stop 5 The lithologic features and carbonate cycle of the Jxy_3

Location: The stop 5 is located about 400 m above the slope of Qingshan Picking Garden. GPS: E 117.443833°, N 40.122186°.

Contents of investigation: (1) A sedimentary cycle formed by clot-like dolomite (or limestone) and layered stromatolite dolomite (or limestone). (2) Vein structure and its origin. (3) Sedimentary environment analysis.

Description: The gray and dark gray carbonate cycle in the Jxy_3. It consists of clot-like fine crystalline dolomite or limestone (subtidal), middle wavy and layered stromatolite microcrystalline dolostones or limestone (tidal flat) (Fig.43) from the bottom to the top which represents the sedimentary sequence from the subtidal to the tidal flat.

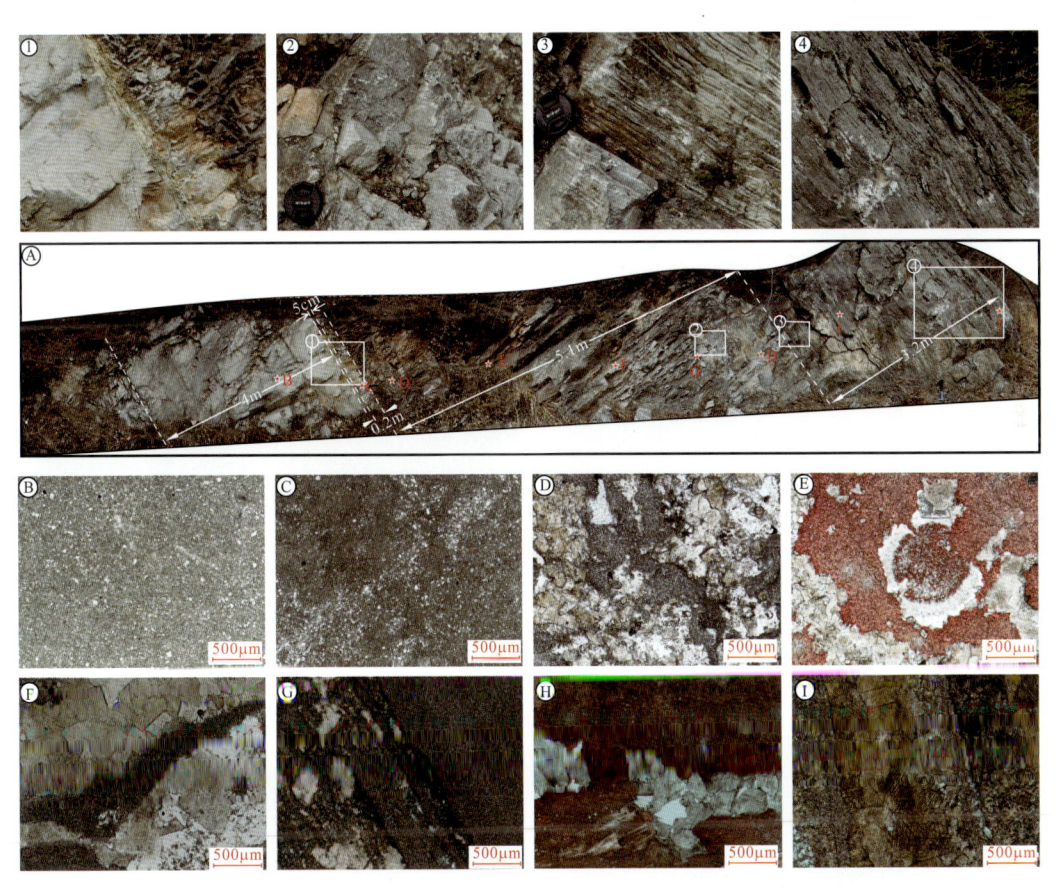

Fig.43 Typical carbonate cycle of the Jxy_3 and micrographs of their rocks

Figure ①②③④ are the enlarged view of the white frame in Figure (A); (A) The typical carbonate rock cycle of the Jxy_3, from the bottom to top, there are thick massive microcrystalline dolostones (4 m), gray-green thin bedded shale (5 cm), medium bedded horizontal laminar dolomite (0.2 m), and clot-like algal limestone (5.1 m) and laminar stromatolite limestone (3.2 m), Stop 5; (B) microcrystalline dolostones, plane polarized light; (C) Laminar stromatolit microcrystalline dolostones, plane polarized light; (D) Siliceous patchy dolomite, plane polarized light; (E) Siliceous-dolomite patchy limestine, corss polarized light; (F) Siliceous dolomite patchy limestone, cross polarized light; (G) Siliceouy patchy dolomite, plane polarized light; (H) Lamiar dolomitic limefone, corss polaiized light; (I) Laminar dolomite limestone, plane polarized light; (J) Lamina siliceous-dolomitic limestone, polarized light Figures (B) — (J) micrographs correspond to the corresponding marked ★ in B—J of figure (A) positions

Gray–white veins are commonly found in wavy and laminar stromatolite dolomite or limestone, filled with mud–crystal dolomite, a few millimeters wide, several to several tens of centimeters long, parallel, oblique or vertical to the rock surface. These veins are formed by the filling of the mud in the algae.

Stop 6　The boundry between the Jx*y* and Jx*w*

Location : The stop 6 is located about 500 m above the slope of Qingshan Picking Garden. GPS : E 117.443030°, N 40.120781°

Contents of investigation : The boundry between the Jx*y* and Jx*w*. The lithological changes above and below the boundry.

Description : There is the boundary between the Yangzhuang Formation and the Wumishan Formation, and they are conformity. Below the boundary there are gray–white silty muddy dolomite; above the interface there are gray thick bedded fine crystalline dolostones with chert nodules and ribbons, and stromatolite dolomite. The topography of the Wumishan formation is a steep cliff.

Stop 7　The boundry between the Jx*y*$_2$ and Jx*y*$_3$

Location : The stop 7 is located at Huaguoyu Village, about 920 m southeast along the Maping Highway. GPS : E 117.464169°, N 40.118511°.

Contents of investigation : The boundry between the Jx*y*$_2$ and Jx*y*$_3$. The lithological changes above and below the boundry.

Description : Stop 7 located on the Maping Highway, the Jx*y*$_2$ and Jx*y*$_3$ are well exposed, and the phenomena are obvious (Fig.44C and D) . The Jx*y*$_2$ is characterized by purple–red dolomitic

Fig.44　Boundary between Yangzhuang Formation (Jx*y*) and Wumishan Formation (Jx*w*) in halfway of the mountain of picking garden in Qingshan Village, and typical rocks of Yangzhuang Formation (Jx*y*) aside the Maping Road near Huaguoyu Village

(A) The boundary between the Yangzhuang Formation and the Wumishan Formation, and they are conformity. Below the boundry there are gray–white silty muddy dolomite; above the interface there are gray thick bedded fine crystalline dolostones with chert nodules and ribbons, and stromatolite dolomite, Stop 6; (B) The boundary between the Yangzhuang Formation and the Wumishan Formation, Stop 6; (C) Purple–red dolomitic mudstone, Stop 7; (D) The boundary between the Jx*y*$_2$ and Jx*y*$_3$, Stop 7 mudstone intercalated in gray dolomitic mudstone (marble meat mudstone). The Jxy_2 is dominated by marble meat mudstone intercalated in gray and dark-gray carbonate rocks with chert nodules.

SECTION Ⅷ Investigation Route of the Wumishan Formation (Jx*w*) of Jixian System along Qingshan Village–Mopanyu Village–Ershilipu Village, Luozhuangzi Town

The Wumishan Formation is the thickest unit within the Proterozoic strata in Jizhou area, its thickness is up to 3416 m. This formation is mainly composed of carbonate rocks, accounting for 80%–90% of its total thickness. The sedimentary rhythm is very well developed, and the huge Wumishan Formation are superimposed by multiple sedimentary rhythm layers, shown in Fig.45 and Fig.46.

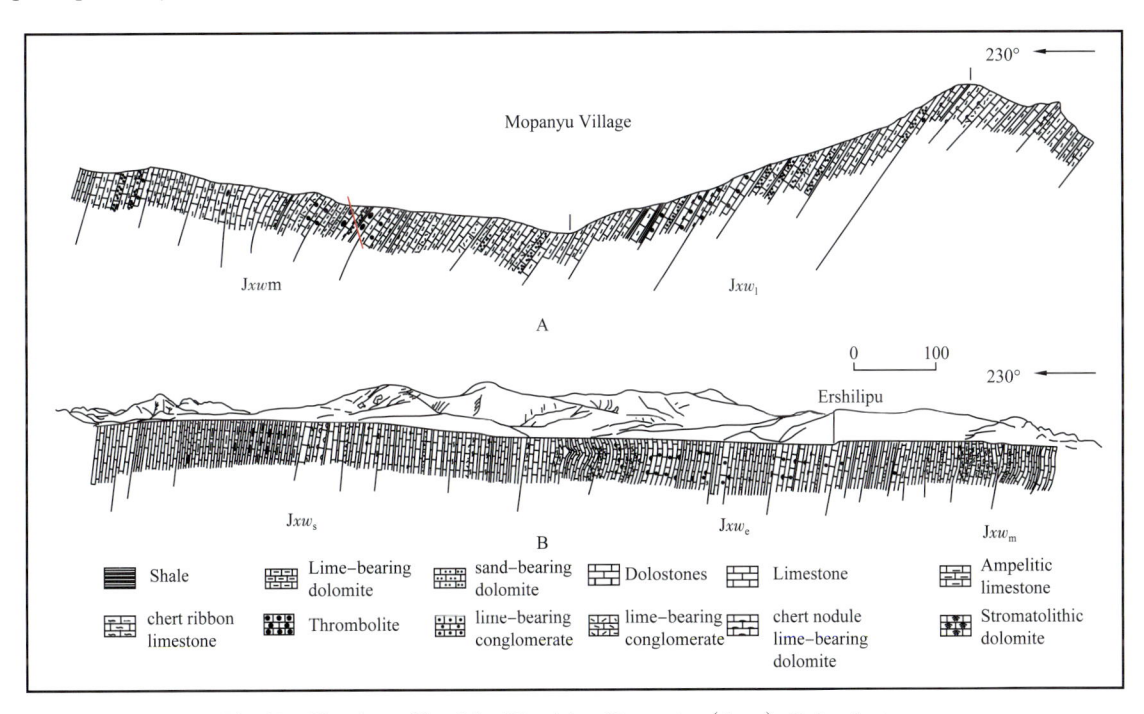

Fig.45 Sketch profile of the Wumishan Formation (Jx*w*), Jixian System,
Neoproterozoic from Ershilipu to Hongshuizhuang in Jizhou District, Tianjin

(A) The sections of the Jxw_1 (Luozhuang Sub-formation) and the Jxw_2 (Mopanyu Sub-formation) from Luozhuang Qingshan to South Mountain of Mopanyu; (B) The sections of the Jxw_3 (Ershilipu Sub-formation) and the Jxw_4 (Shanpoling Sub-formation) from Luozhuang Ershilipu to Shanpoling

1. Route location

The route is located in Ershilipu village and Mopanyu village in Jizhou District, Tianjin. The starting point of the route is located in the eastern of Ershilipu village (E 117.418064°, N 40.121807°), and the ending point is located at the road nameplate of Mojin Road in the southeastern of the Mopanyu village (E 117.438914°, N 40.114359°). The total length of the route is about 4 km. The distribution of the specific investigation stops and route are shown in Fig.47. The stop 1 is located in Ershilipu, which is far

Formation	Member	Thickness (m)	Lithological column	Description
Hongshuizhuang (Jxh)		0		Green shales intercalated in grayish green muddy dolostones
				— Conformity contact —
Wumishan Formation (Jxw)	Jxw₄	3000		Gray-white dolomite quartz sandstones at bottom; chert nodules microcrystalline dolostones intercalated in gray-white calcite dolostones, light-gray chert ribbons calcite dolostones, algal dolostones, thick stromatolite dolostones
	Jxw₃	2000		Purple-red sand-bearing muddy dolostones at bottom; Gray-white microcrystalline dolostones, algal dolostones and oolitic siliceous rocks intercalated with dolomitic shales
	Jxw₂	1000		Gray thick massive algal dolostones, laminar stromatolite dolostones, chert nodules or ribbons microcrystalline dolostones, dolomitic shales formed many cycles
	Jxw₁			Many cycles consisting of massive algal dolostones, laminated algal dolostones, and grayish green shales. Chert nodules and ribbons common.
Yangzhuang F.m. (Jxy)		0		— Conformity contact — purple-red muddy dolostones

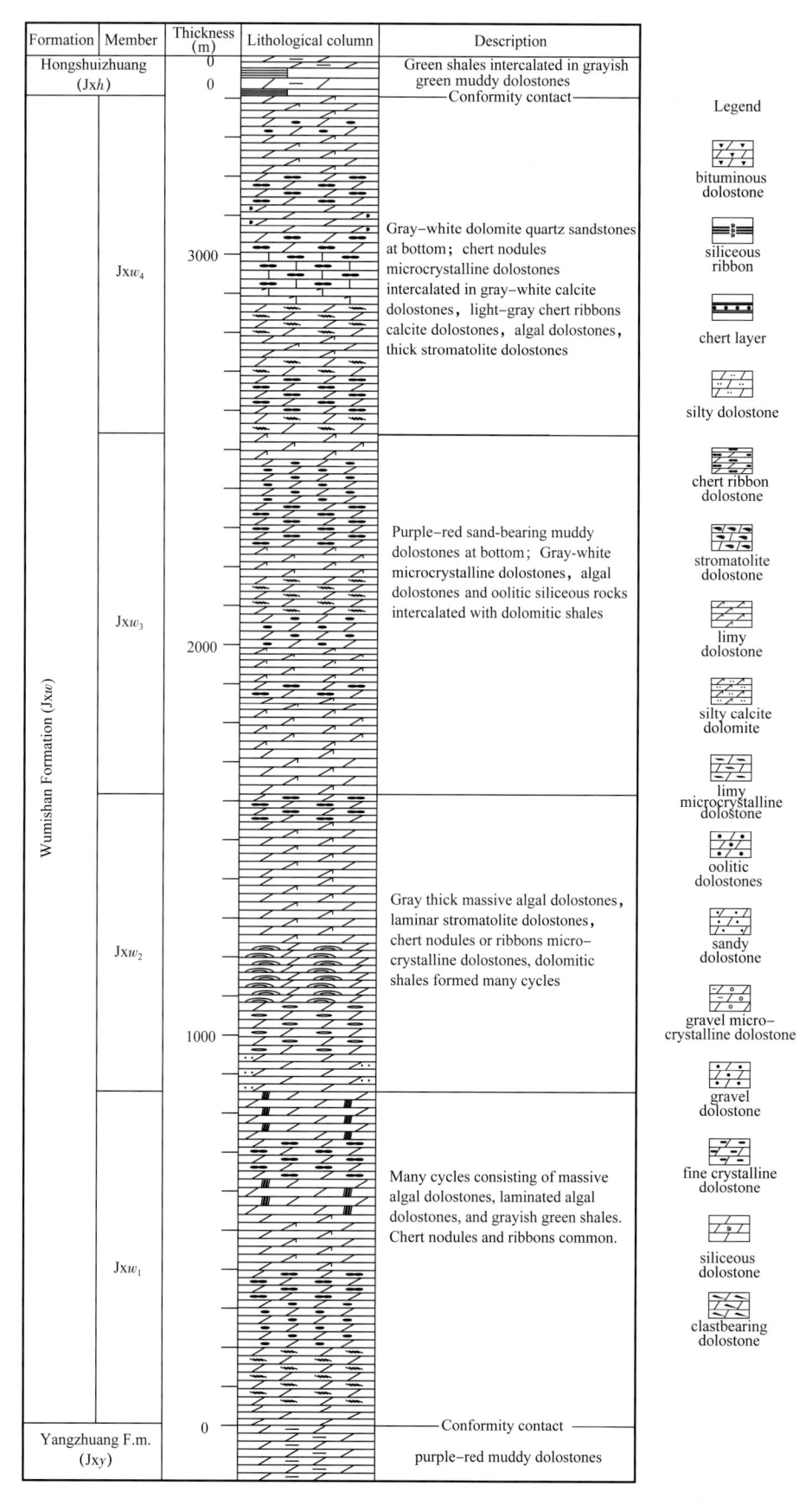

Legend: bituminous dolostone; siliceous ribbon; chert layer; silty dolostone; chert ribbon dolostone; stromatolite dolostone; limy dolostone; silty calcite dolomite; limy microcrystalline dolostone; oolitic dolostones; sandy dolostone; gravel microcrystalline dolostone; gravel dolostone; fine crystalline dolostone; siliceous dolostone; clastbearing dolostone

Fig.46 Stratigraphic column of the Wumishan Formation (Jxw), Jixian System, Neoproterozoic from Ershilipu to Hongshuizhuang in Jizhou District, Tianjin

Fig.47　Distribution of investigation stops and route of the Wumishan Formation (Jxw),
Jixian System, Neoproterozoic in Jizhou District, Tianjin

from Stop 2 to 4. It is suggested that the visitors should ride to the Mopanyu village after the observation of stop 1, and the observation point 2 to 3 should walk into the valley on the north side of the village. Finally, the stop 4, in the southeast direction, is about 400m distance to drive along the Mojin Road.

2. Objectives of investigation

(1) To observe the lithological characteristics and stratigraphic sequence of the Mesoproterozoic Jixian System Wumishan Formation in Jizhou District, Tianjin.

(2) To observe the basic stacking pattern of the sedimentary-cycle layer in the lower part of the Jxw_2.

The north-south S101 highway from Jixian to Huangyaguan Great Wall cuts through the Wumishan Formation in the direction almost vertical to the strike of the strata. It should be a great outcrop, but the fresh surfaces of the strata exposed by newly widening of the road has be covered by dirt and plants due to perennial weathering. Thus, the geologic phenomena are not clear now. In addition, the section is closed to the road with busy traffic, so it's unsafe and unsuitable for the investigation of big team.

The current investigation route is located at the east of the S101 highway along the Ershilipu village-Mopanyu village-Guoyuanxi village-Xinglongbao village-Siwa village-S302 (Provincial Highway). This route is a tortuous asphalt road with width enough for two bus to pass together. Along this route, there are intermittent exposed outcrops with clear geologic phenomena. Meanwhile, this road has little traffic. However, due to the high tortuosity of the road, the moving direction is sometimes to the new strata, but sometimes to the old strata, which is easy to make confused and is hard to build a complete sequence in the minds of vistors, until helped with a map all the time.

However, the Wumishan Formation is superimposed by numerous similar sedimentary cycles, so we can make a good knowledge of the whole formation by carefully observing one part of the outcrop. Along this route, the best investigation stop is near Mopanyu village.

(3) To interpret of the depositional environments and analyze the origin of the sedimentary cycles.

3. Contents of investigation and description of stops

From Ershilipu to Mopanyu, the Jxw_1 and the Jxw_2 are investigated in turn, and four specific investigation stops are selected along the route, including the stop of lithology and sedimentary rhythm of the lower part of the Jxw_1 (P1 in Fig.47), the stop of the middle part of the Jxw_1 (P2 in Fig.47), the stop of the upper part of the Jxw_1 (P3 in Fig.47) and the stop of continuous meter-scale cyclic deposits at the bottom of the Jxw_2 (P4 in Fig.47). Detailed descriptions of the stops are as follows:

Stop 1 Rock Types and Sedimentary Rhythm of the Lower Part of the Jxw_1

Location: This stop is located at the intersection of the small road in the Eastern Ershilipu village. GPS: E 117.418064°, N 40.121807°.

Investigation Content: Typical rock types and sedimentary rhythm of the lower part of the Jxw_1.

Descriptions: Typical rock types and sedimentary rhythms in the lower part of the Jxw_1 are observed at this point. The rock types mainly include grey-white microcrystalline dolostones, dark-gray massive algal fine crystalline dolostones and pale-green-gray dolomitic shales. The strata occurrence is 188°∠48°. Among them, the ripple marks often develops on the top of grey-white microcrystalline dolostones (Fig.48A). Conical stromatolites are common in dark gray agglomerated (mottled) fine crystalline

Fig.48 The lithology and sedimentary structures of the lower part of the Jxw_1 in Jizhou District, Tianjin
(A) Ripple marks in gray-white microcrystalline dolostones, stop 1; (B) Sequence consisting of massive algal dolostones, laminar stromatolite dolostones and dolomitic shales, stop 1; (C) Massive algal dolostones, stop 1; (D) Laminar stromatolite dolostones, Jxw_1, stop 1

dolostones, with light and dark mottles mixed (Fig.48B and C). Light green grey dolomitic shale develops horizontal beddings. From the bottom to top, the sedimentary cycle consists of the light green-gray dolomitic shales (subtidal zone of confined clastic rocks), coagulated dolostones (subtidal zone of carbonate rocks), and stratified stromatolite dolostones (with chert ribbons and nodules)(tidal flat of carbonate rocks)(Fig.48B).

Stop 2　Rock Types and Sedimentary Rhythm in the Middle Part of the Jxw_1

Location : The stop is located in the almost north-south valley at the north side of the Mopanyu village. It takes about 400 meters to walk from the village to the northeast. GPS : E117.443598°, N40.120346°.

Investigation Content : Typical rock types and sedimentary rhythm in the middle part of the Jxw_1.

Description : The typical rock types and sedimentary rhythms in the middle of the Jxw_1 are mainly observed here. The rock types mainly include massive microcrystalline dolostones, laminar and wavy stromatolite dolostones containing chert, which are deposited alternately. These stromatolites have a great variety of forms : wavy, domal (Fig.49A), and/or laminar (Fig.49B), which show obvious alternated light and dark laminations under microscope caused by variation of crystal sizes (Fig.49C). The siliceous nodules and ribbons protuberate to the surface after weathering (Fig.49D). The grey-white microcrystalline dolostones is generally characterized by massive structure and blurred bedding (Fig.49E and F).

Stop 3　Rock Types and Sedimentary Rhythm of the Upper Part of the Jxw_1

Location : The stop is still located in the valley, about 200 m to the south of the previous stop. GPS : E 117.441724°, N 40.117973°.

Investigation Content : Typical rock types and sedimentary rhythm in the upper part of the Jxw_1.

Description : The main rock types and sedimentary rhythms of the upper part of the Jxw_1 are observed here. The main rock types at this stop include grey-black clot-like fine-crystalline dolostones (Fig.50A), grey laminar stromatolite microcrystalline dolostones bearing with cherts (Fig.50B and C) and light-green-grey dolomitic shales (Fig.50D). From the bottom to top, the typical rhythms consist of light green-grey dolomitic shales (subtidal flat), grey-black clot-like fine crystalline dolostones (subtidal flat), and grey laminar stromatolite microcrystalline dolostones bearing with cherts (intertidal-supratidal flat), indicating a shallowing-upward sedimentary sequence, which is similar to that in the stop 2.

Stop 4　Sedimentary Rhythm of the Jxw_2

Location : The stop is located at the road nameplate of the Mojin Road in the southeastern of Mopanyu village. GPS stop : E 117.438914°, N 40.114359°.

Investigation Content : Typical sedimentary rhythm of the lower part of the Jxw_2.

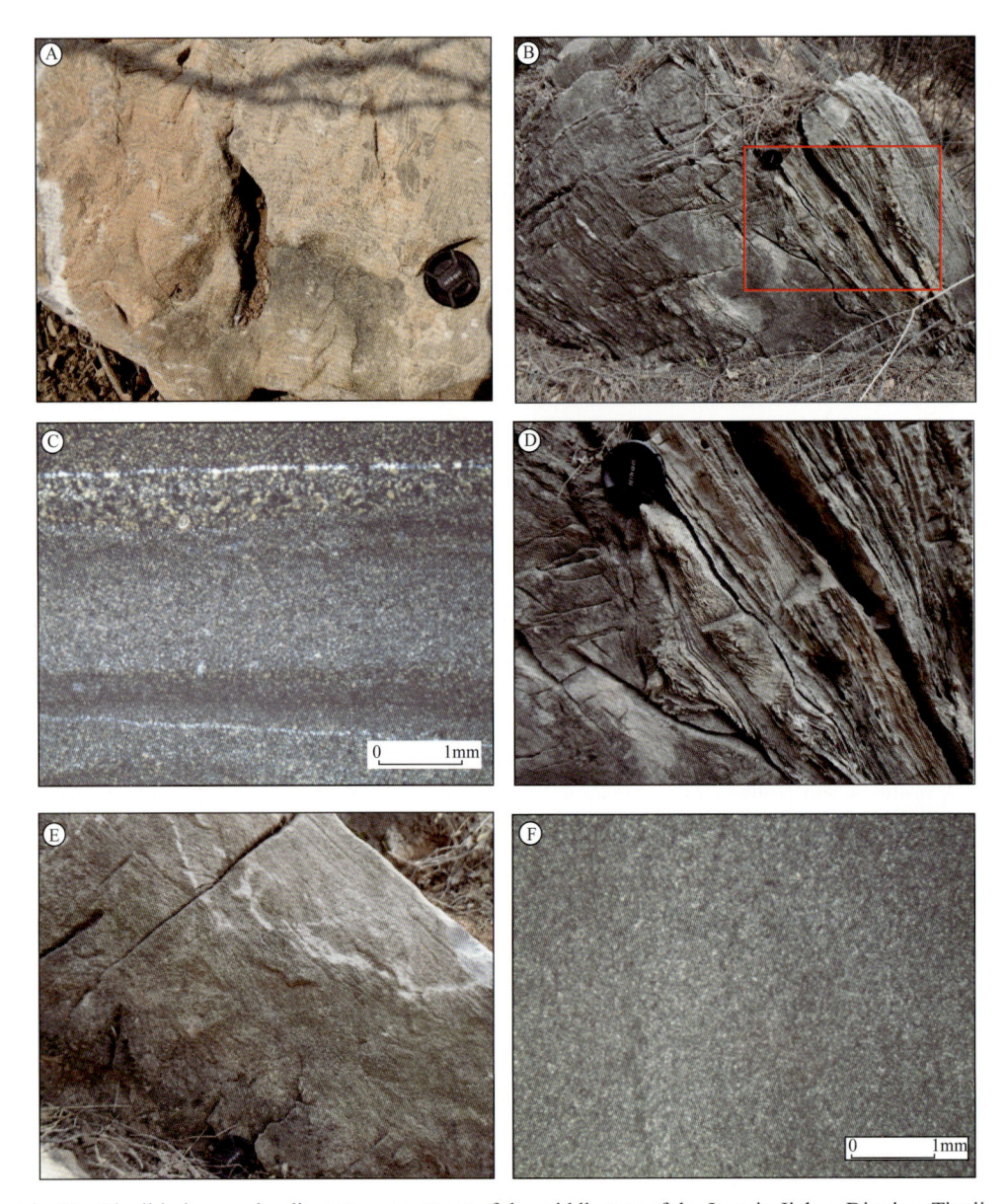

Fig.49 The lithology and sedimentary structures of the middle part of the Jxw₁ in Jizhou District, Tianjin
(A) Wavy stromatolite dolostones, the Jxw₁, stop 2; (B) Laminated stromatolite dolostones with banded cherts, the Jxw₁, stop 2;
(C) Microscopic characteristics of the laminated stromatolite dolostones, showing obvious alternated light and dark laminations, cross-polarized light, the Jxw₁, scale is 1 mm, stop 2; (D) Banded chert, the Jxw₁, stop 2; (E) Microcrystalline dolostones, the middle part of the Jxw₁, stop 2; (F) Microscopic characteristics of microcrystalline dolostone, plane polarized light, scale is 1 mm, the middle part of the Jxw₁, stop 2

Description : This stop is selected for the sedimentary cycle of the the Jxw₂. Moving southeast along the Mojin Road, the characteristics of the lithologic cycle in the lower part of the Jxw₂ can be continuously observed. The sedimentary cycle is well-developed, consisting of light green-gray dolomitic shales (subtidal flat), grey-black clot-like fine crystalline dolostones (subtidal flat), and grey laminar stromatolite microcrystalline dolostones bearing with cherts (intertidal-supratidal flat) (Fig.51).

Fig.50 The lithology and sedimentary structures of the upper part of the Jxw₁ in Jizhou District, Tianjin
(A) Grey–black clot–like dolostones, the upper part of the Jxw₁, stop 3; (B) Grey laminated stromatolite dolostones, the upper part of the Jxw₁, stop 3; (C) Grey–white laminated stromatolite dolostones with banded and nodular cherts, the upper part of the Jxw₁, stop 3; (D) Light green–grey dolomitic shales, the upper part of the Jxw₁, stop 3

Fig.51 The sedimentary features of the high-frequency cyclic of the Jxw₁ in Jizhou District, Tianjin

SECTION IX Investigation Route of the Hongshuizhuang Formation (Jx*h*) of Jixian System in Xiaolingzi Village, Yuyang Town

The Mesoportoterozoic Jixian System Hongshuizhuang Formation is a set of grey–black, grey–green shales intercalated with dolostones and thin bedded siltstones, with a thickness of 131 m. This formation can be divided into two parts : the lower 1st member of the Hongshuizhuang Formation, consisting of medium dolostones with intercalated grey–green, blue–green and grey shales ; and the upper 2nd member of the Hongshuizhuang Formation, mostly consisting of grey-black and grey-green shales with intercalated microcrystalline dolostones lenses (algal reef) and thin bedded siltstones in its top grey-green shales. The Hongshuizhuang Formation shales are rich in micropaleophytes (Sun, 2000). Except some stratigraphic gaps and/or unconformity contact with the underlying Wumishan Formation in the eastern part of Yanshan Mountains, this formation has a conformable contact relationship with the overlying Tieling Formation and the underlying Wumishan Formation in this and other areas (Yuan *et al.*, 1994; Chen and Wu, 1997; Wang *et al.*, 2014).

The Hongshuizhuang Formation are well exposed in Hongshuizhuang village and its southeast area in the Luozhuangzi Town, Jizhou District, Tianjin City, especially along the stone-slab road around the entrance of Xiaolingzi village, but only with the Jxh_2 exposed. Meanwhile, its boundary with the overlying Tieling Formation can be observed here. The boundary between the Hongshuizhuang Formation and the Wumishan Formation can be seen around the roadside about 2km east of Fujunshan Park and 1km south of Siwa village. The Sketch profile and stratigraphic column of the Jxh_2 in Xiaolingzi village are shown in Fig.52 and Fig.53, respectively.

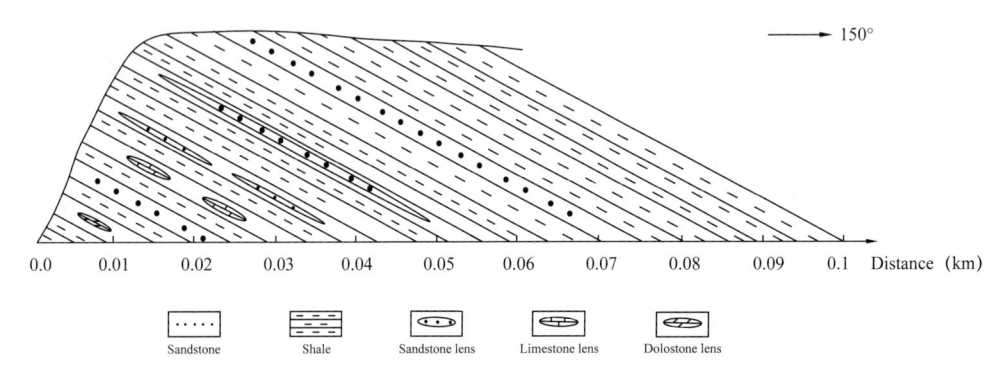

Fig.52 Sketch profile of the Hongshuizhuang Formation in Xiaolingzi village, Jizhou District, Tianjin

1. Route location

The route is located at Xiaolingzi village, Yuyang Town, Jizhou District, Tianjin. The starting point of the route is at the east of Jinwei highway and the bifurcation road between the Beitaoyuan village and the Nantaoyuan village (GPS: E 117.388025°, N 40.084112°), and the ending point is about 100 m eastward (GPS: E 117.389207°, N 40.083879°). The total length of the route is about 110 m. This route is located around the country road with width enough for vehicles of all sizes to pass. The distribution of the specific route and investigation stops is shown in Fig.54.

2. Objectives of investigation

(1) To observe the lithological characteristics and stratigraphic sequence of the Mesoproterozoic Jixian system Hongshuizhuang Formation in Jizhou District, Tianjin;

(2) To observe the typical characteristics of the black shales, and lenses of carbonate and quartz sandstone;

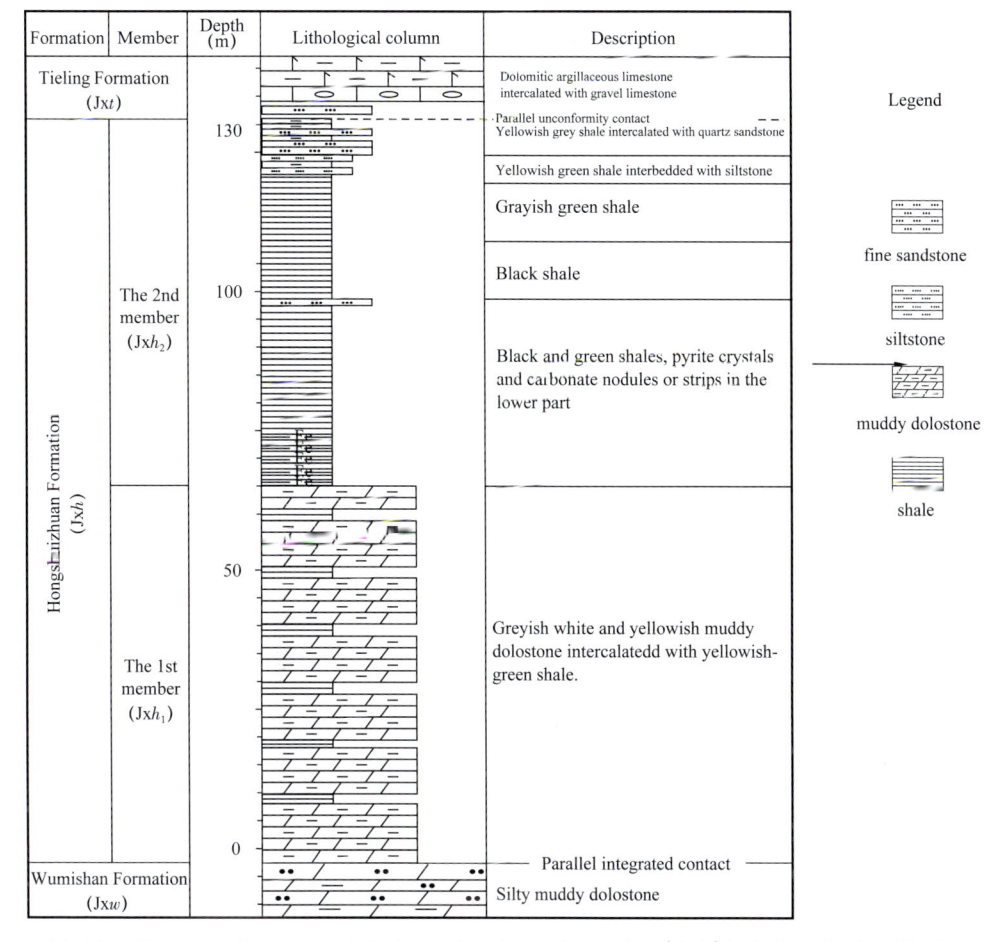

Abstract　Field Guide of Stratigraphy and Sedimentary Facies of the Proterozoic in Jizhou District, Tianjin

Formation	Member	Depth (m)	Lithological column	Description
Tieling Formation (Jxt)				Dolomitic argillaceous limestone intercalated with gravel limestone
		130		Parallel unconformity contact
				Yellowish grey shale intercalated with quartz sandstone
				Yellowish green shale interbedded with siltstone
				Grayish green shale
Hongshuizhuan Formation (Jxh)	The 2nd member (Jxh₂)	100		Black shale
				Black and green shales, pyrite crystals and carbonate nodules or strips in the lower part
	The 1st member (Jxh₁)	50		Greyish white and yellowish muddy dolostone intercalatedd with yellowish-green shale.
		0		Parallel integrated contact
Wumishan Formation (Jxw)				Silty muddy dolostone

Legend:
- fine sandstone
- siltstone
- muddy dolostone
- shale

Fig.53　Stratigraphic column of the Hongshuizhuang Formation (Jxh) in Jixian District, Tianjin

Fig.54　Distribution of investigation stops and route of the Hongshuizhuang Formation (Jxh) in Jizhou, Tianjin

- 213 -

(3) To interpret the depositional environments.

3. Contents of investigation and description of stops

The lithological characteristics of the Hongshuizhuang Formation can be observed at the bifurcation road between the Beitaoyuan village and the Nantaoyuan village and the east of Jinwei highway. Due to the limitation of exposure conditions, only two detailed investigation stops were selected, including the stop of the black shales of the Hongshuizhuang formation, and the stop of quartz sandstone lens. Detailed descriptions of the stops are as follows:

Stop 1   Black Shales of the Hongshuizhuang Formation (Jx*h*)

Location: The stop is located at the east of Jinwei highway and the east of the bifurcation junction between the Beitaoyuan village and the Nantaoyuan village, GPS: E 117.388025°, N 40.084112°.

Investigation contents: (1) The color change of the shales from the bottom to top and its corresponding change in depositional settings; (2) Dolostones lenses (algal reef).

Description: Deep grey and grey green shales with dolomite (or limestone) lenses with 1.5–3 m in length and 0.5 m in thickness (Fig.55). Horizontal laminae of the shales in the Hongshuizhuang Formation are straight and clear, with stable lateral distribution, indicating a shale-shelf setting with low hydrodynamic energy.

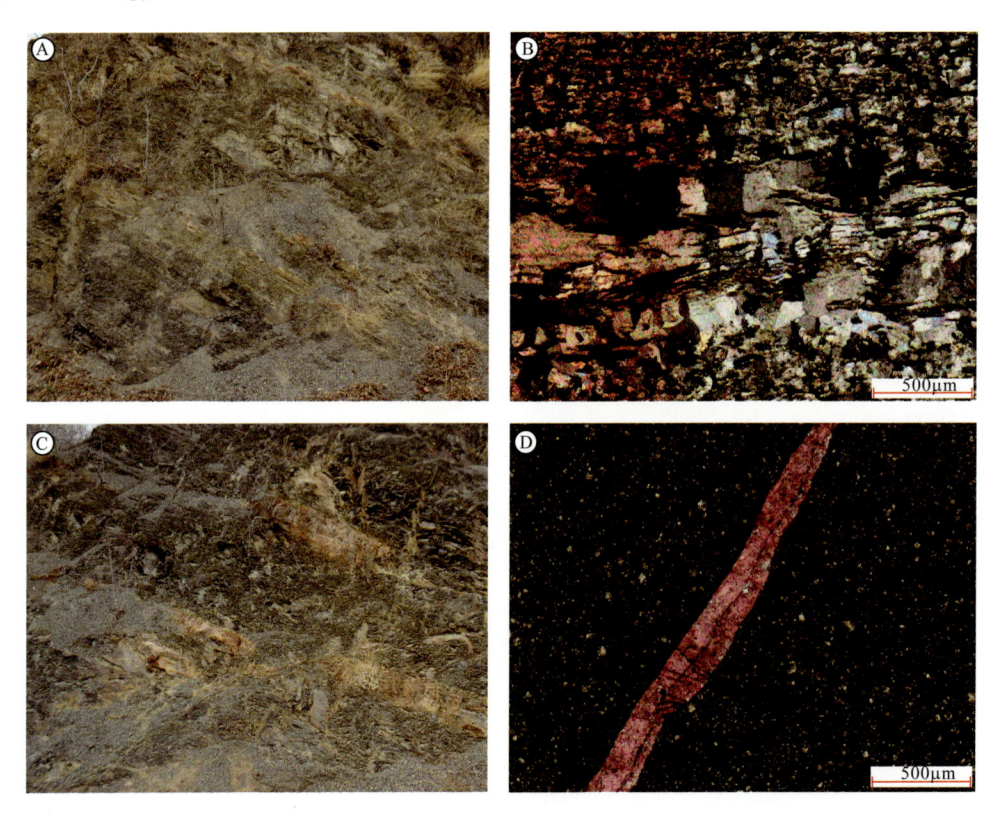

Fig.55   Black shales and dolostones of the Hongshuizhuang Formation at Jixian System

(A) Deep gray shales of the Hongshuizhuang Formation, stop 1; (B) Photomicrograph of limestone lenses, cross-polarized light, and scale is 500 um in length, stop 1; (C) Shales with carbonate lenses (algal reef), stop 1; (D) Photomicrograph of the dolostones lens, cross-polarized light, scale is 500mm in legth, stop 1

Stop 2 Fine Quartz Sandstone Lenses in the Hongshuizhuang Formation

Location: The stop is about 10 m to east of the last stop 1; GPS: E 117.408010°, N 40.084112°.

Investigation Content: Fine quartz sandstone lenses and analysis of its origin.

Description: Shales with intercalated thin lenses of fine quartz sandstones which are 1.5–2 m in length and 0.1 m in thickness (Fig.56).

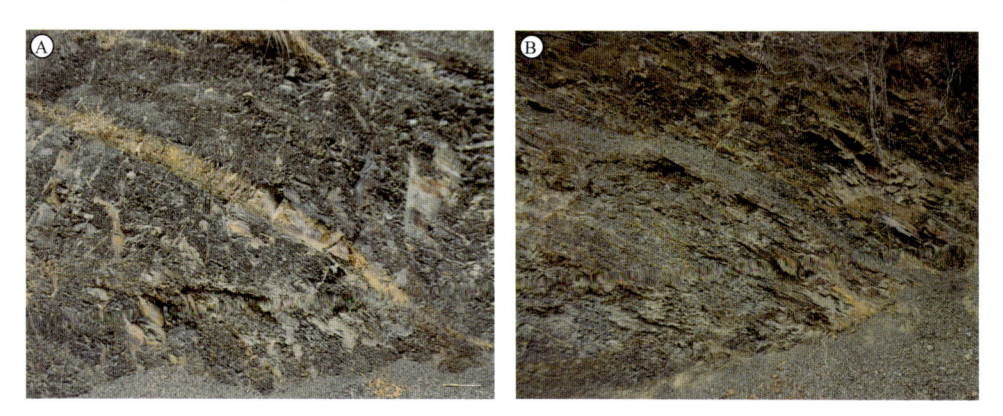

Fig.56 Yellowishgray shales with quartz sandstone lenses of the Hongshuizhuang Formation
(A) Grey-green shales with intercalated sandstone lenses, the width of the field of view is 1m stop 2; (B) Yellow-green shales, the width of the field of view is 1m, stop 2

SECTION X Investigation Route of the Tieling Formation (Jxt) of Jixian System from Xiaolingzi Village in Yuyang Town to Tielingzi Village in Luozhuangzi Town

The Mesoproterozoic Jixian System Tieling Formation in the mostly consist of carbonate rocks with intercalated purple and blue-green shales and stromatolites developed extremely at its upper part (Fig.57, Fig.58). The total thickness of the Tieling Formation is about 303 m. According to the difference of lithological assemblage, the Tieling Formation was previously divided into two members (including Daizhuangzi Sub-formation and Laoting Sub-formation) (Chen et al., 1980), while three members are divided in this book.

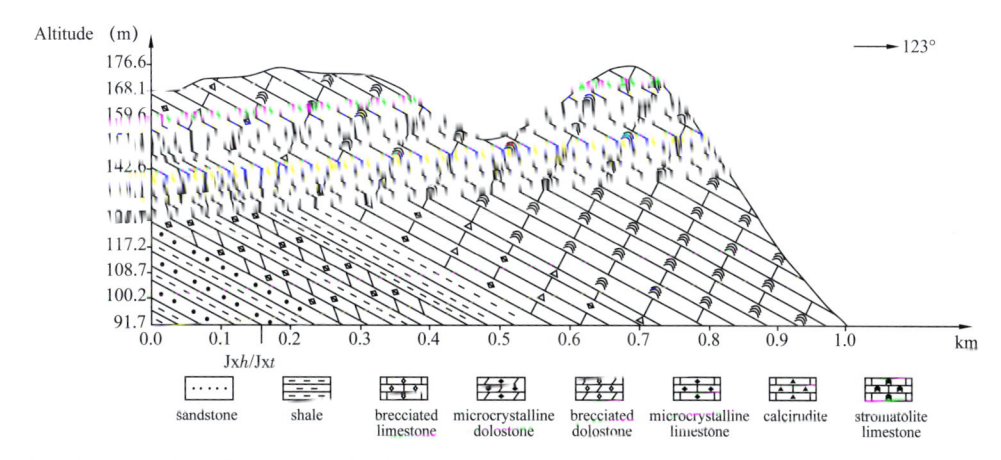

Fig.57 Sketch Profile of the Tieling Formation (Jxt) from the Xiaolingzi village to Tieling village in Jizhou District, Tianjin

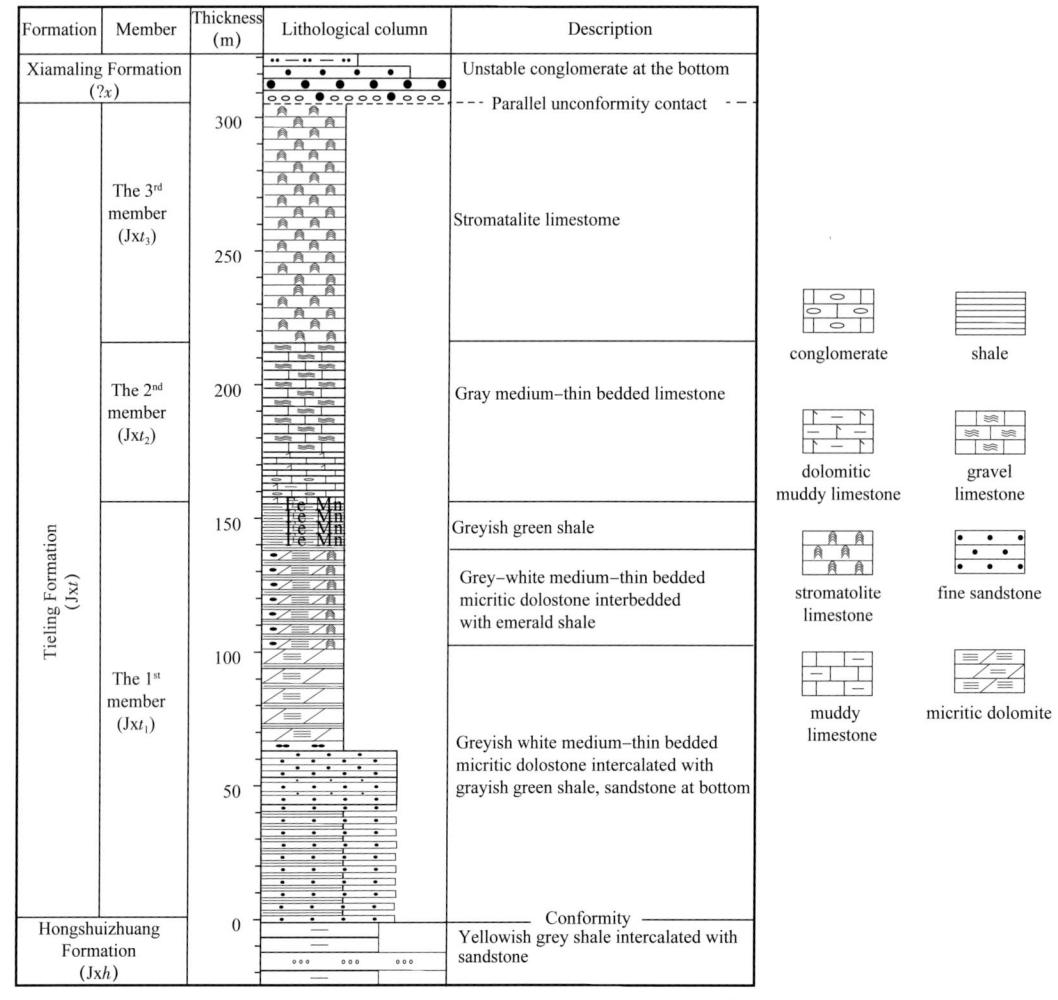

Fig.58 Stratigraphic column of the Tieling Formation (Jxt) from the Xiaolingzi village to the Tieling village in Jixian District, Tianjin

The thickness of the Jxt_1 is 55 m, with grayish-white medium-thin bedded quartz fine sandstone at the bottom and gray manganese-bearing algal microcrystalline dolostones (algal reef) at the top. The main body is medium-thick bedded grey stromatolite dolostones with intercalated grey-green and blue-green shales, and the top is blue-green glauconitic shales and ferromanganese-bearing dark brown shales.

The Jxt_2 is about 60 m in thickness, consisting of dolomud stripped micritic limestones with intercalated wormkalk calciruditeand a small amount of stromatolite limestones, and with local nodular cherts.

The Jxt_3 is 88m in thickness and consists of stromatolite limestones with various types and shapes, including laminar, wavy, columnar and wall-like stromatolites, which is interval in this area with the most abundant and well-developed Mesoproterozoic stromatolites. At the top of this section is a few meters thick dolomud stripped limestones.

This formation is in conformable contact with the underlying Hongshuizhuang Formation, and has a parallel unconformity contact with the overlying Xiamaling Formation of the Qingbaikou System.

Tieling Formation is well exposed in the southern of the Xiaolingzi village and Tielingzi village in Jizhou District, Tianjin City, which is the best route for the investigation of the Tieling Formation.

1. Route location

The route is located in the area from the Xiaolingzi village to Tieling village in Jizhou District, Tianjin. The starting point of the route is near the Xiaolingzi village, east of Jinwei highway (GPS: E 117.389208°, N 40.083879°), and the ending point is in the south of the Tielingzi village (GPS: E 117.397112°, N 40.079933°). The total length of the route is about 1000 m. The first half of the route is cement road, which is wide and accessible to small and medium-sized vehicles. However, when it reaches the stromatolite park, the terrain slope is too steep to drive, only with walking steps. The investigation stops are mostly distributed along the walking stops, which is suitable to visit by walking. Vehicles can detour to and wait at the intersection between the investigation route at the south of the Tielingzi village and the Jinyan Road. The specific route and investigation stops are shown in Fig.59.

2. Objectives of investigation

(1) To observe the lithological characteristics and stratigraphic sequence of the Mesoproterozoic Jixian System Tieling Formation in Jizhou District, Tianjin City;

(2) To observe the typical rock characteristics of ferriferous dolostones, limestones, blue-green shales and stromatolites;

(3) To interpret the depositional environments.

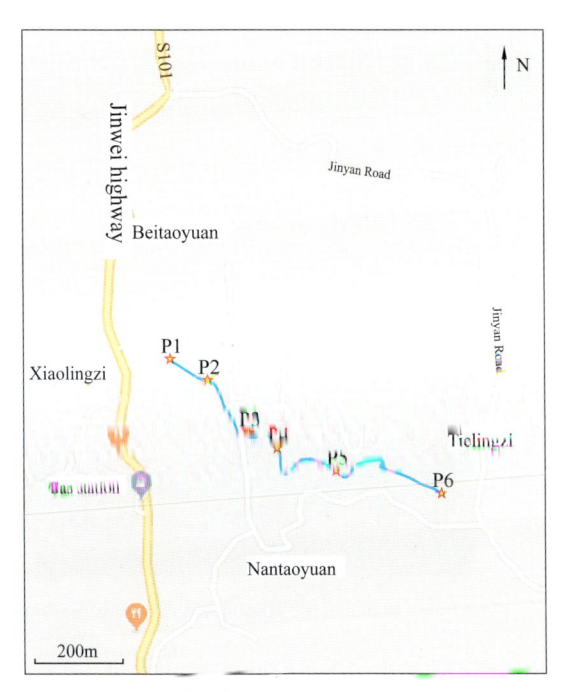

Fig.59 Distribution of investigation stops and route of the Tieling Formation (Jxt) in Jixian District, Tianjin

3. Contents of investigation and description of stops

In this section, Tieling Formation is well exposed and the transportation is convenient. There are six detailed investigation stops in this route, including the stop of the boundary between the Tieling Formation and the Hongshuizhuang Formation (P1 in Fig.59), the stop of quartz sandstone lens (P2 in Fig.59), the stop of grey medium bedded microcrystalline dolostones with intercalated grey-green shales (P3 in Fig.59), the stop of blue-green shales (P4 in Fig.59), the stop of the boundary between the Jxt_1 and Jxt_2 (P5 in Fig.59), and the stop of stromatolites in the Jxt_3 (P6 in Fig.59). Detailed descriptions of the stops are as follows:

Stop 1 Boundary between the Tieling Formation (Jx*t*) and the Hongshuizhuang Formation (Jx*h*)

Location: The stop is located at the boundry tablet of the Hongshuizhuang Formation and the Tieling Formation on the divergent path between the Beitaoyuan village and the Nantaoyuan village, east of Jinwei highway. GPS: E 117.389207°, N 40.083879°.

Investigation contents: (1) the boundary between the Tieling Formation and the Hongshuizhuang Formation; (2) sedimentary sequence and depositional environment analysis of sandstones.

Description: This stop is the boundary between the Tieling Formation and the Hongshuizhuang Formation, located at the bottom of a large set of sandstones. There was a conformable contact relationship between the two formations (Fig.60A and B). Under the boundary, it is the top of Hongshuizhuang Formation, consisting of gray-yellow shales with intercalated quartz sandstone lenses, and above the boundary, it is the bottom of the Tieling Formation, consisting of a large set of gray-white medium-thin bedded fine quartz sandstones. Plane beddings are well developed in sandstone and a small normal fault is observed.

Fig.60 Boundary between the Hongshuizhuang Formation (Jx*h*) and the Tieling Formation (Jx*t*), and photomicrograph of the quartz sandstone near the boundary

(A) The boundary between the Hongshuizhuang Formation (Jx*h*) and the Tieling Formation (Jx*t*), which is conformable contact, stop 1;
(B) Photomicrograph of the grayish-white fine quartz sandstones of the Tieling Formation above the boundary between the Hongshuizhuang Formation and the Tieling Formation, cross-polarized light, stop 1

Stop 2　Grey-white medium-thin bedded microcrystalline dolostones with intercalated grey-green shales in the Jxt_1

Location: It is about 90 m to southeast of the last stop 1. GPS: E 117.390059°, N 40.083879°.

Investigation contents: (1) Rock types and characteristics at the lower part of the Jxt_1; (2) Characteristics of sedimentary cycles; (3) Types and forms of stromatolites in dolostones; (4) analysis of sedimentary environment.

Description: The grey medium-thin bedded microcrystalline dolostones with intercalated grey-green shales in the middle part of the Jxt_1 are observed at this stop, with a thickness of about 50 m. Upwards, the shale layers becomes thinner and less, while the dolostone layers increases gradually (Fig.61). Stromatolites are common in dolostones, and the sequences consisting of deposits from the subtidal to tidal flat are well developed. The intercalated shales represent a short term transgression events and are shelf deposits.

Fig.61　Gray-white medium-thin bedded microcrystalline dolostones with intercalated gray-green shales in the Tieling Formation

(A) The grey medium bedded microcrystalline dolostones with intercalated grey-green shales of the Jxt_1, stop 2; (B) Photomicrograph of the grey medium bedded microcrystalline dolostones in the Jxt_1, cross-polarized light, scale is 100μm in length, stop 2

Stop 3　Blue-green Shales of the Jxt_1

Location: The stop is about 260 m to southeast of the last stop 2.

GPS: E 117.391667°, N 40.081389°.

Investigation content: Characteristics and origin of the blue-green shales.

Description: The grey-white medium bedded ferriferous microcrystalline dolostones interbedded with blue-green shales at the upper part of the Jxt_1 (Fig.62). The blue-green shales is deposits of shelf, and its color is linted by glauconite.

Stop 4　Boundary between the Jxt_1 and the Jxt_2 and lithological characteristics of the Jxt_2

Location: The stop is about 53 m to the southeast of the last stop 3. GPS: E 117.392237°, N 40.081207°.

Investigation contents: (1) the boundary between the Jxt_1 and the Jxt_2 of the Tieling Formation; (2) the characteristics of parallel unconformity interface; (3) the sedimentary environment of brown shales; (4) the rock type and sedimentary environment of the Jxt_2.

Fig.62 Interbedding of gray medium-thin bedded ferriferous microcrystalline
dolostones and blue-green shales in the Tieling Formation

(A) Interbedding of ferriferous gray medium bedded microcrystalline dolostones and blue-green shales in the upper part of the Jxt_1, stop 3;
(B) Photomicrograph of the gray medium bedded ferriferous microcrystalline dolostones, cross-polarized light, in length, stop 3

Description: This stop is the boundary between the Jxt_1 and the Jxt_2, which is a parallel unconformity. The top of the Jxt_1 is composed of grey-green shales, and the bottom of the Jxt_2 is made up of brown ferruginous granules with a thickness of about 0.2 m. The distribution of this interface is stable regionally. After the sedimentation of the Jxt_1, the strata were uplifted, and weathered and eroded, corresponding to the tectonic movement called "Tieling Uplift" (Du and Li, 1980; Chen and Wu, 1997). The Jxt_2 mainly consists of thin bedded dolomud stripped micritic limestone with intercalated wormkalk calcirudites. The limestone were dolomized in variable degrees, which is considered to be related to the evaporative pumping and reflux seepage dolomitization (Fig.63).

Fig.63 Geological phenomena of the Jxt_1 and the Jxt_2

(A) The boundary between the Jxt_1 and the Jxt_2. The top of the Jxt_1 consists of blue-green shales and the bottom of the Jxt_2 consists of brown ferruginous conglomerate, stop 4; (B) Gray calcirudites interbedded with micritic limestones, stop 4; (C) Gray calcirudites interbedded with gray medium-thin bedded micritic limestones, stop 4; (D) Photomicrograph of grayish medium-thin bedded micritic limestones, cross-polarized light, occurred with calcite veins, stop 4

Stop 5 Stromatolites in the JXt_3

Location : The stop is located about 160 m southeast of the last stop 4. GPS : E 117.394179°, N 40.080387°.

Investigation contents : (1) types and characteristics of stromatolites ; (2) types of fillings between stromatolites ; (3) morphological characteristics of wall-like stromatolites in different sections ; (4) analysis of controlling factors of stromatolite morphology ; (5) depositional environments of stromatolites.

Description : A variety of stromatolites in the Jxt_3 are observed at this stop, with the thickness of about 80m (Fig.64). According to the morphological characteristics of stromatolites, columnar, wavy, laminar and wall-like stromatolites can be observed. Different forms of stromatolites represent different

Fig.64 Geological phenomena of the Jxt_3

(A) The Jxt_3 is columnar stromatolites, on which dissolution pores are found, stop 6; (B) wavy stromatolites, stop 6; (C) Top of stromatolites, stop 6; (D) Columnar stromatolites with dissolution pores on the surface, stop 6; (E) Laminar stromatolites, stop 6; (F) Photomicrograph of stromatolites showing light and dark laminae, cross-polarized light, scale is 500μm in length, stop 6

sedimentary environments, which are indicator of the sedimentary facies. According to the change of hydrodynamics from weak to strong, laminar stromatolites are mainly deposited in the upper part of the intertidal and supratidal flats, wavy stromatolites are mainly deposited in the middle part of the intertidal flat, and columnar stromatolites are mostly deposited in the lower part of the intertidal and subtidal flats (Jin, 2013). Wall-like stromatolites occur in the middle and lower parts of the intertidal zone, extending in parallel with the direction of tidal currents. The stromatolites pillars are filled with limemud (usually dolomitized) or blue-green mud without grains, suggesting a setting with low energy.

SECTION XI Investigation Route of the Xiamaling Formation (?x) of System to be estabilished from the South of Tielingzi Village to Xiazhuangzi Village, Luozhuangzi Town

The previous classified the Xiamaling Formation into the "Daijian System", while the author thinks that the term of "Daijian System" is very awkward and it should be classified into the Jixian System until some new discoveries. In this area, the Xiamaling Formation is dominated by gray-black, gray-green shales and silty shales, with argillaceous siltstones and fine sandstones. The formation is exposed within a small area around the limbs of the Fujunshan syncline near its core, especially well-exposed in Luotuoling area. The formation can be divided into two members: The lower member, composed of shales with more intercalated fine sandstones; and the upper member, mostly consisting of shales. Here, the sketch profile and stratigraphic column of this formation are shown in Fig.65 and Fig.66, respectively.

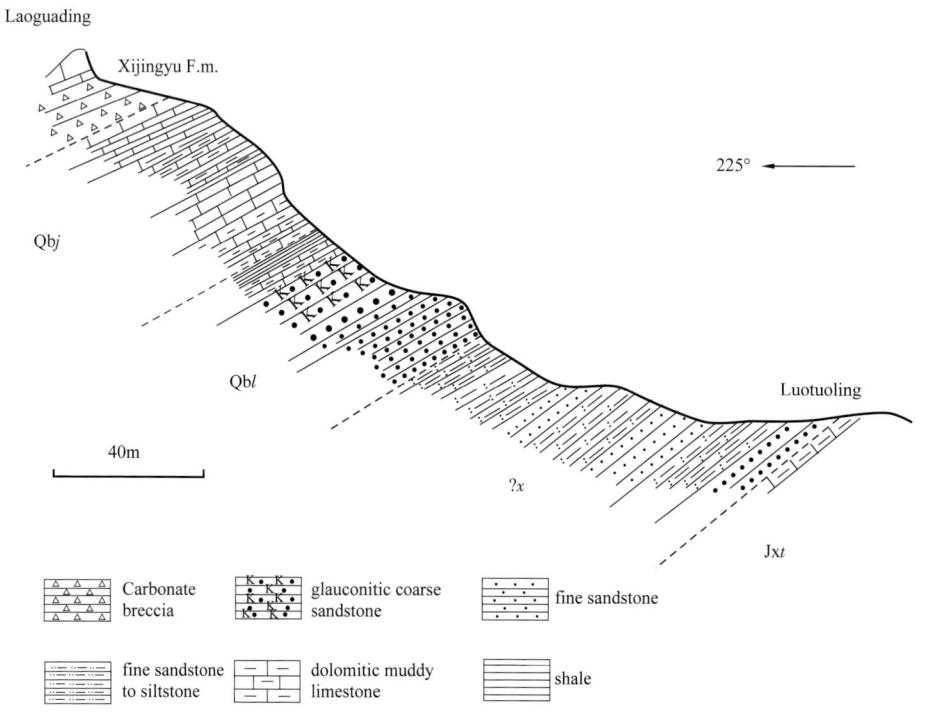

Fig.65 Sketch profile of the Mesoproterozoic Daijian System and the Neoproterozoic Qingbaikou System from Luotuoling to Laoguading in Jizhou District, Tianjin (Modified from the Tianjin Institute of Geology and Mineral Resources, 1964)

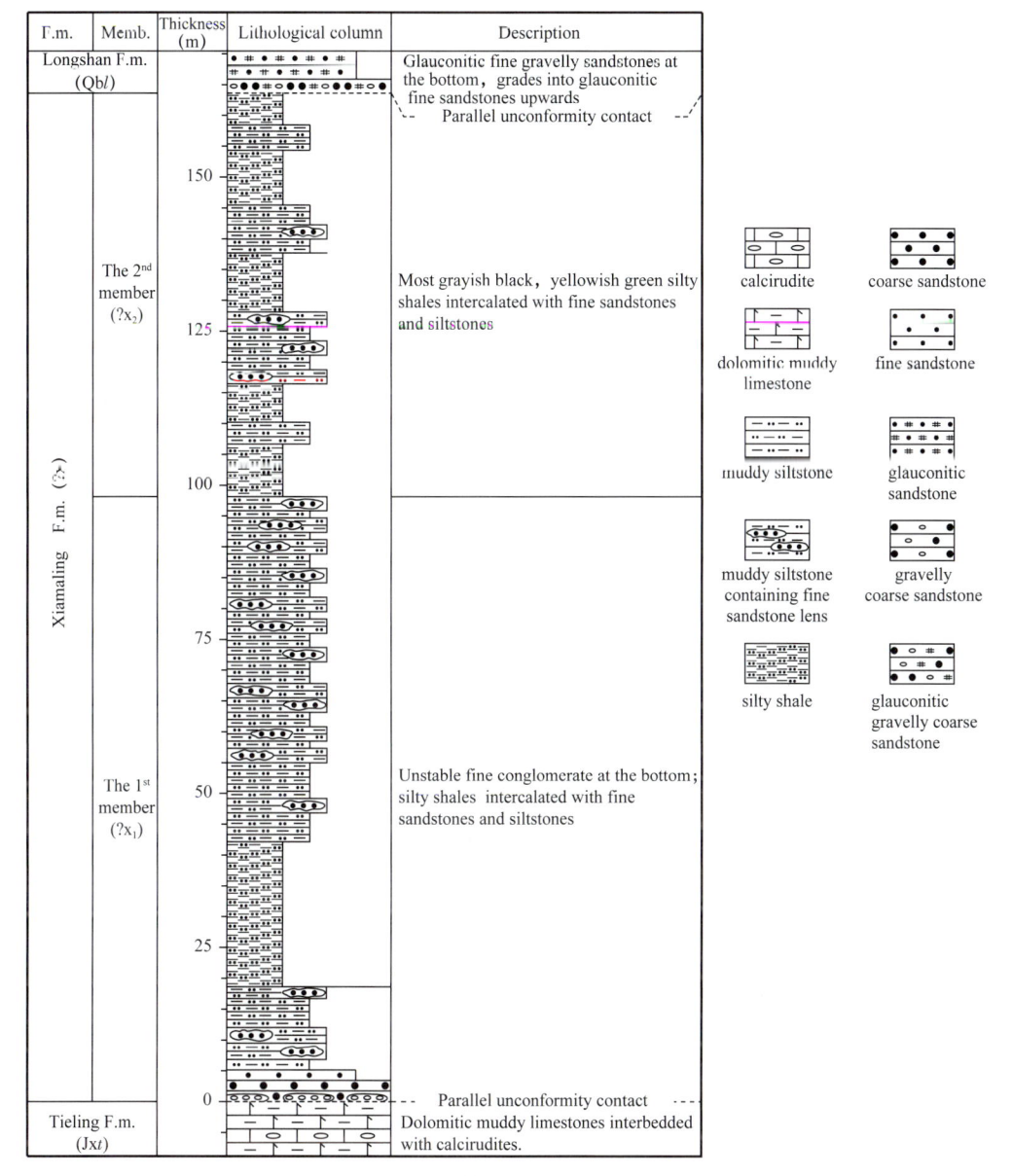

Fig.66 Stratigraphic column of the Mesoproterozoic the Daijian System
Xiamaling Formation in Luotuoling, Jizhou District, Tianjin

1 Route location

This investigation route is located in the area of Tielingzi Village, Luozhuangzi Town, Jizhou District, Tianjin. The starting point is about 650 m to the south of the Tielingzi Village and near the Jinyan Road (GPS: E 117.397112°, N 40.079933°). The ending point is about 250 m to the southeast of the Xiazhuangzi Village closed to the Jinyan Road (GPS: E 117.403186°, N 40.075445°). The total route is about 788 m. The distribution of specific routes and investigation stops is shown in Fig.67. This route extends mainly along the Jinyan Road which is in good condition to allow large or small vehicles to pass. Some steps for walking and sign boards are set near the investigation stops.

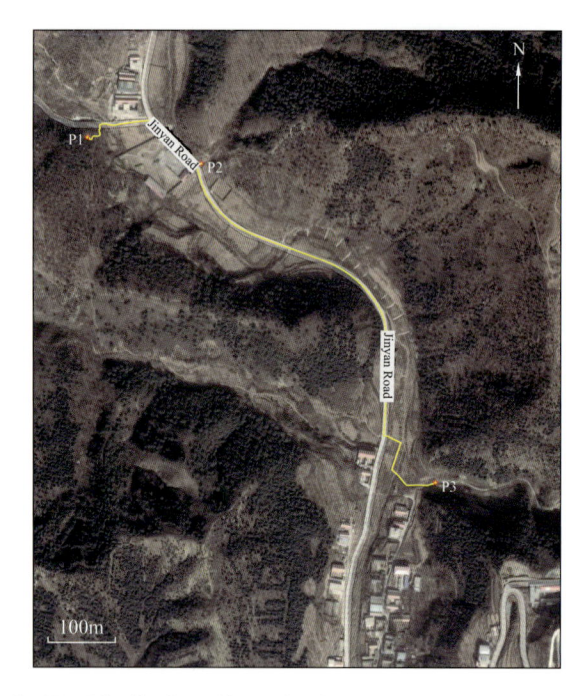

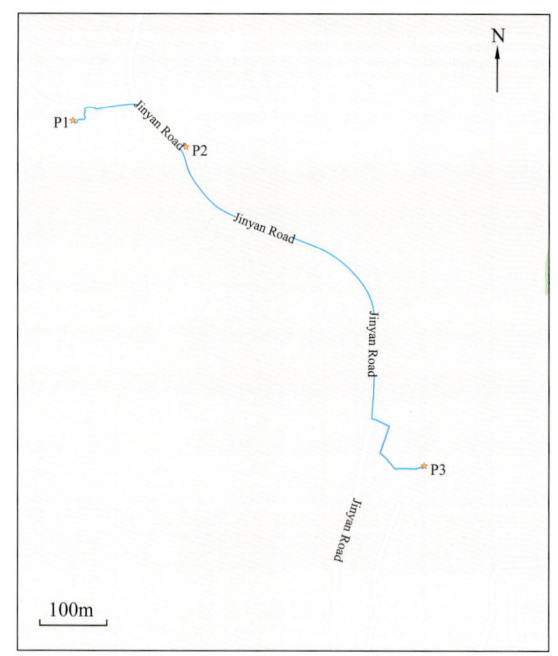

Fig.67　Distribution of investigation stops and route of the Daijian System Xiamaling Formation in Jizhou District, Tianjin

2. Objectives of investigation

(1) To observe the lithologic characteristics and stratigraphic sequence of the Mesoproterozoic Daijian System Xiamaling Formation;

(2) To observe the parallel disconformity between the Xiamaling Formation and the Tieling Formation;

(3) To interpret the depositional environments.

3. Contents of investigation and description of stops

From the Tielingzi village to Xiazhuangzi village, we will observe the lithologic changes and stratigraphic sequence of the Xiamaling Formation from the bottom to top, walking along this route. Due to the limit of exposure condition, only to specific investigation stops are selected, including the stop of the boundary between the Xiamaling Formation and Tieling Formation (P1 in Fig.67) and the stop of typical lithology of $?x_1$ (P2 in Fig.67), The detailed description of each stop is as follows:

Stop 1　Boundary between the Xiamaling Formation and the Tieling Formation

Location: This stop is located at the boundary maker of the Xiamaling Formation and the Tieling Formation, about 650m south of Tielingzi Village next to the Jinyan Road. GPS: E 117.520037°, N 40.219031°.

Contents of investigation: (1) The boundary between the Xiamaling Formation and the Tieling Formation; (2) Rock types and characteristics above and below the boundary; (3) Weathering crust.

Description: This stop is the boundary between the Xiamaling Formation and the Tieling Formation, and there is a parallel disconformity between them (Fig.68A), corresponding to a tectonic movement called "Qinyu uplift" (Chen *et al.*, 1980). Below the boundary, it is the top of the Tieling Formation consisting of thin dolomud striped micritic limestones with intercalated wormkalk calcirudites (Fig.68B), indicating – tidal–flat deposits. The wormkalk calcirudites occurs as lenses, and its rudites are in radial arrangement (Fig.68C), indicating a storm origin. Above the boundary it is the bottom of the Xiamaling Formation consisting of rust–colored ferruginous fine sandstones developed with basal conglomerates, and locally containing hematite lenses (Fig.68D and E), which is considered as the deposits of the initial stage of transgression.

Fig.68 Geological phenomena near the boundary between the Jixian System Tieling Formation and the Daijian System Xiamaling Formation

(A) Boundary between the Tieling Formation and the Xiamaling Formation, a parallel disconformity contact between them, stop 1; (B) Thin bedded dolomud striped micritic limestones, the top of the Tieling Formation, stop 1; (C) Lenses of wormkalk calcirudites, the top of the Tieling group, note that the pebbles are arranged in a radial shape, stop 1; (D) Rust–colored ferruginous argillaceous fine sandstone, bottom of the Xiamaling Formation, stop 1; (E) Ferruginous, argillaceous fine sandstone plane–polarized light, stop 1; (F) Thin grayish green shales sandwiched with thin argillaceous siltstones, the $?x_1$, stop 2

Stop 2 Lithologic characteristics of the ?x_1

Location: This stop is located about 140m southeast of the Jinyanxiang Farmhouse in Tielingzi village along the Jinyan Road. GPS: E 117.399128°, N 40.079632°.

Contents of investigation: (1) Rock types and characteristics of the ?x_1; (2) Depositional environment interpretation.

Description: This stop is an exposure site of typical lithology of the lower part of the ?x_1, which consists of rust-colored fine sandstones and siltstones at the bottom gradually changing into thin gray, gray-green shales and argillaceous siltstones (Fig.68F). The siltstones are intermittently distributed in the layer and partially lenticular. In general, it indicates a shale shelf setting.

SECTION Ⅻ Investigation Route of the Longshan Formation (Qb*l*) of Qingbaikou System in Housigou Valley in Yuyang Town

The Longshan Formation of the Qingbaikou System in the Neoproterozoic is a set of clastics, mainly consisting of pebbly arkoses, quartz sandstone, glauconitic sandstone and variegated sandstone. The grains in this formation are coursing-upward with a total thickness of about 118 m. According to the lithological combination change, the Longshan Formation can be divided into two members: the lower 1st member, a sandstone interval, with gray-green, gray-yellow, glauconitic, -gravelly, feldspar-quartz sandstone developed with tabular and wedge cross beddings and intercalated with gray-yellow argillaceous siltstones at the bottom.; and the upper 2nd member, mostly consisting of gray-green and gray-black shales with intercalated thin bedded fine glauconitic sandstones and with purple-red mudstones at the top. The Longshan Formation has parallel unconformity contacts both with the underlying Daijian System Xiamaling Formation (?x) and the overlying Qingbaikou System Jingeryu Formation (Qb*j*).

The Longshan Formation is widely exposed around the Luotuoling and Housigou Valley area in Jizhou District. Moreover, the Longshan Formation is exposed excellently around the Housigou area with great continuity and integrity, thus it is the best invigation route for this formation. The sketch profile and the stratigraphic column of the Longshan Formation here are shown in Fig.65 and Fig.69.

1. Route location

The route is located at Housigou Valley in Luozhuangzi Town, Jizhou District, Tianjin. The starting point is about 120 m to the north of Xiazhuangzi village next to the Jinyan Road (GPS: E 117.409593°, N 40.076888°), and the ending point is about 350 m south of the Housi village (GPS: E 117.416372°, N 40.075982°). The total length of the route is about 703m. The distribution of specific route and investigation stops is shown in Fig.70. The investigation route is located in the valley, in which there is a stone road suitable to walk. There are steps and indicators around the stops.

Abstract Field Guide of Stratigraphy and Sedimentary Facies of the Proterozoic in Jizhou District, Tianjin

Fm.	Memb.	Thickness (m)	Lithological column	Description
Jingeryu F.m. (Qbj)				Purplishred thin limestones intercalated with gray muddy limestones
Longshan Formation (Qbl)	Qbl$_2$	90–110		Grayish green shales; Shales at top are purplish red; glauconitic sandstone occurs at bottom
	Qbl$_1$	60–90		Yellowish green glauconite feldspar quartz silt –fine sandstones and glauconite feldspar sandstones
		50–60		Grayish white siltstones and medium sandstone lens intercalated in the yellowish green shales
		10–50		Micaceous siltstone intercalated in the grayish green feldspathic sandstones
		0–10		Yellowish green pebbly sandstones and conglomerates, and the gravels are quartzites and chert
Xiamaling F.m (? x)				Ferric sandstone intercalated in illite shales

Legend:
- Feldspathic sandstone
- Pebbled sandstone
- Ferric sandstone
- Muddy limestone
- Shale
- Feldspar–quartz sandstone
- Siltstone
- Glauconitic sandstone

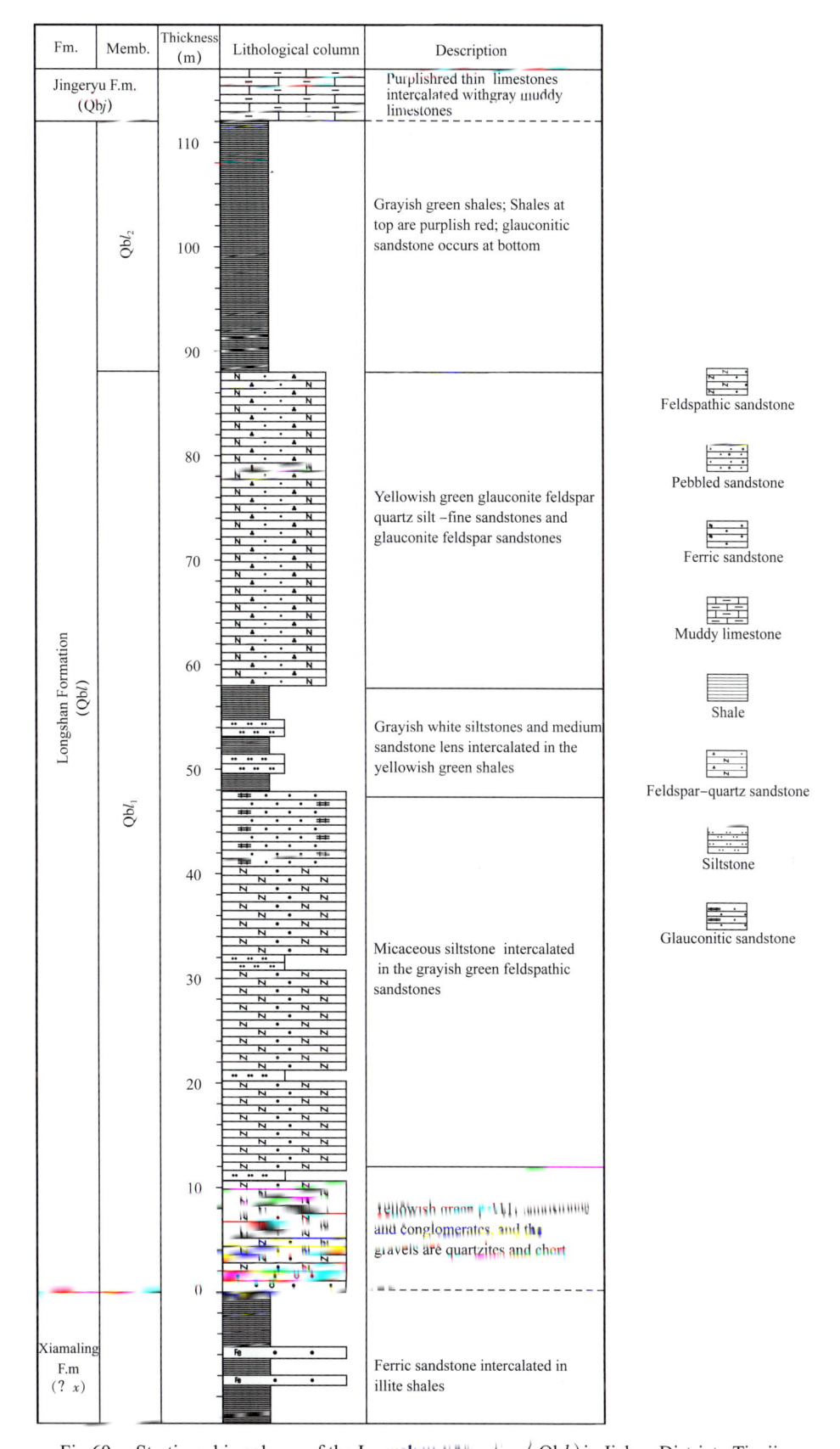

Fig.69 Stratigraphic column of the Longshan Formation (Qbl) in Jizhou District, Tianjin

Fig.70 Distribution of investigation stops and route of the Longshan Formation (Qb*l*) in Jizhou District, Tianjin

2. Objectives of investigation

(1) To observe the lithologic characteristics and stratigraphic sequence of the Neoproterozoic Qingbaikou System Longshan Formation in Jizhou District, Tianjin;

(2) To observe the parallel unconformity between the Xiamaling Formation and the Longshan Formation, as well as the sedimentary structures such as wave marks and cross beddings;

(3) To analyze the sedimentary environment.

3. Contents of investigation and description of stops

From the Xiazhuangzi village to the south of the Housi village, the lithologic change and stratigraphic characteristics of the Longshan Formation would be observed from the bottom to top along this investigation route. Only 4 detailed investigation stops are selected, including the stop of the boundary between the Longshan Formation and the Xiamaling Formation, (P1 in Fig.70), the stop of glauconitic sandstone (P2 in Fig.70), the stop of glauconitic shales (P3 in Fig.70), and the stop of sedimentary structures such as flaser beddings and cross beddings (P4 in Fig.70). The detailed descriptions of each investigation stop are as follows:

Stop 1 Boundary between the Longshan Formation (Qb*l*) and the Xiamaling Formation (?*x*)

Location: This stop is located at the boundary tablet of the Xiamaling Formation and the Longshan Formation, about 120 m north of the Xiazhuangzi village next to the Jinyan Road. GPS: E 117.409593°, N 40.076888°.

Contents of investigation: (1) the characteristics of the boundary between the Longshan Formation and the Xiamaling Formation; (2) the types and characteristics of the rocks above and below the boundary, and interpretation of their sedimentary environments.

Description: this stop is the boundary between the Ximaling Formation (?*x*) and the Longshan

Formation (Qb*l*), as well as the boundary between the Daijian System and the Qingbaikou System. The yellow-grey (weathering color) basal conglomerate at the bottom of the Longshan Formation underlies the blue-green glauconitic shales at the top of the Xiamaling Formation, and there is abrupt contact between them, i.e. erosional surface (Fig.71A). In the late sedimentary period of the Xiamaling Formation, some strata denudation occurred due to the crustal uplift, which could be compared with the strata around the Qingbaikou village in the West Mountain of Beijing. In Jizhou District, the middle and upper parts of the Xiamaling Formation are missing, suggesting a relatively long-term sedimentary hiatus to form a parallel unconformity, which corresponds to the tectonic movement called "Yuxian uplift". (Chen *et al.*, 1980).

The compositions of the grains in the gravelly coarse sandstones at the bottom of the Longshan Formation are mostly include quartz, chert and a small amount of feldspar, containing glauconite, and the pore-fillings are mud and ferruginous cements. There are some uneven erosional surfaces (Fig.71C and D); and the grains are of poor sorting and relatively high roundness changing from sub-angular to sub-rounded, and trough cross beddings and scour surfaces are found within the sandstones (Fig.71B), which both suggest a set of channel deposits of braided river.

Fig.71 Geological phenomena near the boundary between the Xiamaling Formation (?*x*) and the Longshan Formation (Qb*l*)
(A) Parallel unconformity between the Xiamaling Formation and the Longshan Formation, with thick bedded gravelly coarse sandstones of the Longshan Formation above the boundary, and blue-green shales of the Xiamaling Formation below the boundary, stop 1;
(B) Trough cross beddings and scour surfaces in the yellowish grey gravelly coarse sandstones in the bottom of the Longshan Formation, stop 1; (C) Photomicrograph of the yellowish gray gravelly coarse sandstones, plane polarized light, the bottom of the Longshan Formation; (D) Photomicrograph of the yellowish gray gravelly coarse sandstones, the same view with C, cross polarized light

Stop 2　Lithological features and sedimentary structures of the Qbl_1

Location: The stop is about 200 m away from the last stop 1 uphill to the east. GPS: E 117.412086°, N 40.076710°.

Contents of investigation: (1) Rock types and characteristics in the Qbl_1; (2) Sedimentary structures and paleo-current direction in sandstone; (3) Analysis of sedimentary environment.

Description: The lithology and sedimentary structures of the Qbl_1 are observed at this stop. The Qbl_1 mainly consists of grayish green feldspar-quartz sandstone, commonly containing glauconites. The grains of the sandstones is in various sizes, ranging from fine to coarse sand, and locally with granule; the roundness of the grains is generally high, and mostly are sub-rounded; and the pore-fillings of the sandstones is siliceous cements (Fig.72C and D). Some wave marks are found on the bedding surfaces (Fig.72A), and some cross beddings are also found in the sandstones (Fig.72B). These sandstones are typical coast deposits.

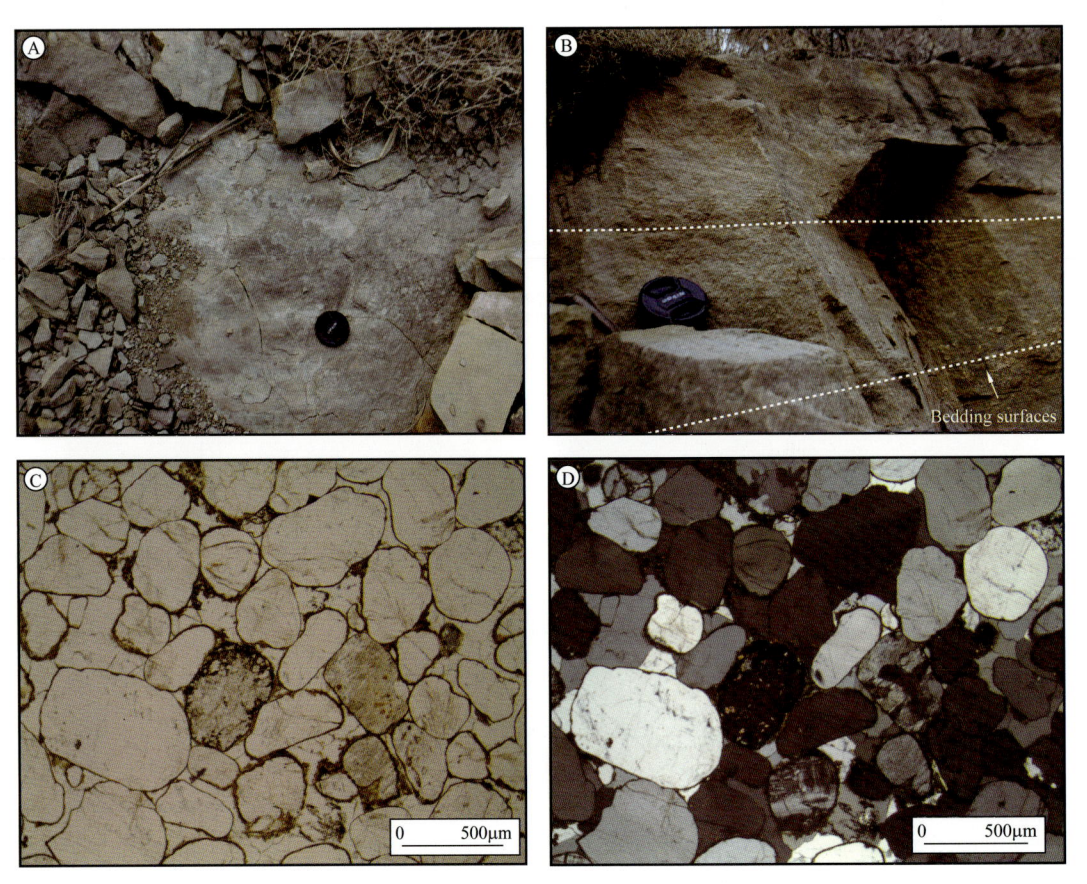

Fig.72　Arkose sandstone and its sedimentary structures in the Qbl_1
(A) Wave-mark structures in the feldspar quartz sandstone, the Qbl_1, stop 2; (B) Tabular crossbedding, the Qbl_1, stop 2;
(C) Photomicrograph of the feldspar-quartz sandstone in the Qbl_1, plane polarized light; (D) Photomicrograph of the feldspar-quartz sandstone in the Qbl_1, the same view with Figure C, cross polarized light

Stop 3　Glauconitic shales and its sedimentary structures in the Qbl_1

Location: This stop is located at the sign of the glauconitic shales, GPS stop: N 40.076815°, E 117.414159°.

Investigation content: characteristics of glauconitic shale.

Description: Glauconitic shales and its sedimentary structures in the Qbl_1 are observed at this stop. The fresh surface of the glauconitic shales is blue-green and develops lamellations (Fig.73); and the interbeddings between the glauconitic shales and the gray-white siltstones are common (Fig.73B), which suggest a shelf setting.

Fig.73　Glauconitic shales and its sedimentary structure in the Qbl_1
(A) Glauconitic shales, the middle part of the Qbl_1, stop 3; (B) Thin interbedded of blue-green glauconitic shales and siltstones, the middle part of the Qbl_1, stop 3

Stop 4　Glauconitic sandstones and its sedimentary structures at the top of the Qbl_1, and the purplish red shales in the Qbl_2

Location: This stop is about 85 m to the east from the last stop 3. GPS: N 40.076499°, E 117.415360°.

Contents of investigation: (1) Flaser beddings in the glauconitic sandstones; (2) fuchsia and variegated shales.

Description: The glauconitic sandstones and its sedimentary structures at the top of the Qbl_1, and the variegated shale of the Qbl_2 are observed at this stop. The fine glauconitic sandstone is yellowish green after weathering, and its main minerals include quartz and glauconite. The glauconite is orientated, and the quartz particles are well sorted and sub-rounded (Fig.74C and D). The herringbone cross beddings (Fig.74A) and flaser beddings occur in the sandstones (Fig.74B), indicating a sandy flat in the intertidal zone. The purplish red shales (Fig.74E) indicates the exposed oxidizing environment, which is the supratidal zone.

— 231 —

Fig.74 Glauconitic sandstones and its sedimentary structure in the Qbl_1 and purplish red shales in the Qbl_2
(A) Herringbone cross beddings in the glauconitic sandstones in the Qbl_1, stop 4; (B) Flaser beddings, the Qbl_1, stop 4;
(C) Photomicrograph of the glauconitic sandstones, plane polarized, scale is 0.5 mm in length, the Qbl_1;
(D) Purplish red shales, the Qbl_2, stop 4

SECTION XIII Investigation Route of the Jingeryu Formation (Qbj) of Qingbaikou System north of Xijingyu Village, Yuyang Town

The Jingeryu formation (Qbj) consists mostly of medium-thin bedded carbonate rocks with intercalated thin bedded glauconitic sandstones and shales, which can be divided into three members: the lower 1st member (Qbj_1), comprised mainly of gray and grayish purple medium-thin bedded micritic bimestone, at bottom of which there is several tens of centimeters of coarse glauconitic sandstones bearing with granules; the middle 2nd member (Qbj_2), consisting mostly of grayish green micritic limestones with intercalated argillaceous limestones; and the upper 3rd member (Qbj_3), made mainly up of gray thin bedded muddy dolostone and dolomitic limestones with intercalated grayish green shales.

The formation is exposed within a small regions around the Fujunshan Syncline, close to its core, especially well-exposed in the Laoguading area. The sketch profile and stratigraphic column of this formation are shown in Fig.65 and Fig.75.

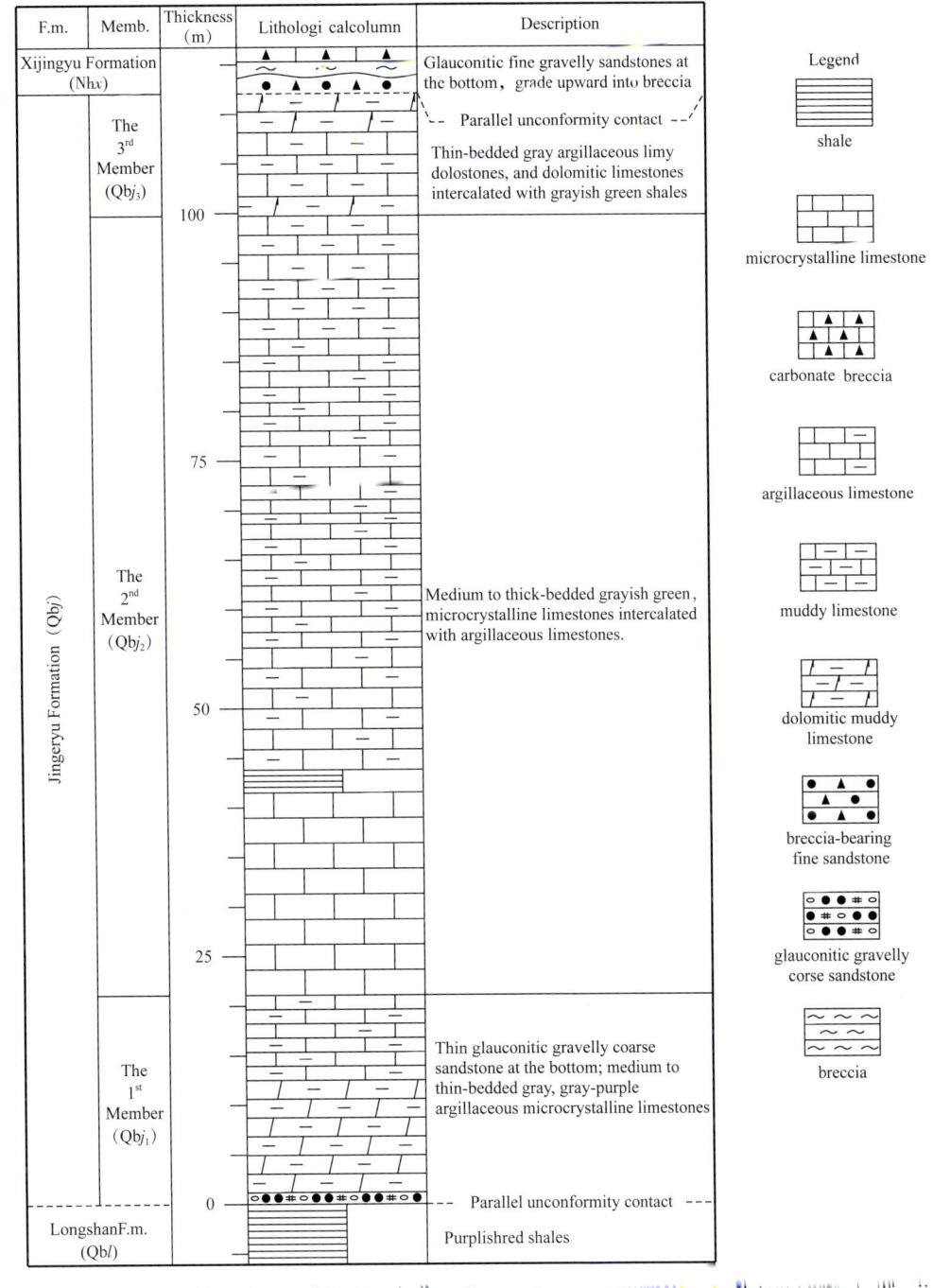

Fig.75 Stratigraphic column of the Qingbaikou System Jingeryu Formation in Jizhou District, Tianjin

1. Route Location

The route is located in the Beiling area of Xijingyu village, Chengguan Town, Jizhou District, Tianjin, and the starting point is next to the road from Xijingyu Village to Housi Village (GPS: E 117.409733°, N 40.074470°). The ending point is located in the quarry pit at the corner of the south slope (GPS: E 117.409780°, N 40.073561°). The total route is about 110 m. The distribution of specific routes and investigation stops is shown in Fig.76. The main investigation route is near the village road, which is in good condition and allows large and small vehicles to pass, so visitors can take bus to the

— 233 —

vicinity of the investigation stops. The investigation route is short, but uphill. However, it is convenient is that the upslope is cement pavement and suitable for walking.

2. Objectives of investigation

(1) Lithologic characteristics and stratigraphic sequence of the Jingeryu Formation;

(2) Boundaries between the Longshan Formation and Jingeryu Formation, and between the Neoproterozoic (the Jingeryu Formation) and the Paleozoic (the Cambrian Fujunshan Formation);

(3) Depositional environment interpretation.

3. Contents of investigation and description of stops

Up along the slope, the lithologic changes and stratigraphic characteristics of the Jingeryu Formation will be investigated from the bottom to top, and three detailed investigation stops are selected, including the stops of boundary between the Longshan Formation and the Jingeryu Formation (P1 in Fig.76), the stop of the typical lithology of 2nd member of the Jingeryu Formation (P2 in Fig.76) and the stop of boundary between the Jingeryu Formation and the Xijingyu Formation (Fig.76).

Fig.76 Distribution of investigation stops and route of the Jingeryu Formation in Jizhou District, Tianjin

Stop 1 Boundary between Longshan formation (Qb*l*) and Jingeryu formation (Qb*j*)

Location: this stop is located on the road from Xijingyu village to Housi Village. GPS: E 117.409733°, N 40.074470°.

Contents of investigation: Boundary between the Longshan Formation (Qb*l*) and the Jingeryu Formation (Qb*j*).

Description: this stop is the boundary between the Longshan Formation and the Jingeryu Formation

(Fig.77A and B), below which it is the top of the Longshan Formation composed of purple-red shales with intercalated gray-green shales; above which it is the Jingeryu Formation consisting of interbedded gray and purple, medium-thin bedded, muddy micritic limestones, with locally developed, granule-bearing, coarse glauconitic sandstones at its bottom (Fig.77B). The two formations are common to be considered as conformable contact, but due to the occurrence of the granules at the bottom of the Jingeryu Formation, a possible temporary sedimentary hiatus cannot excluded (Zhu et al., 2016).

Stop 2 Lithologic characteristics of the Qbj_2

Location: this stop is located in a quarry pit at the corner of the slope in the south of the last stop. GPS: E 117.409780°, N 40.073561°.

Contents of investigation: (1) The medium-thin bedded micritic limestones with intercalated gray-green shales in the lower part of the Qbj_2; (2) Depositional environments interpretation.

Fig.77 Geological phenomena near the boundary between the Qingbaikou System Jingeryu Formation and the underlying Longshan Formation, and the overlying Cambrian Fujunshan Formation (now Proterozoic Xijingyu Fm.)

(A) The boundary between the Jingeryu Formation and the Longshan Formation, below which it is purplish red shales with intercalated grayish green shales. Stop 1; (B) The boundary between the Jingeryu Formation and the Longshan Formation. Note that the bottom of Jingeryu Formation is a layer of glauconitic sandstones (red arrow). Stope 1; (C) Medium-thin bedded micritic limestones in the Qbj_2, Stop 2; (D) The boundary between the Jingeryu Formation and lower part of the Fujunshan Formation (the newly built Xijingyu Formation in this book), below which it is the top part of the Jingeryu Formation consisting of thin bedded gray limestones, and over which it is the lower part of the Fujunshan Formation (now Xijingyu Formation) consisting of breccia limestones, Stop 3

Description: This stop is located in a quarry pit, and the section of typical lithology of the the Qbj_2 is well exposed, which is mainly composed of the egg-white like color, thin-medium bedded micritic limestones with intercalated thin bedded gray-green shales (Fig.77C), and their beddings are flat, indicating a relatively deep-water carbonate ramp setting.

Stop 3　Boundary between the Jingeryu Formation (Qb*j*) and the Xijingyu Formation (Qb*x*)

Location: The stop is located at boundary maker of the Jingeryu Formation and the Xijingyu Formation (lower part of the previous Fujunshan Formation).

Contents of investigation: (1) Boundary between the Jingeryu Formation and the Xijingyu Formation; (2) Lithologic characteristics of the strata overlying and underlying the boundary.

Description: This stop is the boundary between the Jingeryu Formation and the Xijingyu Formation. There is a micro-angular unconformity contact between them (Fig.77D), corresponding to the famous "Jixian Movement" (Sun, 1957; Chen et al., 1980). Under the boundary, it is the top of the Jingeryu Formation consisting of thin bedded, gray, lime-bearing, muddy dolostones, dolomitic limestones, and over the boundary it is composed of carbonate breccia (glacial deposit) of the Xijingyu Formation.

SECTION XIV　Investigation Route of the Xijingyu Formation (Nh*x*) of Nanhua System west of Fujunshan Park in Yuyang Town

The Xijingyu Formation is only exposed in Beiling area of Jizhou District. The section of Xijingyu Formation studied this time is located on the west side of Fujunshan Park. The section is comprised mainly of breccia carbonate rocks, with a total thickness of 155 m. The breccia is generally massive, without beddings. The breccia composition is diverse, and it is with grey-white color and mostly comprised of dolostones, but small amount of limestones in the bottom, and the chert breccia gradually increases upwards. Most of the breccia are angular, various in size, chaotic in arrangement and poorly sorted, suggesting a glacial origin.

The sketch profile and stratigraphic column of the formation are shown in Fig.78.

1. Route Location

This route is located on the west side of the Fujunshan Park in the north of Jizhou District, Tianjin. The starting point of the route is located in the south of the Xijingyu village (GPS: E117.411278°, N 40.069987°). The ending point is in the north of Taoyuan village (GPS: E117.418635°, N 40.062835°). The total route is about 1000 m. The road is generally wide, where large and small sizes vehicles could pass and it is convenient to park. The distribution of specific route and investigation stops are shown in Fig.79.

Abstract Field Guide of Stratigraphy and Sedimentary Facies of the Proterozoic in Jizhou District, Tianjin

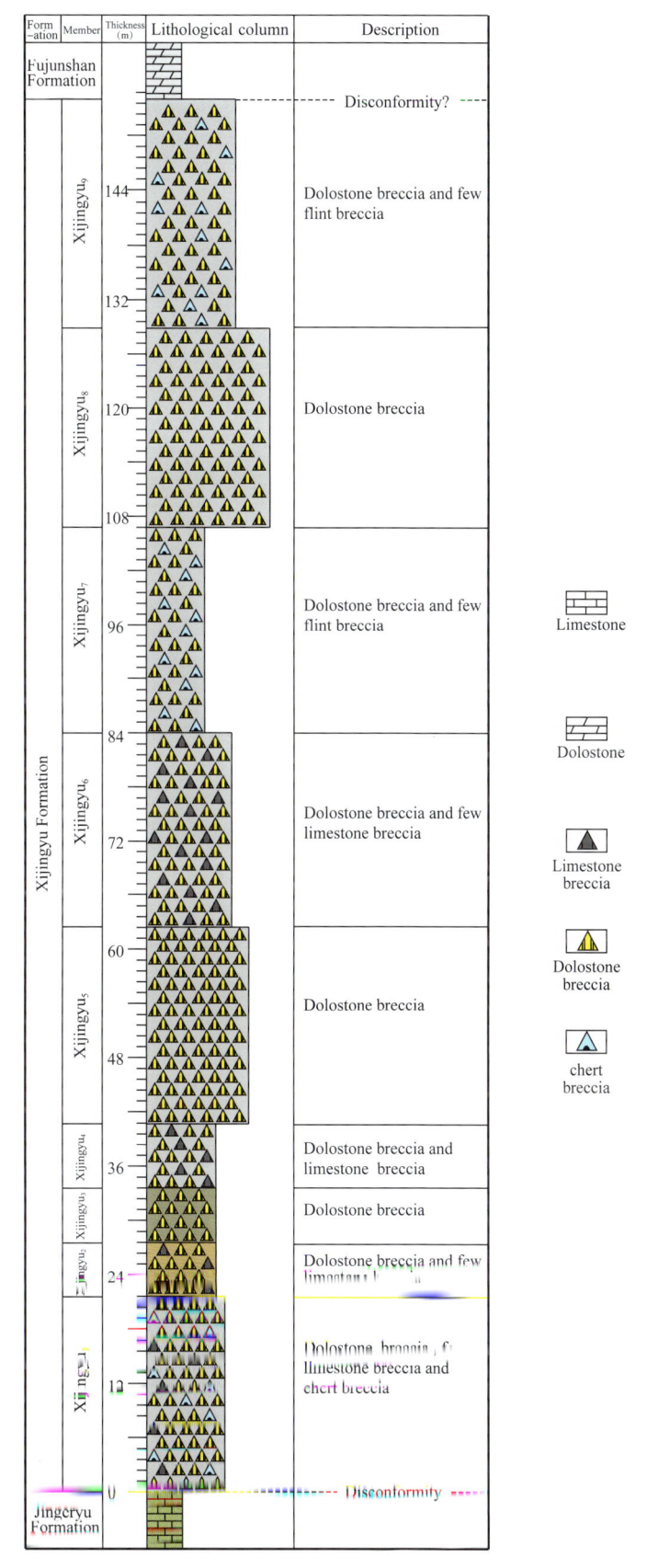

Fig.78 Stratigraphic column of the Xijingyu Formation in Jizhou District, Tianjin

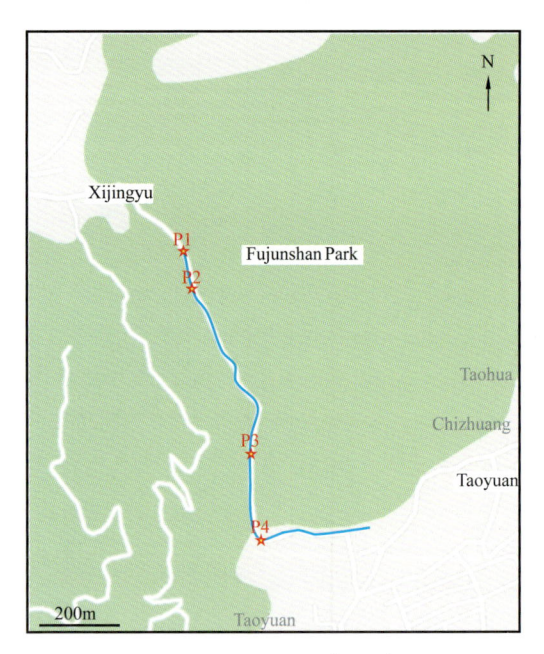

Fig.79　Distribution of investigation stops and route of Xijingyu Formation (Nhx),
Nanhua System, Jixian District, Tianjin

2. Objectives of investigation

(1) To observe the lithologic characteristics and stratigraphic sequence of the Xijingyu Formation in Jizhou District, Tianjin;

(2) To observe the sedimentary characteristics of different types of breccia and carbonate rocks;

(3) To analyze the origin of the breccia.

3. Contents of investigation and description of stops

From the south of the Xijingyu village to the north of Taoyuan village, the road condition is good and suitable to investigate. Only four detailed investigation stops are selected along the route, including the stop of the boundary between the Jingeryu Formation and the Xijingyu Formation (P1 in Fig.79), the stop of the breccia in the Xijingyu Formation (P2 in Fig.79), the stop of the breccia carbonate rock and diabase (P3 in Fig.79) and the stop of the overlying carbonate rock of the Fujunshan Formation (P4 in Fig.79). The detailed descriptions of each stop are as follows:

Stop 1　Boundary between the Jingeryu Formation (Qbj) and the Xijingyu Formation (Nhx)

Location: This stop is located on the west side of the Fujunshan Park, south of Xijingyu village. GPS: E117.412785°, N 40.069987°.

Content of investigation: The boundary between the Jingeryu Formation and the Xijingyu Formation.

Description : This stop is the boundary between the Jingeryu Formation and the Xijingyu Formation, and there is a parallel unconformity contact relationship between the two formations (Fig.80A and B). Under the boundary, it is the medium-thin bedded gray micritic limestones at the top of the Jingeryu Formation; and above the boundary, it is the breccia at the bottom of the Xijingyu Formation.

Fig.80 Typical rocks of the Jingeryu Formation and the Xijingyu Formation

(A) Grey micritic limestones at the top of the Jingeryu Formation, stop 1; (B) Breccia at the bottom of the Xijingyu Formation, note that the dolostone fragments are sporadically scattered on the surface in sizes of 2-4 cm, stop 1

Stop 2 Characteristics of breccia in the lower part of the Xijingyu Formation

Location : This stop is about 150 m southeast from the last stop 1. GPS : E 117.413223°, N 40.069136°.

Content of investigation : The boundary between the Jingeryu Formation and the Xijingyu Formation.

Description : A large number of breccia widely develop between this stop and the Xijingyucun bus stop. The breccia is mainly comprised of dolomite, as well as limestones, cherts and so on (Fig.81).

Stop 3 Characteristics of breccia in the middle part of the Xijingyu Formation

Location : The stop is located at the memorial gate of Dongtianfudi, south of the Xijingyucun bus stop. GPS : E 117.413223°, N 40.069136°,

Content of investigation : Characteristics of breccia in the middle part of the Xijingyu Formation.

Description : A large number of breccia developed between this stop and the Xijingyucun bus stop. The breccia is mainly comprised of dolomites, as well as limestones, cherts and so on (Fig.82).

Fig.81　Breccia of the Xijingyu Formation and their photomicrographs

(A) Breccia dolostones in the Xijingyu Formation with nodular cherts on the surface in the sizes of 2-6 cm, stop 2;
(B) Breccia limestones in the Xijingyu Formation, comprised mainly of calcite, stop 2;
(C) Breccia in the Xijingyu Formation, stop 2; (D) Microphotograph of the Breccia, cross polarized light, stop 2

Stop 4　Boundary between the Xijingyu Formation and the Fujunshan Formation

Location: The stop is located at the corner of the road from the south slope to memorial gate of Dongtianfudi (there is a house). GPS: E 117.415430°, N 40.062772°.

Content of investigation: Characteristics of breccia in the middle part of Xijingyu Formation. The boundary between the Xijingyu Formation and the Fujunshan Formation and the lithology over or below the boundary.

Description: At this stop, the boundary between Xijingyu Formation and Fujunshan Formation can be seen, which an abrupt contact. Over the boundary, it is the Fujunshan Formation, consisting of the medium-thin bedded limestones and dolostones with well-developed beddings; while below the boundary, it is the Xijingyu Formation, consisting of the breccia (Fig.83).

Abstract Field Guide of Stratigraphy and Sedimentary Facies of the Proterozoic in Jizhou District, Tianjin

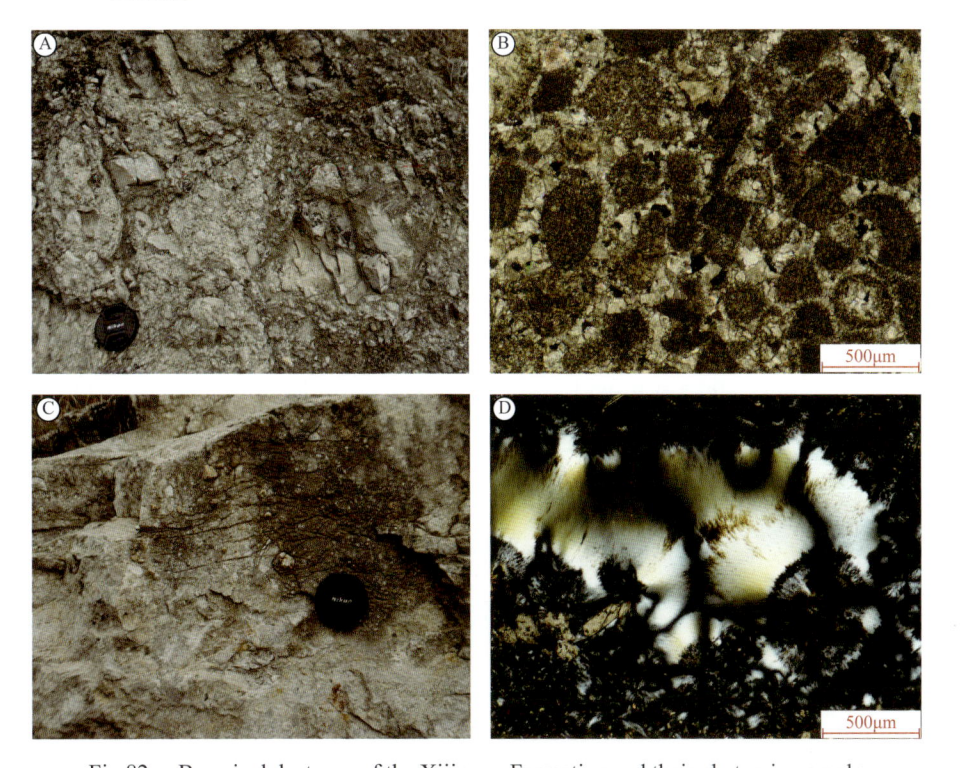

Fig.82 Breccia dolostones of the Xijingyu Formation and their photomicrographs

(A) Breccia, of which most are comprised of dolomite, with a diameter from 1 to 6 cm; (B) Microphotograph of doloarenites, cross polarized light, stop 3; (C) Breccia dolostones containing chert breccia with length up to 3 cm; (D) Microphotograph of chert, cross polarized light, stop 3

Fig.83 Carbonate of the Cambrian Fujunshan Formation and their photomicrographs

(A) Limestones at the bottom of the Fujunshan Formation, stop 4; (B) Photomicrograph of the micritic limestones, cross polarized light, stop 4; (C) Dolostones of the Fujunshan Formation, stop 4; (D) Photomicrograph of the dolostones with calcite veins in the Fujunshan Formation, stop 4

SECTION XV Sedimentary facies of each formation

1. Introduction

In the Proterozoic, the depositional background of the Jizhou area is a NE-SW rift trough (Fig.84). During the depositional period of the Changcheng System, the rift trough is long and narrow, with about 1000 km in length and only 100 km in width, which is considered as one branch of the aulacogen and became wide during the depositional periods of the Jixian System and Qingbaikou System.

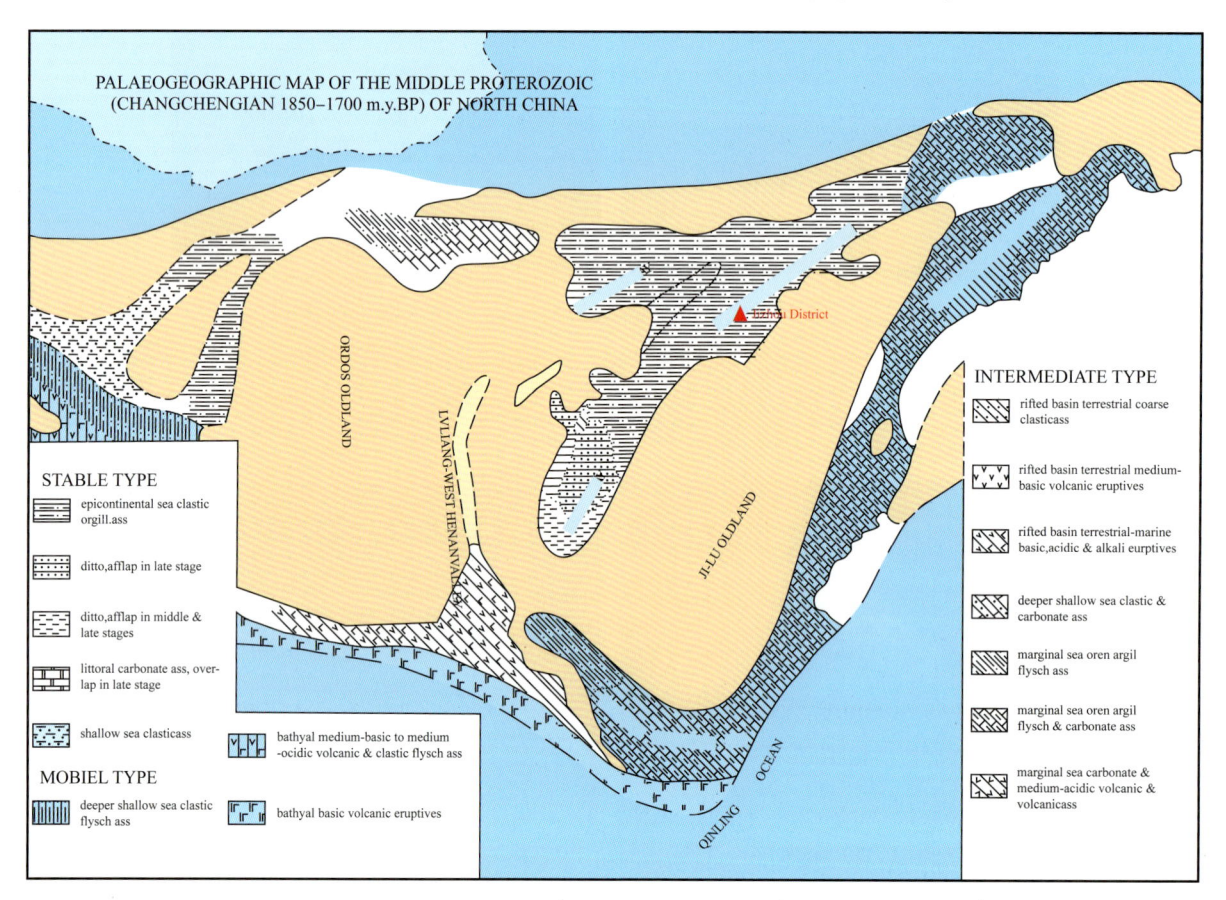

Fig.84 Paleogeographic map of the Mesoproterozoic (now Paleoproterozoic) Changcheng period (modified from Wang *et al.*, 1985). Red triangle represents Jixian area

Based on the data of outcrop observation and thin-section identification of the Proterozoic in Jizhou District, we analyzed the sedimentary facies of each formation in Proterozoic. The facies indicators used in the facies analysis include rock types, colors, mineral compositions, textures (i.e., grain size, roundness, sorting, pore-filling types), sedimentary structures (e.g., cross beddings, parallel beddings, wavy beddings, flaser beddings, horizontal beddings, ripple marks, mud cracks, stromatolites), sedimentary sequences, paragenetic associations. There are no calcareous paleontological fossils in the Proterozoic.

The sedimentary types of the Proterozoic can be divided into 4 types: clastic deposits, carbonate deposits, clastic-carbonate mixed deposits, and glacial deposits.

The strata dominated by clastics include the Changzhougou Formation, the Chuanlinggou Formation, the Xiamaling Formation, and the Longshan Formation. There are rare carbonate in these formations, and their sedimentary model is shown in Fig.85. These formations were deposited in one or several facies belts of the sedimentary model. The Chc_1 was deposited in the braided plain, the Chc_2 was deposited in the upper part of coast, and the Chc_3 was deposited in the lower part of coast. The Chch_1 was formed in shale shelf, while the Chch_2 and Chch_3 were deposited in gentle slop-basin. The Hongshuizhuang Formation, the Xiamaling Formation and the Longshan Formation were mainly deposited in the coast and shale shelf.

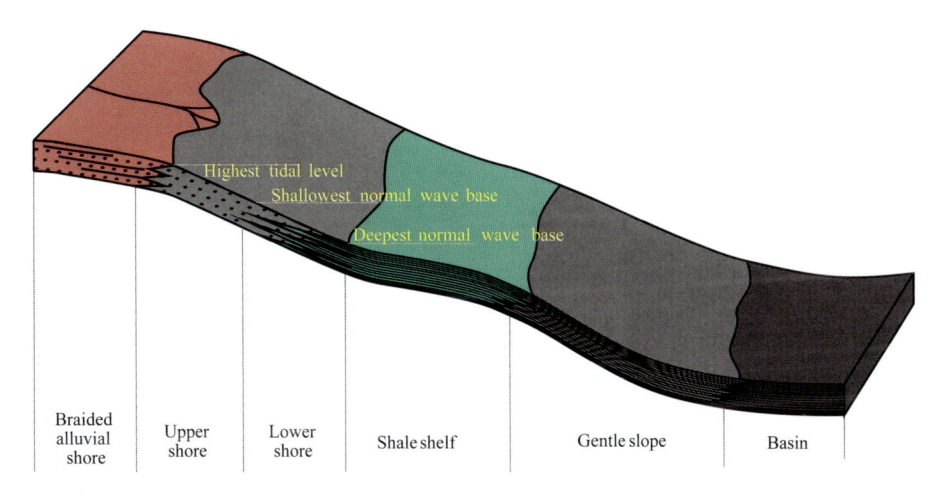

Fig.85 Depositional model of the Proterozoic Changzhougou Fm., Chuanlinggou Fm., Xiamaling Fm. and Longshan F.m. in Jixian area

Carbonates are dominant in the Gaoyuzhuang Formation, the Wumishan Formation, the Tieling Formation and the Jingeryu Formation, although there are also some clastics. The strata consist of clastic-carbonate mixed deposits, mostly including the Tuanshanzi Formation, the Dahongyu Formation, the Yangzhuang Formation, and the Hongshuizhuang Formation. The clastic-carbonate mixed depositional model is shown in Fig.86. Among them, the facies belt was completely developed in the Tuanshanzi Formation, including the clastic supratidal mudflat, intertidal mixed flat, subtidal sandflat, subtidal mixed flat, shale restrict sea, carbonate tidal flat (supratidal flat and intertidal flat), open platform, carbonate gentle slope, and basin.

In respect to the strata consisting of interbedded coastal sandstones and carbonates, the carbonate platform directly contacted with the sandy coast, and no shale restricted sea occurred between them.

In respect to the strata consisting of intebedded tidal-flat mudstones and carbonates, the muddy tidal flat directly contacted with the carbonate platform, and there are generally no high-energy coastal sandy facies belts and shale restricted shallow sea (Fig.87). The Yangzhang Formation and the Wumishan Formation belong to this model. In the Yangzhuang Formation, red mudstones and dolomitic mudstones represents the strongly oxidizing supratidal mudflat, while the light gray mudstones represents the

- 243 -

intertidal mudflat. Although the intertidal mudflat is exposed frequently, it is also frequently submerged by the tidal water, and the sediments are always wet, so it is weakly reducing in their interiors to alter the red mudstones into light grayish gray and gray mudstones. This kind of mudstone is different from the grayish green shales, and the later is deposited in shelf.

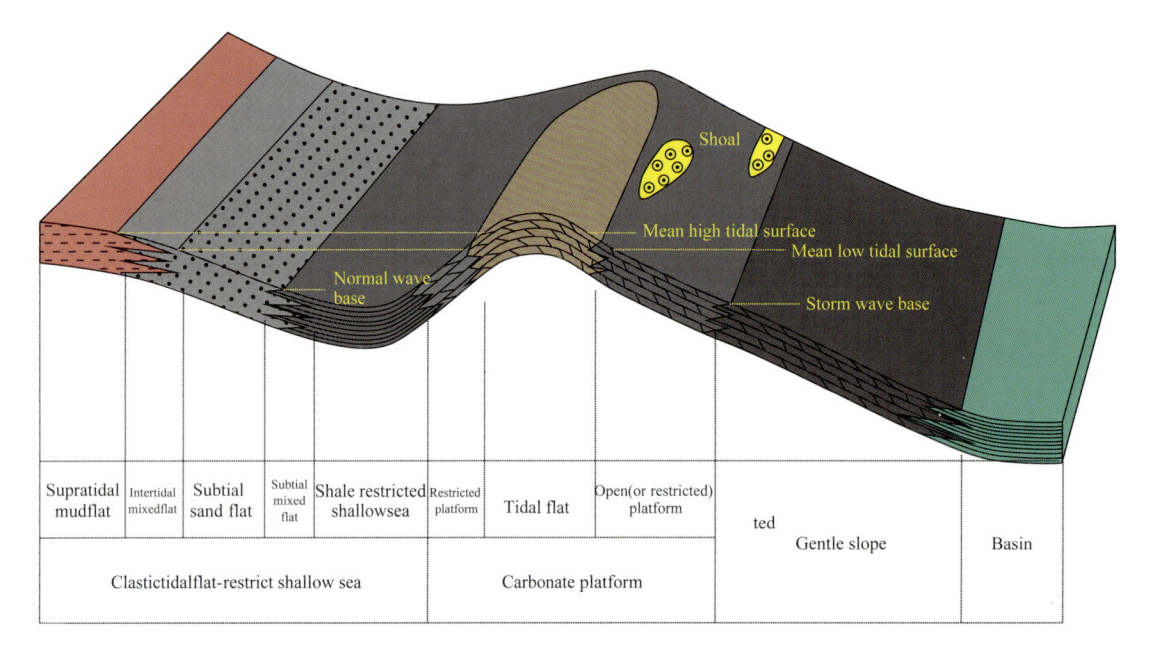

Fig.86　Depositional model of Proterozoic Tuanshanzi Fm., Dahongyu Fm., Gaoyuzhuang Fm., Hongshuizhuang Fm., Tieling Fm. and Jingeryu Fm. in Jixian area

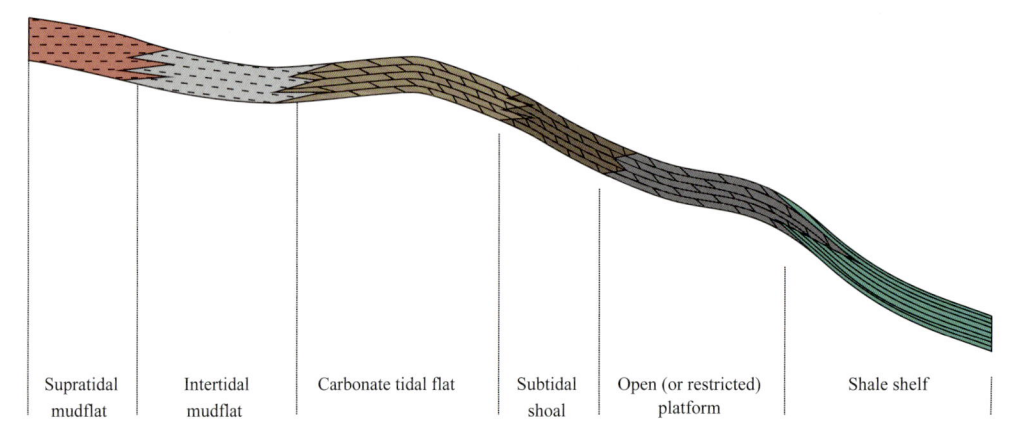

Fig.87　Depositional model of Proterozoic Yang zhuang Fm. and Wumisham Fm. in Jixian area

2. Sedimentary facies interpretation of the Changzhougou Formation and the Chuanlinggou Formation

1) Sedimentary facies interpretation of the Changzhougou Formation

The Changzhougou Formation is a set of thick (859m) sandstone-dominated strata, which is divided into 3 members: the 1st member (Chc_1), the 2nd member (Chc_2), and the 3rd member (Chc_3), from the bottom to top.

(1) Sedimentary facies interpretation of the Chc_1.

Is the Chc_1 a fluvial facies or a marine coast facies? Due to the lack of fossils, it can only be distinguished by the diagnostic indictors, such as color, mineral compositions, sedimentary structures, depositional sequences, paragenetic associations, and so on.

The marine coastal sandstones are mostly weak-reducing color, due to deposition mostly under the water, and commonly contain glauconites (but no glauconite has been found in the Changzhougou Formation, while it is common in Longshan Formation). Its grains is generally well sorted due to the relatively stable hydrodynamic conditions, and the roundness is often high, commonly sub-rounded, as a result of transportation of the sands mostly by rolling (especially in swashing zone). Due to mostly under the water, the ripple marks in this setting are symmetric, and the variable directions of the current, forming the ripple marks, result in the variable dip direction of the laminae in cross beddings. In addition, the characteristic swash cross beddings can be formed here. Because the coastal beach and bar can not erode underlying deposits, the scouring surface is rare, and the beach bar shows either a homogeneous sequence or a coursing-upward sequence. The transgression can form a generally fining-upward sequence, but this is rare in a single beach or bar.

River sandstones are mostly oxidizing color, because it is mostly subaerial, and contain no glauconites. Because of the relatively larger hydrodynamic fluctuation during the flooding and the inter-flooding periods, the grains are generally moderately sorted. Because the sand are transported mainly by saltation (suspension transport maybe occurs during the flooding period), its roundness is generally low, mainly sub-angular. Due to the overall stability of the river direction, the dip directions of the laminae in the cross bedding forming in the river channels are relatively stable. In addition, it can also develop distinct scouring surfaces and fining-upward sequences.

According to the above analysis, the sandstones, gravel-bearing sandstones, gravelly sandstones, and fine conglomerates in the Chc_1 in this area were deposited by sandy braided rivers, moreover the braided rivers were located in deep and large rifting valleys. The evidence is as follows:

① The color is red, oxidizing color (Fig.7 and Fig.9). This member is generally purplish red, including occasional mudstone gravels and mudstone beds, which are oxidizing color, indicating exposure setting.

② Relatively low roundness and moderate sorting (Fig.10), Most of the sand grains in sandstones are sub-angular. Even for granules, their roundness is not high. The sorting is medium. All of these are the sedimentary characteristics of the fluvial facies.

③ Stable dip direction of beddings. From the bottom to top, the dip direction of bedding in the strata consisting of sandstones, gravel-bearing sandstones, gravelly sandstones, and fine conglomerates, with a thickness of 260 m, is NNW, and very stable, especially in the upper part of this member, which reflects the stability of the flow direction.

④ Common scouring surfaces and fining-upward sequences (Fig.8 and Fig.9), which are the characteristics of fluvial facies.

⑤ Rip-up mudstone clasts (Fig.7). Such angular rip-up mudstone clasts cannot be found in coastal deposits, as the repeated washing will destroy them, while these rip-up mudstone clasts are common in fluvial deposits.

⑥ The strata with a thickness of about 260m mostly consist of sandstones, gravel-bearing sandstones, gravelly sandstone and fine conglomerates, occasionally residual mudstone interbed is present, which suggests that the floodplain did not develop around the rivers during this period. Why not? It should be caused by the frequent switching of the channels, so that the deposits of the floodplain is difficult to be preserved. According to modern deposits observation, only the braided rivers in deep rift valleys have such feature, and can form very thick deposits of sandstones but without mudstone interbeds.

In summary, the Chc_1, consisting of sandstones, gravel-bearing sandstones, gravelly sandstone and fine conglomerates in this area, are sandy braided river deposits in deep large rifting valleys, rather than marine coastal deposits, which are different from the overlying Chc_2 and Chc_3.

(2) Sedimentary facies interpretation of the Chc_2 and Chc_3.

The Chc_2 and Chc_3 mainly consist of milk-white fine sandstones. Compared with the Chc_1, the color of sandstones in the Chc_2 and Chc_3 is more reducing, and especially its roundness and sorting are generally better. The comprehensive analysis shows that the Chc_2 is deposited in the upper part of coast, and the Chc_3 is deposited in lower part of coast. The evidence is as follows:

① The grains of sandstones are well rounded and sorted (Fig.11). Most of the quartz grains are sub-rounded. The coast in the stable passive continental margin is the only environment, where the quartz grains can be abraded into a round shape. Herein, under the influence of waves, swashing currents and longshore currents, the sands are transported in a rolling way, and can be swashed and abraded for a long time due to the slow rate of subsidence, so their roundness is generally high. However, no matter how far the sands are transported, it is hard to abraded them into a round shape in the river, because the sands are transported mostly by saltation, even by temporay suspension. The roundness of the sands deposited in the Yangtze River mouth is still low, although they have been transported for thousands of kilometers. In coastal lake, although the sand can be transported in a rolling way, but it is difficult for them to be abraded, due to the instability and frequent fluctuation of the lake level. Therefore, all sands with high roundness (sub-rounded or rounded) are deposited in marine environments. Although marine sandstones after weathering may provide well rounded sands for rivers, but there must be other poorly rounded sands provided by other parent rocks. As a result, the grains of the fluvial sandstones are always poorly-rounded, and the poorly rounded sands are usually in the majority, although some sands could be well rounded. The roundness is a major indicator to distinguish the marine and continental settings. Good sorting is also an important feature of marine sandstones, because the waves, swashing currents and longshore currents in the coast are relatively stable to transport sands with similar sizes during normal weather.

② The pore fillings is cements. Siliceous cements are the main pore fillings in the sandstones in the Chc_2 rather than matrix, which suggests that the depositional setting is continuously agitated to prevent muds from being deposited. The water depth should be above the normal wave base, because only in this condition, there is no quiet stage for mud to deposit.

③ The sandstone is reducing color. Why are the sandstones in the Chc_1 purplish red, while the sandstones in the Chc_2 and Chc_3 are milk-white (gray)? The author considers that the sandstones in the Chc_2 and Chc_3 were deposited in subaqueous setting, which is more reducing, so these sandstones is of reducing color.

④ The sandstones are mostly massive and without beddings. The author believes that this is caused by bioturbation. Although no fossil has been found in the Proterozoic, it does not mean that no organism existed. It is likely that the creatures are without skeleton or shells to preserve as fossils. In respect to the sedimentary structures, if it was not destroyed by bioturbation, there should be beddings preserved in the sandstones. Thus, if there is no bedding, it can be speculated that there should be creatures.

⑤ Compared with the Chc_2, the grayish green mudstone interbeds increase upwards and sandstones become thinner in the Cht_3. And Cht_3 gradually changes into the grayish green shales in the Chuanlinggou Formation deposited in the offshore shelf, which reflects that the water depth gradually increased during the depositional periods of the Chc_2 to the Chc_3. The thin mudstone beds represent the deposits of quiet periods. There is no mudstone beds in the Chc_2, which indicates that the seawater depth is shallow and above the normal wave base, suggesting upper part of the coastal setting where the water depth is above the shallowest normal wave base. There are mudstone beds in the 3rd member, indicating that it was deposited in intermittently agitated environments. Sand is deposited during the agitated period while mud is deposited in quiet period, and sometimes waves can affect the bottom of the sea and sometimes cannot. This is the feature of the wave-fluctuation belt. Because the deposits are still dominated by sand, suggesting that the depositional water is most of the time agitated, and the quite period is short and not frequent, so they were deposited in the lower part of the coast, where the water depth is between the shallowest and the deepest normal wave base. If the water depth was below the deepest normal wave base, it would be dominated by muddy deposits.

⑥ The undulating bedding surfaces of the sandstones were wave marks formed by storms. The uneven and undulating bedding surfaces of the sandstones in the Chc_2 and Chc_3 are actually large-scale wave marks. The ripples are roughly symmetrical, with wavelengths of tens of centimeters and waves heights ranging from several centimeters to ten centimeters, which are formed by large-scale storm waves.

In summary, it is a complete transgression sequence from the Chc_1 to the Chc_3, evolving from the continental facies to the marine facies, and the water depth increased gradually.

2) Sedimentary facies interpretation of the Chuanlinggou Formation

The Chuanlinggou Formation is a set of thick shales (889 m), which shows gradual contact with the underlying Changzhougou Formation and no sedimentary hiatus occurred between them.

This formation is divided into 3 members: the 1st member ($Chch_1$), the 2nd member ($Chch_2$) and the 3rd member ($Chch_3$), from the bottom to top.

The shales in the $Chch_1$ is grayish green, intercalated with thin bedded fine sandstones and siltstones in the lower part, and microcrystalline dolostone lenses (grayish white after weathering) at the top,

(Fig.13). The Chch_2 is a large set of black shales, with diabase intrusions in the middle part. The Chch_3 consists of black shales with intercalated thin bedded (the thickness of single bed is about 1cm) muddy-silty microcrystalline dolostones (Fig.15 and Fig.18D). Dolostones is ferriferous, and show brown color after weathering. There are diabase intrusions at the top of this member.

The whole lithology of the formation is shales, which is deposited in quiet and low-energy environment. The color is reducing, suggesting that it was deposited subaqueously and the water depth is below the normal wave base. Were the shales deposited in lacustrine or marine settings? Since there were no fossils at that time, it was not possible to distinguish its depositional setting with fossils. However, because there is a large set of marine carbonate rocks in the overlying Tuanshanzi Formation, There are some microcrystalline dolostone lenses intercalate in the shales, and the regional sedimentary backgrounds are taken into consideration, it is believed that the shales of the Chuanlinggou Formation should be deposited in marine rather than lacustrine settings.

How deep was the sea? The author puts forward that the Chch_1 was deposited in the outer shelf, while Chch_2 and Chch_3 were formed in the deep-water gentle slope-basin. The evidence is as follows :

① The shales in the Chch_1, with intercalated thin bedded fine sandstones and siltstones with ripple marks and small-scale cross beddings (Fig.17D, E), is lighter in color and is grayish green (Fig.17C, E), reflecting a weakly reducing environment. These thin bedded fine sandstones and siltstones should be formed by storm, which suggests its water depth should be above the storm wave base. In the middle part of this member, where the garyish green shales are sandwiched with the thin-bed siltstones (the stop 3 of the Chuanlinggou Formation), some symmetrical ripple marks and the suspected mud cracks are found locally (Fig.17F). Some scholars believe that they are mud cracks, but the author thinks they are not mud cracks, because their shapes and the fillings (silts) in the cracks are not similar with the typical mud cracks and are associated with shelf shales. In author's opinion, these "mud cracks" are actually formed by the uneven weathering of the wave crests and the wave troughs of the wave marks in the underlying thin siltstones. The thin siltstones with fish-scale small ripple marks (multi-directional interference ripple marks), overlain by thin shales. During weathering, the muddy covers above the wave troughs were preserved, while that on the wave crests were errodded and exposed, with the appearance like muddy fillings. If they were really mudcracks, the sea level should have fallen dramatically and rapidly to expose the sea bottom, which is unlikely to happen.

② The shales of the Chch_2 are black, and darker than the Chch_2, suggesting that they were deposited below redox interface. Moreover, there is no interbeds of fine sandstones and siltstones, and the shales are more pure, indicating that the depositional water was much deeper than the storm wave base. In particular, some small-scale folds are found in the lower part of this member (the stop 4 of the Chuanlinggou Formation), which is likely to be penecontemporaneous folds formed by slumping and indicates a slope setting. However, no turbidite has been found, which suggests that it is a gentle slope rather than a steep slope. Therefore, it is considered that this member was deposited in a deep-water gentle slope-basin setting.

③ The shale in the Chch_3 is still black, and its depositional environment is the same as that of the Chch_2. However, there are thin muddy and silty microcrystalline dolostones beds in this member. These dolostones were formed by mechanical transportation of the dolomuds, terrigenous muds and silts from nearby carbonate platform to the deep-water environment in the form of low-density turbidity current.

In summary, from the bottom to top, the color of shales gradually changes from grayish green to black, siltstones and fine sandstones thin bed gradually disappear, which indicate that the water depth increased gradually, and the depositional setting changed from offshore shelf to gentle slope-basin. The sea in the period when the Chch_2 was deposited is the deepest in the whole Proterozoic (deeper than that during the period when the Hongshuizhuang Formation was deposited), corresponding to the max flooding surface of the Proterozoic.

3. Sedimentary facies interpretation of the Tuanshanzi Formation

The Tuanshanzi Formation is 518m thick (Fig.19 and Fig.20), and is divided into 4 members: the 1st member (Cht_1), the 2nd member (Cht_2), the 3rd member (Cht_3), and the 4th member (Cht_4), from the bottom to top.

1) Sedimentary facies interpretation of the Cht_1

The Cht_1 consists of dark gray medium-thin bedded muddy microcrystalline dolostones interbeded with silty microcrystalline dolostones and dark gray dolomitic shales. After weathering, the dolostones are yellowish brown, reflecting they are ferous.

The most obvious features of this member are that the bedding surfaces of all rocks are flat, and the horizontal beddings are extremely developed (Fig.22). Their dark color and horizontal beddings suggest that they were deposited in a subaqueous reducing environment, where the water was deep, far below the storm wave base, and the water bottom was quiet. According to the law of the facies sequence proposed by Walther, the Cht_1 should be deep-water deposits, because it is overlying the Chch_3 consisting of black shales deposited in ramp-basin settings. In view of the fact that this member mostly consist of dolostones, its depositional water depth should be shallower than that of the Chch_3, and the depositional setting should be slope. Due to the lack of high-density turbidites and slumping structures, the slope should be a very gentle.

In respect to the dolostones in this member, the author considers that they were formed by mechanical transportation rather than primary or secondary precipitation origin. The microcrystalline dolomites, formed by primary precipitation or evaporation pumping dolomitization on the adjacent carbonate platform, were stirred up by storm waves and transported to the deep-water environment in the form of low-density turbidity currents and pelagic suspension. The evidence is as follows:

① The dolostones were deposited in a deep-water environment, and it is not possible to explain its origin with the dolomitization mechanisms that have been proposed so far. The evaporation pumping dolomitization is not possible, because it was not deposited in the tidal flat environment; the reflux seepage dolomitization is not possible, because they are comprised of microcrystalline dolomites rather than fine or even coarser crystals of dolomite; and the dolostones formed by dorag, burial and hydrothermal dolomitization should all consist of saccharoidal dolomite crystals.

② The dolostones are generally argillaceous and silty (Fig.22), which suggests that they were deposited in a turbid water body. Especially, the existence of silts is the proof of mechanical transportation.

If these dolostones were mechanically deposited, it should be called a "dolomudstone", similar to the limemudstone.

2) Sedimentary facies interpretation of the Cht_2

The Cht_2 is dominated by gray medium-thick bedded microcrystalline dolostones with intercalated thin bedded muddy microcrystalline dolostones. Because of the high iron content of these dolostones, most of the weathered surface is yellowish brown.

Compared with the Cht_1, the medium-thick bedded microcrystalline dolostones are common in the Cht_2, with uneven bedding surfaces and symmetrical ripple marks (Fig.88). The interior of the dolostones is massive or with blurry horizontal beddings, and locally with normal graded beddings in thickness of several millimeters to about 2 cm.

Fig.88　Symmetrical ripple marks on the bedding surface of the Cht_2, west of the Tuanshanzi village, Jizhou District, Tianjin

The bedding is not obvious, which may be caused by bioturbation. As to what organism, it is not clear. It is likely to be the algae or worms, and both have not been preserved as fossils.

The horizontal beddings in the argillaceous microcrystalline dolostones are similar to that in the Cht_1.

Generally speaking, the water body where the Cht_2 were deposited is shallower than the Cht_2, which should be the upper part of the gentle slope, below the normal wave base, but near the storm wave base. The normal graded beddings should be formed by the low-density turbidity current.

3) Sedimentary facies interpretation of the Cht_3

The Cht_3 is obviously divided into three parts: the lower, the middle and the upper. The lower part is mainly dark gray thin bedded microcrystalline dolostones (yellowish brown after being weathered) intercalated with gray thin bedded fine dolomitic quartz sandstones. Some small-or medium-scale crossbeddings commonly occur in the sandstones. The middle part consists of interbedded thin bedded dark gray shales and fine sandstones (single bed thickness is several centimeters), and some small-or medium-scale crossbeddings commonly occur in the sandstones. The upper part is comprised of medium-thin bedded microcrystalline dolostones interbedded with medium bedded dolomitic quartz sandstone. Some laminar and wavy stromatolites and mud cracks commonly occur in these dolostones, and crossbeddings are common in these sandstones. The Cht₃ is deposits of the carbonate-clastic mixed platform. The common occurrence of the laminar and wavy stromatolites and mud cracks suggest that they were deposited in the intertidal and supratidal flats. The dolostones consist of microcrystals, which should be formed by the evaporation pumping dolomitization.

The sandstones are generally thin-medium bedded, interbedded with tidal-flat dolostones. Some small-scale cross beddings are found in the sandstones, and the dip direction of the laminae in the crossbeddings are inverse in the upper and lower layers, which suggest that they were deposited in a bidirecation-current setting, i.e., tidal flat.

The thin interbedding of dark shales and fine sandstones occurs in the middle part, and the dark shales were deposited in deep reducing environment; while the sandstone with wave marks and crossbeddings were deposited in shallow water. Thus, the middle part of the Cht₃ was a set deposits of restricted subtidal environments.

4) Sedimentary facies interpretation of the Cht₄

The Cht₄ consists of thin bedded purplish red silty shales intercalated with thin bedded fine sandstones. Some asymmetrical wave marks are found on the bedding surfaces of the sandstones, and small-scale crossbeddings are common. From the bottom to top, the lithology changes from purplish red shales into thin bedded fine sandstones, and finally changes into grayish white thick bedded fine quartz sandstones of the overlying Dahongyu Formation.

The shale has an oxidizing color and is deposits of supratidal mudflat. The dip of the small-scale crossbeddings in the upper and lower sandstone layers is opposite, which indicates a bidirection current, so the thin interbeddings of the sandstones and shales are deposits of intertidal mixed flat.

From the bottom to top, the depositional environment changes from the supratidal mudflat to the intertidal mixed flat, and then to the subtidal shoals of the Dahongyu Formation, which is a transgression sequence.

In general, the Tuanshanzi Formation is carbonate-clastic mixed platform deposits.

4. Sedimentary facies interpretation of the Dahongyu Formation

The Dahongyu Formation is divided into 3 members: the 1ˢᵗ member (Chd₁), the 2ⁿᵈ member (Chd₂), and the 3ʳᵈ member (Chd₃), from the bottom to top.

The Chd₁ is dominated by medium-thick bedded milky white quartz sandstones, intercalated with purple red siltstones, sandy dolostones bearing with cherts, dolomitic sandstones and bluish green tuffs. The ripple marks on the bedding surfaces of quartz sandstones are common, some parallel beddings and cross beddings are developed, and the grains are well sorted, rounded, and cemented with siliceous cement, suggesting a high-energy subtidal shoal deposit.

The Chd₂ is mainly a set of volcanic lava and volcanic breccia, intercalated with a small amount of quartz sandstones and tuffs. According to the associated strata, the eruption environment should be shore.

The Chd_3 is mostly comprised of medium- thick bedded grayish white dolostones containing cherts, and its bedding surfaces are undulating. Conical stromatolites are common in dolostones. Stromatolite is the indicator of tidal flat facies. Therefore, the sedimentary environment of this member is mainly the tidal flat on the carbonate platform. The chert is secondary, and replaces dolomites. The replacement occurred very early, not long after burial, that is, the penecontemporaneous stage. The acidic microenvironment formed by decaying of the algae is conducive to dolomite dissolution and siliceous precipitation, which is proved by the prior siliceous replacement of the stromatolites.

5. Sedimentary facies interpretation of the Gaoyuzhuang Formation

The Gaoyuzhuang Formation can be divided into six members: the 1st member (Jxg_1), the 2nd member (Jxg_2), the 3rd member (Jxg_3), the 4th member (Jxg_4), the 5th member (Jxg_5), and the 6th member (Jxg_6), from the bottom to up.

1) Sedimentary facies interpretation of the Jxg_1

The Jxg_1 is grey medium-thick beded stromatollites microcrystalline dolostones with chert ribbons and nodules, intercalated with grayish green thin bedded shales, and a quartz sandstone bed about 3m thick at the bottom, which shows parallel unconformity contact with the Dahongyu Formation. Small- and medium-scale symmetrical ripples are found in sandstones.

The stromatolites is mainly laminar, wavy and semi-spherical, which are indicators of the tidal flat environments. The bedding surfaces of the dolostones is undulating, some of which are ripple marks, and some are caused by wave-like and semi-spherical stromatollites. Therefore, the medium-thick bedded stromatolites microcrystalline dolostones with chert ribbons and nodules were mainly deposited in the tidal flat setting on the carbonate platform. The origin of the chert is the same as that in the Dahongyu Formation. The quartz sandstones at the bottom are well sorted, rounded, and cemented by silia, and have symmetrical ripple marks, indicating that they are subtidal shoal deposits. The sandy shoals are directly overlain by carbonate tidal flats, suggesting that there are sandy beach in the coast, which changed rapidly into carbonate platform in the direction of the sea, similar to the present east coast in Austrialia.The grayish green shale bed are shelf deposit, which represents a short transgression event, indicating that the carbonate platform was often submerged in deep water.

2) Sedimentary facies interpretation of the Jxg_2

The Jxg_2 consists of interbeddings of grayish brown manganese-bearing silty shales, manganese-bearing silty-dolostones and manganese-bearing siltstones, with well-developed horizontal beddings. The lower-middle parts of this member contains the "Jixian-type manganese (boron) ore".

This member was deposited in a deep shelf. Muddy deposits are dominant, indicating that the water body is quiet and low-energy, and the horizontal beddings is developed, indicating that the water depth is below the storm wave base. The color is grayish brown caused by manganese. Manganese is formed in subaqueous reducing environment, similar to siderite.

3) Sedimentary facies interpretation of the Jxg$_3$

The Jxg$_3$ is interbeddings of medium-thick bedded massive microcrystalline dolostones and microcrystalline dolostones with horizontal beddings, intercalated with few shales. The dolomite is ferous and show brown color after being weathered. The chert nodules and ribbons are rare, and are concentrated in the microcrystalline dolostones with horizontal beddings. Stromatolites are undeveloped.

The bedding surfaces of the strata in this interval are flat, commonly with horizontal beddings and rarely with stromatolites, which indicates that the deposition water body of these dolostones is deeper than that of the Jxg$_1$ but shallower than that of the shales in the Jxg$_2$, and belongs to the open (or restricted) platform below the the normal wave base. The water depth of the dolostones with horizontal beddings is deeper than that of the massive dolostones. The shale interbeddings represents transgression events, similar to that in the Jxg$_1$. The dolomite crystals are very fine, and contains Fe^{2+}, which is most likely to be formed by penecontemporaneous dolomitization or primary precipitation induced by microorganisms.

4) Sedimentary facies interpretation of the Jxg$_4$

The Jxg$_4$ consists of dark gray medium bedded limy dolostones and dolomitic limestones intercalated with thin bedded argilliferous limy dolostones and dolomitic limestones. Although there are some chert nodules and ribbons, but are rare.

Medium bedded limy microcrystalline dolostones and dolomitic micritic limstones are massive, without beddings, grains and stromatolites, and their bedding surfaces are relatively flat (Fig.89), which both suggest that the depositional environment is quiet and low-energy, and the water depth is below the normal wave base. The environments should be open platform. Lack of bedding may be due to bioturbation.

Fig.89   Massive dolomitic micritic limestones bearing chert nodules and argilliferous dolomitic micritic limestones with horizontal beddings in the Jxg$_4$

Dark gray argilliferous limy dolostones and dolomitic limestones show not only horizontal beddings (Fig.89), but also some small-scale symmetrical ripple marks which made the thin beds look like pod or chain. The mud content is high and the horizontal bedding is developed, which indicates that they were deposited on gentle slope below the storm wave base, while the argilliferous limy dolostones and dolomitic limestone with ripple marks were formed near the storm wave base.

In general, this formation is alternately deposited between open platform and gentle slope.

The nodular micritic limestones at the bottom of this member may be related to the microorganisms. The "molar" limestones or "chicken-claw" limestones at the top should be formed by filling incompletely developed mud cracks. This indicates that at the end of the member sedimentation, the water body was shallow, and overlain by the tidal flat deposits in the Jxg_5.

5) Sedimentary facies interpretation of the Jxg_5

The lower part of the Jxg_5 is the laminar bituminous fine crystalline dolostones and fine crystalline limestone. The upper part consists of coarse crystalline dolostones with oncolites, interbedded with laminar medium-fine crystalline dolostones and the microcrystalline dolostones. The laminae are laminar stromatolites.

The existence of a large number of laminar stromatolites and oncolites indicates that the depositional environment is the tidal flat on the carbonate platform. Fine crystalline dolostones is hydrothermal dolostones, which is related to the deep fault nearby. Limestones are generally recrystallized, which is also the result of the reaction between rocks and hydrothermal fluids.

6) Sedimentary facies interpretation of the Jxg_6

There are gray medium-thick bedded microcrystalline dolostones with chert nodules and ribbons. The laminar and wavy stromatolites are common, which are carbonate tidal flat deposits.

In summary, the whole Gaoyuzhuang Formation is very thick, and has undergone many environmental changes. The Jxg_1, Jxg_5, and Jxg_6 are mainly tidal-flat deposits on the carbonate platform. The Jxg_2 is deep-water shale shelf deposits. The Jxg_3 is open platform deposits. The Jxg_4 is mainly open platform and gentle slope deposits. The transgression started from the Jxg_1 to Jxg_2, then the regression started from the Jxg_2 to Jxg_3, then the transgression started again from the Jxg_3 to Jxg_4, then the regression started again from the Jxg_4 to Jxg_6. The maximum flooding surface of the Gaoyuzhuang Formation is in the Jxg_2.

6. Sedimentary facies interpretation of the Yangzhuang Formation

The Yangzhuang Formation is characterized by distinctive purplish red and grayish white mudstones, looking like the "streaky pork". The mudstones often contains silts and microcrystalline dolomite. This formation is in sharp contact with the dolostones in the underlying Gaoyuzhuang Formation, which may be a parallel unconformity.

The formation can be divided into 3 members: the 1st member (Jxy_1), the 2nd member (Jxy_2), and the 3rd member (Jxy_3), from the bottom to top. The Jxy_1 consists of interbeddings of red and white mudstones, which commonly contains silts and microcrystalline dolomite and is 209 m in thickness. The Jxy_2, with a thickness of 434 m, is the red intercalated with white mudstones, which also often contains silts and microcrystalline dolomites and contains a tens of centimeters thick coarse sandstone bed in the upper part. The Jxy_3 is comprised of interbeddings of the red and white mudstones intercalated with gray massive algae dolostones, limestones and stromatollites dolostones, with a thickness of 130 m.

The purplish red mudstone represents the exposed oxidizing environment, and mud cracks can be seen in some places, suggesting tidal flat in arid climate.

Grayish white mudstone represents a low-energy subaqueous reducing environment, such as a restricted sea or lagoon.

Occasional sandstone is subtidal shoal deposits.

The gray massive algal dolostones and limestones were deposited in the subtidal zone, and the laminar stromatolites microcrystalline dolostone is tidal-flat deposits.

Generally speaking, Yangzhuang Formation is dominated by deposits of muddy tidal flat, which alternates frequently with shallow restricted sea or lagoon, controlled by the frequent fluctuations of sea level.

7. Sedimentary facies interpretation of the Wumishan Formation

The Wumishan Formation is of huge thickness, but the lithology is monotonous, and especially developed are the meter-scale cycles. The typical cyclic sequence consists of greenish gray dolomitic shales, gray clot-like (massive) fine microcrystalline algae dolostone, laminar stromatolite microcrystalline dolostones with cherts, from the bottom to top.

The shales are in weakly reducing color, and were deposited in subaqueous low-energy setting, representing shallow water restricted sea, located between carbonate platform and land.

The gray clot-like fine microcrystalline algae dolostones are deposits of subtidal zone, and the laminar stromatolite microcrystalline dolostones with cherts are tidal flat deposits.

The depositional water is gradually shallowing from grayish green dolomitic shales to gray clot-like fine microcrystalline algae dolostones, and laminar stromatolite microcrystalline dolostones with cherts.

A small amount of purplish red mudstones are found in the upper part of the Wumishan Formation, which represents muddy tidal flat deposits (supratidal zone).

There are occasionally grayish white quartz fine sandstones in the Wumishan Formation, which are well sorted and rounded and are deposits of subtidal shoal.

8. Sedimentary facies interpretation of the Hongshuizhuang Formation

The Hongshuizhuang Formation is mainly a set of shales, which is in conformity contact with the Wumishan Formation.

From the bottom to the top, it can be divided into the 1st member (Jxh_1) and the 2nd member (Jxh_2). The Jxh_1 is the medium bedded microcrystalline dolostones intercalated with grayish green, bluish green and gray shales. Stromatolites can always be found in the dolostones, which are deposits of carbonate platform, while the shales are shelf deposits. The frequent interbeddings between the two indicates that sea level oscillations are frequent. The Jxh_2 is mainly black and grayish green shales, with muddy microcrystalline dolostone lenses (algal reef) and thin siltstones at the top. The shales are in reducing color, with well-developed horizontal beddings, suggesting a quiet low-energy reducing setting, i.e., shale shelf. Upwards, the deposits of shale shelf change into mouth-bar sandstones in delta front and coastal sandstones of the Tieling Formation.

In a word, after the deposition of the Wumishan Formation, transgression occurred and the sea water gradually deepened. After experiencing the transitional stage of interactive deposition between platform dolostones and shelf shales in the Jxh_1, it was completely transformed into shale shelf in the Jxh_2, and finally changed to delta front mouth-bar sandstones or coastal sandstones in the Tieling Formation.

9. Sedimentary facies interpretation of the Tieling Formation

The Tieling Formation was conformable with the underlying Hongshuizhuang Formation, and can be divided into 3 members: the 1st member (Jxt_1), the 2nd member (Jxt_2), and the 3rd member (Jxt_3), from the bottom to top.

1) Sedimentary facies interpretation of the Jxt_1

The Jxt_1 is 155m in thickness. Its bottom is grayish white medium-thin bedded fine quartz sandstones changing upwards into gray manganese-bearing algal microcrystalline dolostones, and most of Jxt_1 is medium-thick bedded gray dolostones with stromatolites intercalated with grayish green and blueish green shales. Its top is blueish green glauconitic shales and ferromanganese-bearing dark brown shales.

On the outcrop of Xiaolingzi village in the northwest of Fujunshan Park, Jizhou District, quartz fine sandstones are well sorted, with parallel beddings, and have gradual contact with the shelf shales of the Hongshuizhuang Formation, showing a coarsening-upward sequence and belonging to mouth-bar deposits of the delta front. On the outcrop of about 2 km east of Fujunshan Park and 1 km south of Siwa Village, there are symmetrical wave marks on the bottom surfaces of quartz sandstones, which has an abrupt contact with the underlying black shale, and should be coastal sandstones.

The medium-thick bedded gray stromatolite dolostones are tidal flat deposits on the carbonate platform, while the thin grayish green and blue-green shales represent the temporary flooding.

The relatively thin bedded blue-green glauconitic shales and ferromanganese-bearing dark brown shale at the top represent temporary shale shelf deposits.

The tectonic uplifting occurred after the deposition of the Jxt_1, and the strata were subjected to weathering and denudation, forming a parallel unconformity between the Jxt_1 and Jxt_2.

2) Sedimentary facies interpretation of the Jxt_2.

During the deposition of the Jxt_2, the area subsided to accept deposits mainly composed of dolomud ribboned micritic limestones intercalated with wormkalk calcirudites and few stromatolite limestones, and chert nodules locally occurred. The existence of stromatolites and wormkalk calclrudites indicate that the water body is not deep and they are mainly tidal flat deposits on carbonate platform. The wormkalk calcirudites were deposited on the tidal flat during storms. Locally, some deposits of small tidal channels can be found. The thin bedded microcrystalline dolostones, showing brown color after being weathered, should be formed by evaporative pumping dolomization on the tidal flat.

3) Sedimentary facies interpretation of the Jxt₃

The Jxt₃ consists of stromatolite limestones, which are the most abundant and well-developed in this area in the Proterozoic.

There are many types and shapes of stromatolites, including laminar, wavy, columnar and wall-like stromatolites. A large number of stromatolites indicate that the Jxt₃ was deposited in a tidal flat and shallow subtidal zone. The fillings between the columnar stromatolites are limemud or grayish green mud, without grains, which indicates that the energy of water body is not high. The fillings are often dolomitized. Among them, the wall-like stromatolites can indicate the direction of tidal current, which is parallel to the extension direction of the stromatolites.

The stromatolites body of the Jxt₃ is a large stromatolite reef. It is thickest in Jizhou area, and becomes thin or disappear in other areas.

After the deposition of the Jxt₃, tectonic uplifting occurred and the strata suffered from weathering and denudation, forming a parallel unconformity between it and the Xiamaling Formation.

10. Sedimentary facies interpretation of the Xiamaling Formation

The Xiamaling Formation is in parallel unconformity contact with the Tieling Formation. It mainly consists of grayish green shales. At the bottom there are several meters thick gray shales intercalated with sandstones. Gravels appeared locally.

Sandstones are coastal deposits and shales are shelf deposits. The basis of analysis is the same as above.

The whole Xiamaling Formation is a transgression sequence.

11. Sedimentary facies interpretation of Longshan Formation

The Longshan Formation has an abrupt contact with the underlying Xiamaling Formation, which is an uneven erosion surface and is likely a parallel unconformity.

The Longshan Formation can be divided into 2 members: the 1st member (Qbl_1) and the 2nd member (Qbl_2).

The Qbl_1 is medium-thick bedded glauconitic sandstones, with tabular and wedge crossbeddings. Glauconite is a typical marine mineral. This formation is typical coastal deposits.

The Qbl_2 is mainly grayish green and grayish black shales, intercalated with thin bedded fine glauconitic sandstones, which are mainly shale shelf deposits. The top several meters are purplish red mudstones, and the purplish red is a strongly oxidizing color, representing the exposed oxidizing environment, so they should be tidal mud flat deposits.

The whole Longshan Formation is a sequence of transgression and regression.

12. Sedimentary facies interpretation of the Jingeryu Formation

There are about ten centimeters thick coarse glauconitic sandstones with fine gravels at the bottom of this formation; the lower part is mainly grayish white and grayish purple medium-thin bedded micritic limestones; the middle part is grayish, green medium-thin bedded muddy micritic limestones intercalated in micritic limestones; and the upper is gray thin bedded grayish green shales intercalated in muddy micritic limestones.

The formation is generally relatively deep-water open carbonate platform deposits, because the bed is thin, the bedding surfaces are relatively flat, and the grayish green shales are intercalated in it.

13. Glacial deposits in the Xijingyu Formation

The Xijingyu Formation is a new formation established by the author, which is a set of massive carbonate breccia, with a thickness of 155m. The author believes that it is glacial deposits.

The strata of this formation were previously classified to the Fujunshan Formation of the Lower Cambrian, but its lithologic characteristics are extremely different from those of the midium-thin bedded microcrystalline and fine crystalline dolostones of the overlying Lower Cambriain Fujunshan Formation. Therefore, the author thinks that it is necessary to build a new formation. Because the section of this set of strata is located near the Xijingyu village, it is called "Xijingyu Formation".

The Xijingyu Formation shows abrupt contacts both with the underlying Jingeryu Formation and the overlying Fujunshan Formation. The abrupt contact between the Xijingyu Formation and the Jingeryu Formation is considered as a parallel unconformity or microangular unconformity, corresponding to the tectonic activity called "Jixian Movement".

The abrupt contact between the Xijingyu Formation and the overlying Fujunshan Formation is likely to be a parallel unconformity contact. However, due to the lack of fossils and other dating minerals, it is uncertain whether there is a lack of strata between the two formations.

The age of Xijingyu Formation should be younger than the Jingeryu Formation, and elder than the Fujunshan Formation. Although there is no definite evidence to prove whether it belongs to the Neoproterozoic or the early Cambrian, it is more likely to belong to the Neoproterozoic. We did not find any trilobites or other fossils in this set of breccia. If it belongs to the Cambrian, there should be fossils. Moreover, flint breccia is common in breccia, but there are no flint nodules or ribbons in the Cambrian system in North China. This evidence suggests that the parent rock cannot be Cambrian. According to the composition of breccia, the parent rocks are mainly from the Wumishan Formation, and a small amount are from Tieling Formation. According to this, it is speculated that its age belongs to the Neoproterozoic rather than Cambrian.

1) Textural characteristics of breccia

(1) Particle size, sorting and roundness of breccia.

The breccia size of the Xijingyu Formation is very different, the diameter of the largest breccia is about 1m, and the whole section is dominated by breccia with diameter of 1-5cm. The breccia with large sizes are unevenly distributed laterally and vertically, and the phenomenon of local concentration is common (Fig.90A). The breccia are very poorly sorted (Fig.90B), even should be described as "no sorting". The roundness of breccia is very low, which is generally angular and sub-angular (Fig.90C, D). Breccia has a variety of shapes, such as rhomboid, square, triangular, strip-shaped, or irregular (Fig.90F). Compared with carbonate breccia, flint is mainly strip-shaped in the outcrops. Dolostone breccia and limestone breccia is mainly rhomboid, square and triangular. Dolostone breccia surface shows chopping mark structure (Fig.90E). From as spect of sorting and roundness of breccia, the textural maturity of breccia is very low.

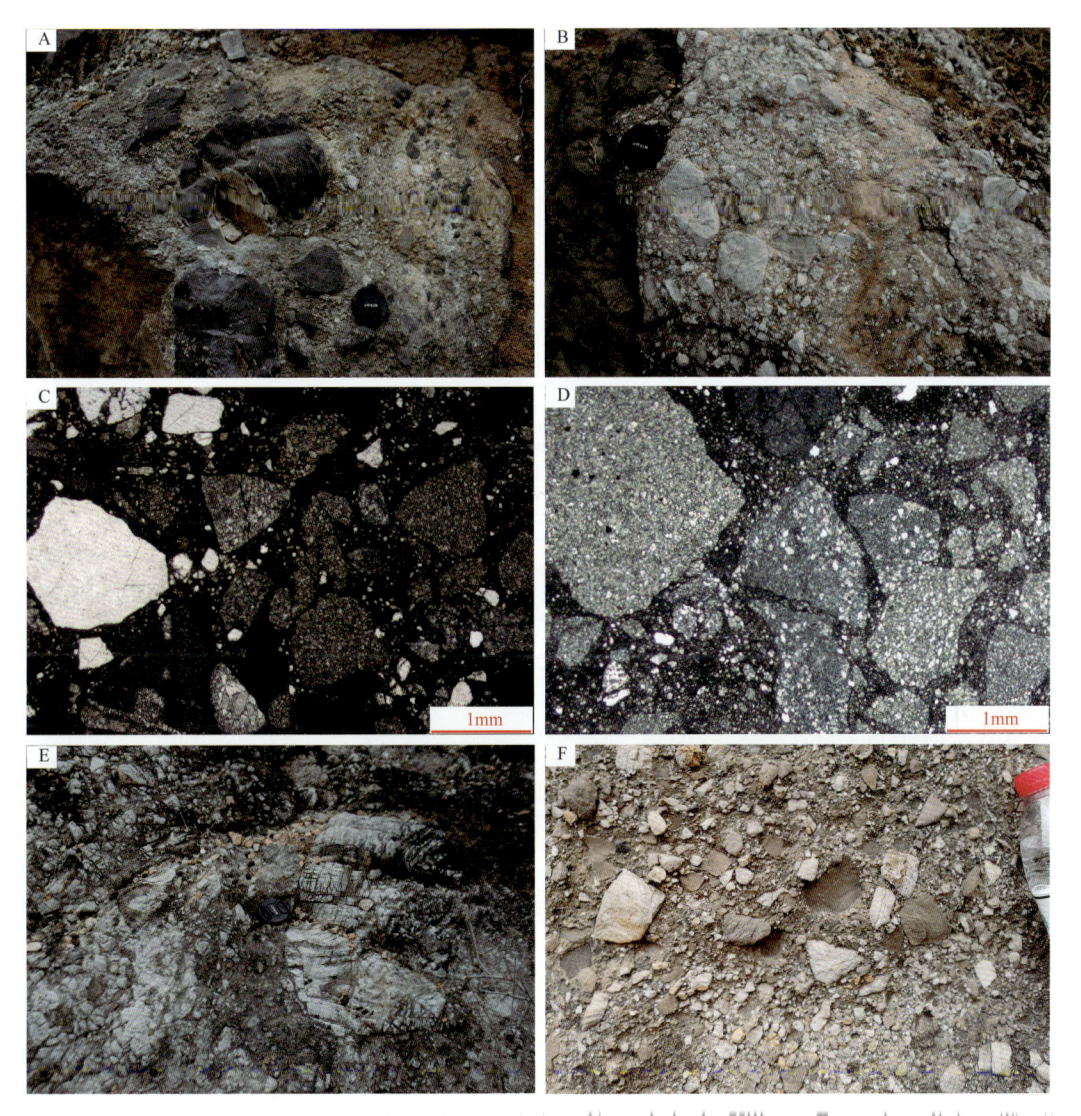

Fig.90 Size, sorting, roundness and shape characteristics of breccia in the Xijingyu Formation, Jixian, Tianjin
(A) Concentrated distribution of large breccia in the Xijingyu Formation; (B) Mixed breccia of the Xijingyu Formation; (C) Characteristics of breccia under microscope. Angular, poorly sorted, plane polarized light; (D) Microscopic characteristics of breccia, generally dolomitic breccia, poor sorting, plane polarized; (E) Dolomite breccia in breccia; (F) Breccia of various shapes, dominated by square, rhomboid, rectangular shapes

(2) Supporting type and characteristics of fillings.

The breccia in the study area is grain-supported, with high content of breccia, about 80%–90% (Fig.90). Most of the grains are point-contact with each other (Fig.91).

The fillings between breccia are the matrix, with content about 10%–20%. The matrix is mainly terrigenous dolomitic (a small amount of limy) muds and silts (Fig.91A, B, D, E), basically undyed (alizalin red), dark under the microscope; locally contain quartz silt (Fig.91C).

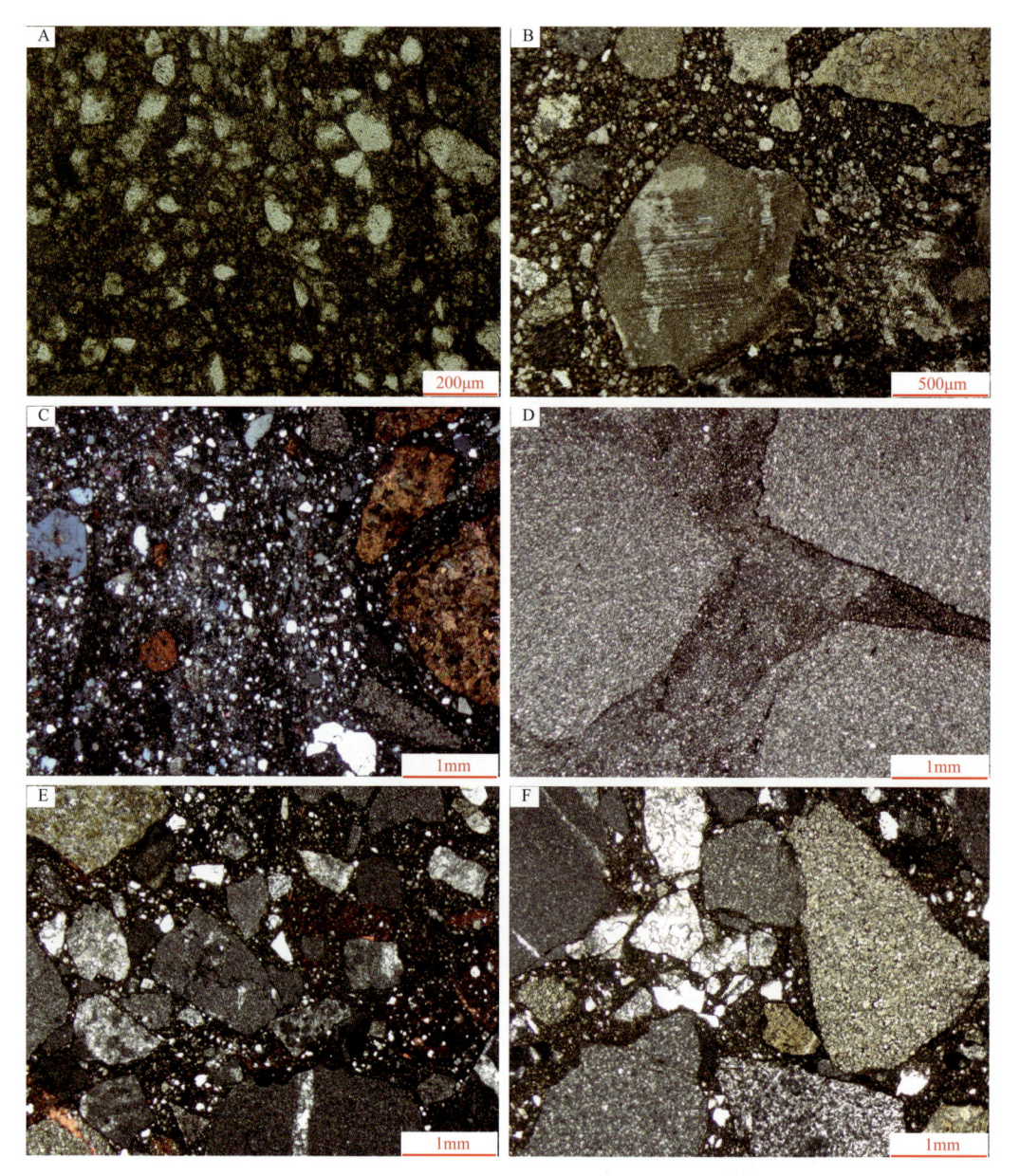

Fig.91　Characteristics of the fillings and supporting type of breccia in the Xijingyu Formation, Jizhou District, Tianjin
(A) Fine silt size dolomite fillings, plane polarized light;(B) Dolomite breccia, containing dolomite muds and silts, and a small amount of clay, plane polarized light;(C) Limestone breccia(dyed red)and various-size quartz particles fillings, cross polarized light;(D) Dolomitic matrix filled between the dolostone breccia, plane polarized light;(E) Dark dolomud and silts matrix in the breccia, containing little limemud, plane polarized light;(F) Grain-supported dolostone breccia with a dark dolomud and silts matrix, plane polarized light

(3) Composition of breccia.

According to the field observation and identification under microscope, the compositions of breccia in the study area are mainly dolostones, some are limestones and flints, and a very small amount of igneous rock fragments are also found. The average content of dolostone breccia is about 80% (Fig.92). Limestone and flint breccia are nonuniformly distributed vertically.

Fig.92 Characteristics of dolostone breccia and limestone breccia in the Xijingyu Formation, Jixian, Tianjin

(A) Microcrystalline dolostone breccia with horizontal beddings, plane polarized light; (B) Microcrystalline dolostone breccia, plane polarized light; (C) Microcrystalline dolostone breccia, plane polarized light; (D) Medium-fine crystalline dolostone breccia, plane polarized light; (E) Coarse crystalline dolostone breccia, cross polarized light; (F) Doloarenite breccia, plane polarized light; (G) Oolitic dolostone breccia, plane polarized light; (H) Coarse crystalline limestone breccia (dyed red), cross polarized light, sample; (I) Microstalline limestone breccia (dyed red), plane polarized light

The types of dolostone breccia in the Xijingyu Formation are various. According to the crystal size, the dolostone breccia consisting of microcrystals to coarse crystals all occurred, but are dominated by microcrystals (Fig.92A-E). According to sedimentary structures, microcrystalline dolostone breccia can be divided into two types, namely microcrystalline dolostone breccia with horizontal bedding (Fig.92A) and homogeneous microcrystalline dolostone breccia (Fig.92B). In addition, two kinds of grain dolostones are developed, namely, sparry doloarenite breccia (Fig.92) and sparry oolitic dolomite breccia (Fig.92G). These dolostone breccia with special grains and sedimentary structures provide a basis for further analysis of their provent rock types.

The limestone breccia of the Xijingyu Formation are relatively few, dominated by coarse crystalline and microcrystalline limestone breccia (Fig.92H and I). Because the lithology of the Meso- and Neoproterozoic carbonate strata underlying the Xijingyu Formation is generally dolostone, and limestones are not common, this results in rare limestone breccia in the Xijingyu Formation. But there are a local concentrated zones of limestone breccia.

Chert breccia are common in the breccia of the Xijingyu Formation (Fig.93). Compared with carbonate breccia, chert has strong weatherproof ability and protrudes out farm the surface of breccia, resulting in uneven surface of breccia. Chert is generally white under plane polarized light (Fig.93A), and some chert contains some dolomite crystals, which show as mottles (Fig.93C). Some cherts are microcrystal line (Fig.93D), while others are cryptocrystal line (Fig.93E and F). The structure of parent rock is often preserved in the process of replacement by, so siliceous oolitic dolostone and flint breccia with horizontal beddings are common in the study area (Fig.93G). The cortex structure of ooids in oolitic dolomite breccia is obvious, and the fillings are commonly replaced by flint, while the ooids are generally dolomitized. The flint breccia with horizontal beddings is the result of replacement of the horizontal-bedding microcrystalline dolostone by flints. The incomplete replacement results in the alternate laminae of the white flint and the residual microcrystal dolomite (Fig.93H). There are very few igneous breccia in the breccia of the Xijingyu Formation, and they are diabase breccia with typical diabasic texture under microscope (Fig.93I). They mainly come from the diabase intrusions in the underlying Meso- to Neo-proterozoic strata.

2) Discussion

(1) The provenance of breccia.

According to the composition analysis, the parent rock of the breccia in the Xijingyu Formation is the Meso- and Neo-proterozoic strata underlying it. The composition of breccia is mainly carbonate rock, which is mainly determined by the Neoproterozoic thick carbonate strata in the underlying strata. Although the Meso- and Neo-proterozoic are generally carbonate rocks, the characteristics of rocks in different formations are also very obvious. The correlation between the types of breccia and the rock characteristics of each formation can provide a information about the provenance of breccia.

The Jingeryu Formation is mainly grayish green thin micritic limestones, and the Changlongshan Formation is mainly glauconitic quartz sandstones and shales. The Xijingyu Formation is in direct contact with the Qingbaikou strata, but there are no quartz sandstone in the breccia except few micritic limestone breccia. The limestone breccia is gray, not grayish green, which suggests that the Qingbaikou System is not the provenance of breccia.

The Xiamaling Formation is a set of strata dominated by grayish green shales, with a few of sandstones at the bottom. However, through field observation and microscopic identification, the breccia of the Xijingyu Formation does not contain shales breccia, and even the fillings of breccia is mainly microcrystalline dolomite mud and silt, while clay minerals are very few. These characteristics suggest that the Xiamaling Formation is not the provenance of the breccia.

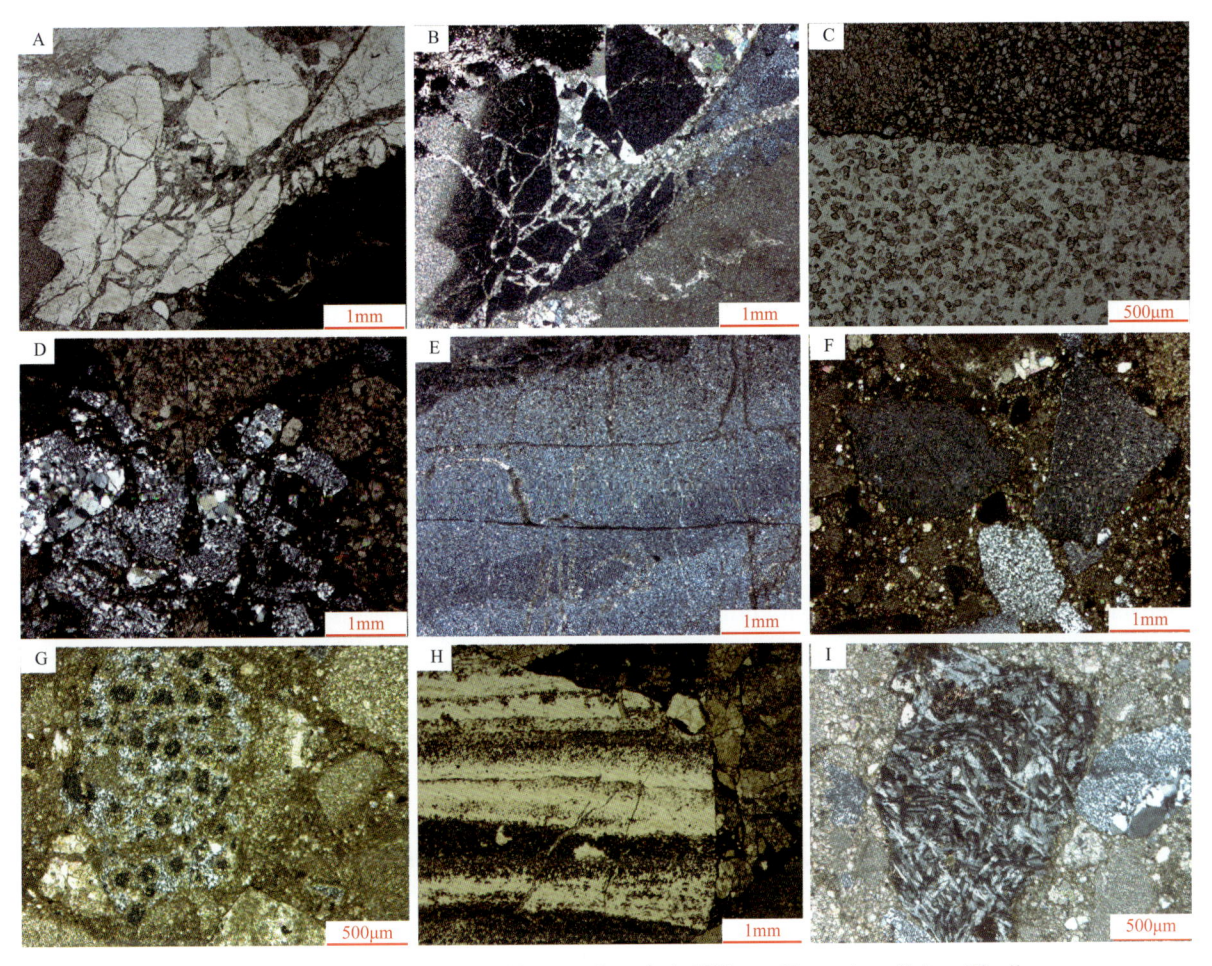

Abstract Field Guide of Stratigraphy and Sedimentary Facies of the Proterozoic in Jizhou District, Tianjin

Fig.93 Characteristics of flint and igneous breccia in Xijingyu Formation, Jixian, Tianjin
(A) Broken flint breccia, white under plane polarized light, plane polarized light; (B) Broken flint breccia, cryptic under cross polarized light, with the same field of view as a cross polarized light; (C) Flint breccia indicating floating powder dolomite particles, plane polarized light; (D) Millet flint, cross polarized light; (E) Cryptic flint breccia, cross polarized light; (F) The chert of cryptocrystalline matter and small rice grain structure, cross polarized light; (G) Chert dolomite breccia, cross polarized light; (H) Horizontal bedding argillaceous dolomite chert breccia, plane polarized light; (I) Diabase breccia, cross polarized light

The Tieling Formation is dominated by limestones, and there are many types of stromatilites, while some huge stromatilite limestone breccia with a diameter of about 1m has been found in the breccia of the Xijingyu Formation. The characteristics of the stromatolite are consistent with those of the Tieling Formation.

The Hongshuizhuang Formation is dominated by grayish green and black shales, and is obviously not the provenance of breccia in the Xijingyu Formation.

There are thousands of meters thick dolostones with chert ribbons in the Wumishan Formation, and dolostones from micro-crystalline to coarse crystalline all exist, including oolitic dolostones, doloarendites and so on. Sandstones are occasionally found locally. The characteristics of dolostone breccia and chert breccia in the Xijingyu Formation are similar to those in dolostones with chert ribbons of the Wumishan Formation, so the Wumishan Formation is one of the important provenances.

Yangzhuang Formation is mainly composed of red dolomitic mudstones interbedded with light gray dolomitic mudstone. However, such breccia in the Xijingyu Formation is rare, so the possibility of this composition as the main source is very low.

Dolostones with chert ribbons in the Gaoyuzhuang Formation occur in the lower and upper part, limestones and limy dolostones are in the middle part, and the lithologic characteristics are similar to those breccia in the Xijingyu Formation, but because of the lack of breccia from the Yangzhuang Formation, and the Yangzhuang Formation is overlying the Gaoyuzhuang Formation, so it can be speculated that it is very unlikely that Gaoyuzhuang Formation will become the main provenance of Xijingyu Formation.

The lower part of the Dahongyu Formation is quartz sandstones, the middle part is volcanic breccia and volcanic rocks, the upper part is dolostones with chert ribbons; the lower part of the Tuanshanzi Formation is dolostones, the middle part is dolomitic sandstone, the upper part is purplish red dolomitic mudstones with intercalated sandstones; the Chuanlinggou Formation is a large set of shales, and the Changzhougou Formation is large set of sandstones. There are neither sandstone breccia nor volcanic breccia in the breccia of the Xijingyu Formation, and these formations are all overlain by the Yangzhuang Formation, so they can not be their provenance.

In summary, the dolostone and chert breccia of the Xijingyu Formation mainly come from the dolostone with chert ribbons of the Jixian System Wumishan Formation in the Mesoproterozoic, and the limestone breccia comes from limestones of the Tieling Formation. This means that the strata exposed in the provenance area are mainly the Wumishan Formation and the Tieling Formation. Because of the parallel unconformity, there are no folds in the provenance area which should be mainly controlled by faults. The valley of the provenance area should bedeep, cutting through the Tieling Formation and into the Wumishan Formation.

(2) Genesis of breccia.

There are two possibilities for the genesis of breccia in the Xijingyu Formation: glacial and debris flow. The authors believe that the breccia in this area is not debris flow deposit, but glacial deposit, i.e. moraine rock.

Although the conglomerate of debris flow deposit is similar to that of moraine, such as the disorderly arrangement of gravel, poor sorting and poor rounding, some evidence in this area show that the breccia of the Xijingyu Formation is not debris flow deposition. The evidence are as follows:

① The pore-filling is not muddy. Debris flow deposits can form conglomerate, but their pore-filling is usually muddy. Mud is formed by mixing mud with water, and its lubrication causes debris flow to flow forward under the gravity. However, the pore-filling of breccia in this area is mainly dolomite mud and silt. This matrix is difficult to form by carbonate weathering, because it will be dissolved. This matrix can only be formed by grinding of carbonate rock, and flow of glacier can grind carbonate bedrock into dolomite mud and silt.

② All gravels are poorly rounded. Although debris flow deposits show poor rounding, most of them are angular or sub-angular, but it is also common to have gravel with good rounding, which is rounded in mountainous rivers. However, there is no rounded gravel in the breccia in this area.

③ There is no graded bedding. Reverse graded bedding is often found in debris flow deposits. The breccia up to 155m thick in this area is all massive and there is no reverse graded bedding.

④ There is no clear bed surface. The thickness of debris flow deposits of one time is mostly several meters. When debris flow deposits of different time are superimposed, there will be obvious bed surface. However, the breccia up to 155m thick in this area is all massive, and no layer surface can be found in the interior. Even at the contact of breccia with different grain sizes, there is no clear bed surface at all.

⑤ The breccia are not associated with other water deposits. Debris flow deposits are often associated with alluvial fans or other current deposits. However, all the strata up to 155m in this area are breccia arranged disorderly.

⑥ The slope is too gentle. Before breccia deposition, the terrain in this area should be very flat, because unconformity is parallel unconformity. In the Xishan area of Beijing, there is also parallel unconformity between the Jingeryu Formation and Fujunshan Formation, which indirectly supports this view. Moreover, strata overlying and underlying the Xijingyu Formation are shallow carbonate platform deposits and there are no steep slopes. It is difficult for debris flow to flow on flat terrain and it is difficult to form thick deposits.

According to the above evidence, the author believes that the breccia in this area is not debris flow deposits. The evidence for the glacial origin are as follows:

① The rock type is breccia, and the large gravels are mixed with small ones. (the diameter of the large one can reach about 1m); the arrangement is disordered, the sorting is very poor, the rounding is very poor, and the pore filling material is dolomite mud and silt. These characteristics are consistent with glacial deposits. Breccia is mainly formed by physical weathering and glacial erosion, and is frozen into glaciers and transported in glaciers. After melting of glaciers, large and small breccia is released and accumulated, resulting in mixed breccia size, disordered arrangement, poor sorting and poor rounding. The matrix as a filling material is mainly formed by the grinding between breccia at bottom of glacier and bedrock.

② It is massive, and the 155m thick strata have no bedding and no bed surface. Although glaciers can deposit in many episodes, there is no bed surface between them, which is significantly different from normal current deposits, eolian deposits or event deposits. However, because the eroded strata in the parent rock area can change with time, the grain size and composition of sediments can be different in each stage. There are more limestone breccia in the breccia of the lower Xijingyu Formation, which may be due to the exposure of the Tieling Formation in the parent rock area. With the denudation of the Tieling Formation, the limestone breccia basically disappeared.

③ No glacial striations were found on the gravel surface in the study, but this is not surprising, because most glacial gravels do not have striations. Only gravels frozen at the bottom of the glacier could have striations, and the amount of such gravels is small and it is not easy to find. Moreover, the breccia is mainly carbonate rock, which is easy to be affected by weathering, so it is difficult to preserve glacial striations.

In conclusion, based on the texture, structure and rock assemblage of breccia, the breccia of the Xijing Formation is tillite.

The discovery of these moraine rocks is of great significance to the restoration of the Neoproterozoic climatic characteristics. As it is shown by the paleomagnetic data that the North China Plate was in low latitude tropics or subtropical regions, and a large amount of carbonate deposits were formed in the Jingeryu and Fujunshan Formations during this period. But why did moraine rocks form in low latitudes? Maybe it was related to snowball earth.

附录

实测地质剖面记录表

日期：　　　　第　页

填表人：

倾角换算表

岩层走向与导线方位间夹角

真倾角＼视倾角	80°	75°	70°	65°	60°	55°	50°	45°	40°	35°	30°	25°	20°	15°	10°	5°	1°
10°	9°51'	9°40'	9°24'	9°5'	8°41'	8°13'	7°41'	7°6'	6°28'	5°46'	5°2'	4°15'	3°27'	2°37'	1°45'	0°53'	0°10'
15°	14°47'	14°31'	14°8'	13°39'	13°34'	12°28'	11°36'	10°4'	9°46'	8°44'	7°38'	6°28'	5°14'	3°33'	2°40'	1°20'	0°16'
20°	19°43'	19°23'	18°53'	18°15'	17°30'	16°36'	15°35'	14°25'	13°10'	11°48'	10°19'	8°45'	7°6'	5°23'	3°37'	1°49'	0°22'
25°	24°48'	24°15'	23°39'	22°55'	22°0'	20°54'	19°39'	18°15'	16°41'	14°58'	13°7'	11°9'	9°3'	6°53'	4°37'	2°20'	0°28'
30°	29°37'	29°9'	28°29'	27°37'	26°34'	25°13'	23°51'	22°12'	20°21'	18°19'	16°6'	13°43'	11°10'	8°30'	5°44'	2°53'	0°35'
35°	34°36'	34°4'	33°21'	32°24'	31°13'	29°50'	28°12'	26°20'	24°14'	21°53'	19°18'	16°29'	13°28'	10°16'	6°56'	3°30'	0°42'
40°	39°34'	39°2'	38°15'	37°15'	36°0'	34°30'	32°44'	30°41'	28°20'	25°42'	22°45'	19°31'	16°0'	12°15'	8°117'	4°11'	0°50'
45°	44°34'	44°1'	43°13'	42°11'	40°54'	39°19'	37°27'	35°16'	32°44'	29°50'	26°33'	22°55'	18°53'	14°30'	9°51'	4°59'	1°0'
50°	49°34'	49°1'	48°14'	47°12'	45°54'	44°17'	42°23'	40°7'	37°27'	34°21'	30°47'	26°44'	22°11'	17°9'	11°41'	5°56'	1°11'
55°	54°35'	54°4'	53°19'	52°18'	51°3'	49°29'	47°35'	45°17'	42°33'	39°20'	35°32'	31°7'	26°2'	20°17'	13°55'	7°6'	1°26'
60°	59°37'	59°8'	58°26'	57°30'	56°19'	54°49'	53°0'	50°46'	48°4'	44°47'	40°54'	36°14'	30°29'	24°8'	16°44'	8°35'	1°44'
65°	64°40'	64°14'	63°36'	62°46'	61°42'	60°21'	58°40'	56°36'	54°2'	50°53'	46°59'	42°11'	36°15'	29°2'	20°25'	10°35'	2°9'
70°	69°43'	69°43'	68°49'	68°7'	67°12'	66°8'	64°35'	62°46'	60°29'	57°36'	53°57'	49°16'	43°13'	35°25'	25°30'	13°28'	2°45'
75°	74°47'	74°47'	74°5'	73°32'	72°48'	71°53'	70°43'	69°14'	67°22'	64°58'	61°49'	57°37'	51°55'	44°1'	32°57'	18°1'	3°44'
80°	79°51'	79°51'	79°22'	78°59'	78°29'	77°51'	77°2'	76°0'	74°40'	73°15'	70°34'	67°21'	52°43'	55°44'	44°33'	26°18'	5°31'
85°	84°56'	84°56'	84°41'	84°29'	84°14'	83°54'	83°29'	82°57'	82°15'	81°20'	80°5'	78°19'	75°39'	71°20'	63°15'	44°54'	11°17'
89°	88°59'	88°58'	88°56'	88°54'	88°51'	88°51'	88°42'	88°35'	88°27'	88°15'	88°0'	87°5'	87°5'	86°9'	84°15'	78°41'	44°15'

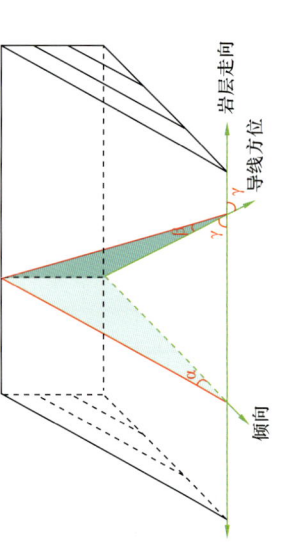

$\beta = \arctan[\tan\alpha \cdot \sin\gamma]$
β—视倾角
α—真倾角
γ—岩层走向与导线方位间夹角

剖面垂直比例尺放大后，岩层倾角大小歪曲结果表

垂向放大倍数 \ 歪曲视倾角 视倾角	1°	5°	10°	15°	20°	25°	30°	35°	40°	45°	50°	55°	60°	65°	70°	75°	80°	85°
×2	2.0°	9.9°	19.4°	28.2°	36.1°	43.0°	49.1°	54.5°	59.2°	63.4°	67.2°	70.7°	73.9°	76.9°	79.7°	82.4°	85.0°	87.5°
×3	3.0°	14.7°	27.9°	38.8°	47.5°	54.4°	60.0°	64.5°	68.3°	71.6°	74.4°	76.9°	79.1°	81.2°	83.1°	84.9°	86.6°	88.3°
×4	4.0°	19.3°	35.2°	47.0°	55.5°	61.8°	66.6°	70.4°	73.4°	76.0°	78.2°	80.1°	81.8°	83.4°	84.8°	86.2°	87.5°	88.7°
×5	5.0°	23.6°	41.2°	53.3°	61.2°	66.8°	70.9°	74.1°	76.6°	78.7°	80.5°	82.0°	83.4°	84.7°	85.8°	86.9°	88.0°	89.0°
×6	6.0°	27.7°	46.0°	58.1°	65.4°	70.3°	73.9°	76.6°	78.8°	80.5°	82.0°	83.3°	84.5°	85.6°	86.5°	87.4°	88.3°	89.2°
×7	7.0°	31.5°	51.0°	61.9°	68.6°	73.0°	76.1°	78.5°	80.3°	81.9°	83.2°	84.3°	85.3°	86.2°	87.0°	87.8°	88.6°	89.3°
×8	7.9°	35.0°	54.8°	65.0°	71.0°	75.0°	77.8°	79.9°	81.5°	82.9°	84.0°	85.0°	85.9°	86.7°	87.4°	88.1°	88.7°	89.4°
×9	8.9°	38.2°	57.9°	67.5°	73.0°	76.6°	79.1°	81.0°	82.5°	83.7°	84.7°	85.6°	86.3°	87.0°	87.7°	88.3°	88.9°	89.4°
×10	9.9°	41.2°	60.5°	69.5°	74.6°	77.9°	80.2°	81.9°	83.2°	84.3°	85.2°	86.0°	86.7°	87.3°	87.9°	88.5°	89.0°	89.5°

$|DB|=t \cdot |AB|$

$|BC|=\dfrac{|AB|}{\tan\alpha}$ $|BC|=\dfrac{|DB|}{\tan\beta}$

$\dfrac{|AB|}{\tan\alpha}=\dfrac{|DB|}{\tan\beta}=t \cdot \dfrac{|AB|}{\tan\beta}$

$\tan\beta=t \cdot \tan\alpha$

$\beta=\arctan(t \cdot \tan\alpha)$

含量估算模版

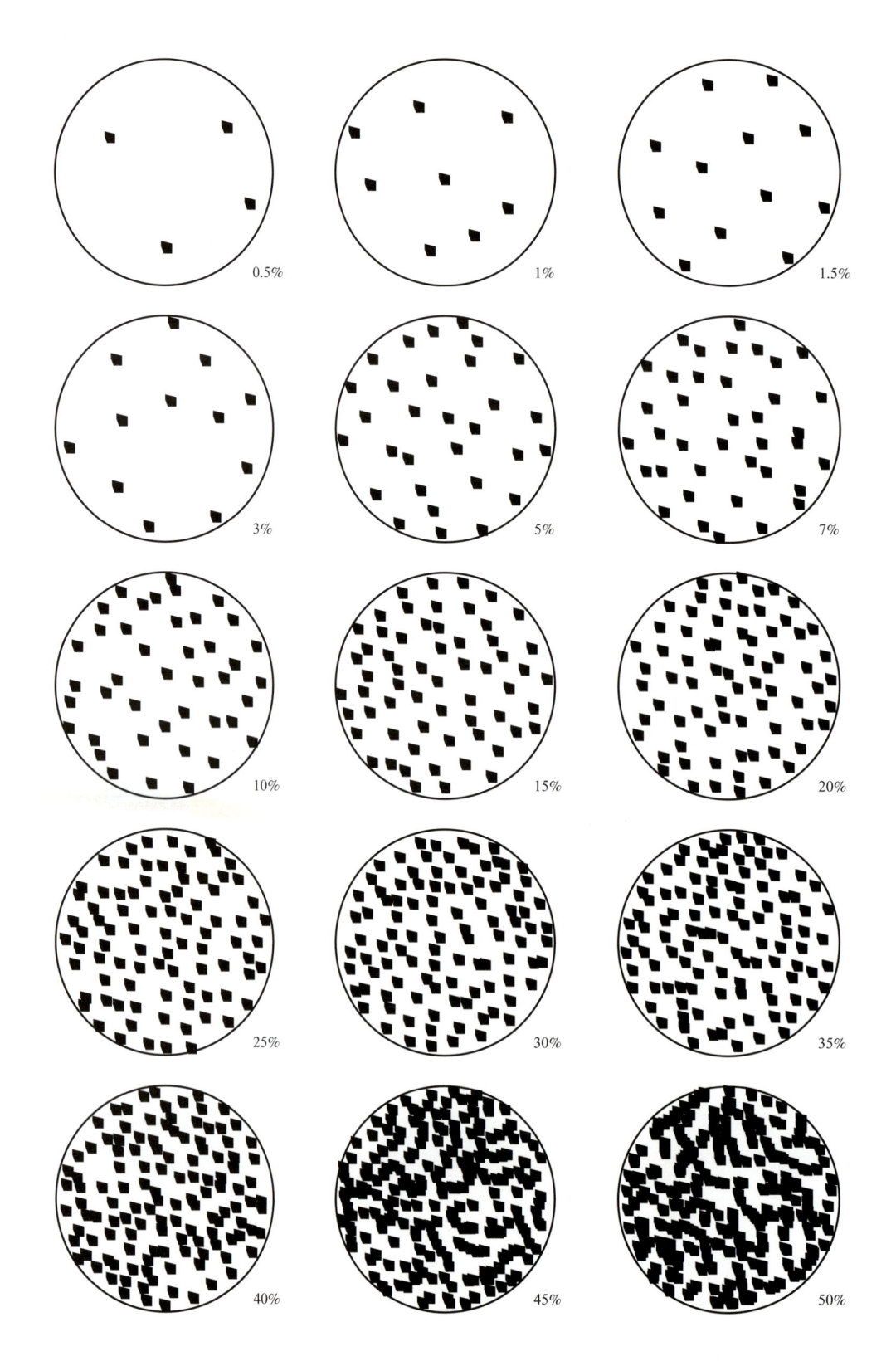

0.5%　1%　1.5%
3%　5%　7%
10%　15%　20%
25%　30%　35%
40%　45%　50%

INTERNATIONAL CHRONOSTRATIGRAPHIC CHART

www.stratigraphy.org — International Commission on Stratigraphy — v 2019/05

IUGS

Eonothem / Eon	Erathem / Era	System / Period	Series / Epoch	Stage / Age	numerical age (Ma)
					present
Phanerozoic	Cenozoic	Quaternary	Holocene	Meghalayan	0.0042
				Northgrippian	0.0082
				Greenlandian	0.0117
			Pleistocene	Upper	0.126
				Middle	0.773
				Calabrian	1.80
				Gelasian	2.58
		Neogene	Pliocene	Piacenzian	3.600
				Zanclean	5.333
			Miocene	Messinian	7.246
				Tortonian	11.63
				Serravallian	13.82
				Langhian	15.97
				Burdigalian	20.44
				Aquitanian	23.03
		Paleogene	Oligocene	Chattian	27.82
				Rupelian	33.9
			Eocene	Priabonian	37.8
				Bartonian	41.2
				Lutetian	47.8
				Ypresian	56.0
			Paleocene	Thanetian	59.2
				Selandian	61.6
				Danian	66.0
	Mesozoic	Cretaceous	Upper	Maastrichtian	72.1 ±0.2
				Campanian	83.6 ±0.2
				Santonian	86.3 ±0.5
				Coniacian	89.8 ±0.3
				Turonian	93.9
				Cenomanian	100.5
			Lower	Albian	~ 113.0
				Aptian	~ 125.0
				Barremian	~ 129.4
				Hauterivian	~ 132.9
				Valanginian	~ 139.8
				Berriasian	~ 145.0

Eonothem / Eon	Erathem / Era	System / Period	Series / Epoch	Stage / Age	numerical age (Ma)
					~ 145.0
Phanerozoic	Mesozoic	Jurassic	Upper	Tithonian	152.1 ±0.9
				Kimmeridgian	157.3 ±1.0
				Oxfordian	163.5 ±1.0
			Middle	Callovian	166.1 ±1.2
				Bathonian	168.3 ±1.3
				Bajocian	170.3 ±1.4
				Aalenian	174.1 ±1.0
			Lower	Toarcian	182.7 ±0.7
				Pliensbachian	190.8 ±1.0
				Sinemurian	199.3 ±0.3
				Hettangian	201.3 ±0.2
		Triassic	Upper	Rhaetian	~ 208.5
				Norian	~ 227
				Carnian	~ 237
			Middle	Ladinian	~ 242
				Anisian	247.2
			Lower	Olenekian	251.2
				Induan	251.902 ±0.024
	Paleozoic	Permian	Lopingian	Changhsingian	254.14 ±0.07
				Wuchiapingian	259.1 ±0.5
			Guadalupian	Capitanian	265.1 ±0.4
				Wordian	268.8 ±0.5
				Roadian	272.95 ±0.11
			Cisuralian	Kungurian	283.5 ±0.6
				Artinskian	290.1 ±0.26
				Sakmarian	293.52 ±0.17
				Asselian	298.9 ±0.15
		Carboniferous — Pennsylvanian	Upper	Gzhelian	303.7 ±0.1
				Kasimovian	307.0 ±0.1
			Middle	Moscovian	315.2 ±0.2
			Lower	Bashkirian	323.2 ±0.4
		Carboniferous — Mississippian	Upper	Serpukhovian	330.9 ±0.2
			Middle	Visean	346.7 ±0.4
			Lower	Tournaisian	358.9 ±0.4

Eonothem / Eon	Erathem / Era	System / Period	Series / Epoch	Stage / Age	numerical age (Ma)
					358.9 ±0.4
Phanerozoic	Paleozoic	Devonian	Upper	Famennian	372.2 ±1.6
				Frasnian	382.7 ±1.6
			Middle	Givetian	387.7 ±0.8
				Eifelian	393.3 ±1.2
			Lower	Emsian	407.6 ±2.6
				Pragian	410.8 ±2.8
				Lochkovian	419.2 ±3.2
		Silurian	Pridoli		423.0 ±2.3
			Ludlow	Ludfordian	425.6 ±0.9
				Gorstian	427.4 ±0.5
			Wenlock	Homerian	430.5 ±0.7
				Sheinwoodian	433.4 ±0.8
			Llandovery	Telychian	438.5 ±1.1
				Aeronian	440.8 ±1.2
				Rhuddanian	443.8 ±1.5
		Ordovician	Upper	Hirnantian	445.2 ±1.4
				Katian	453.0 ±0.7
				Sandbian	458.4 ±0.9
			Middle	Darriwilian	467.3 ±1.1
				Dapingian	470.0 ±1.4
			Lower	Floian	477.7 ±1.4
				Tremadocian	485.4 ±1.9
		Cambrian	Furongian	*Stage 10*	~ 489.5
				Jiangshanian	~ 494
				Paibian	~ 497
			Miaolingian	Guzhangian	~ 500.5
				Drumian	~ 504.5
				Wuliuan	~ 509
			Series 2	*Stage 4*	~ 514
				Stage 3	~ 521
			Terreneuvian	*Stage 2*	~ 529
				Fortunian	541.0 ±1.0

Eonothem / Eon	Erathem / Era	System / Period	numerical age (Ma)
			541.0 ±1.0
Precambrian — Proterozoic	Neo-proterozoic	Ediacaran	~ 635
		Cryogenian	~ 720
		Tonian	1000
	Meso-proterozoic	Stenian	1200
		Ectasian	1400
		Calymmian	1600
	Paleo-proterozoic	Statherian	1800
		Orosirian	2050
		Rhyacian	2300
		Siderian	2500
Precambrian — Archean	Neo-archean		2800
	Meso-archean		3200
	Paleo-archean		3600
	Eo-archean		4000
Hadean			~ 4600

Units of all ranks are in the process of being defined by Global Boundary Stratotype Section and Points (GSSP) for their lower boundaries, including those of the Archean and Proterozoic, long defined by Global Standard Stratigraphic Ages (GSSA). Italic fonts indicate informal units and placeholders for unnamed units. Versioned charts and detailed information on ratified GSSPs are available at the website http://www.stratigraphy.org. The URL to this chart is found below.

Numerical ages are subject to revision and do not define units in the Phanerozoic and the Ediacaran; only GSSPs do. For boundaries in the Phanerozoic without ratified GSSPs or without constrained numerical ages, an approximate numerical age (~) is provided.

Ratified Subseries/Subepochs are abbreviated as U/L (Upper/Late), M (Middle) and L/E (Lower/Early). Numerical ages for all systems except Quaternary, upper Paleogene, Cretaceous, Triassic, Permian and Precambrian are taken from 'A Geologic Time Scale 2012' by Gradstein et al. (2012), those for the Quaternary, upper Paleogene, Cretaceous, Triassic, Permian and Precambrian were provided by the relevant ICS subcommissions.

Colouring follows the Commission for the Geological Map of the World (www.ccgm.org)

CCGM CGMW

Chart drafted by K.M. Cohen, D.A.T. Harper, P.L. Gibbard, J.-X. Fan

To cite: Cohen, K.M., Finney, S.C., Gibbard, P.L. & Fan, J.-X. (2013; updated) The ICS International Chronostratigraphic Chart. Episodes 36: 199-204.

URL: http://www.stratigraphy.org/ICSchart/ChronostratChart2019-05.pdf

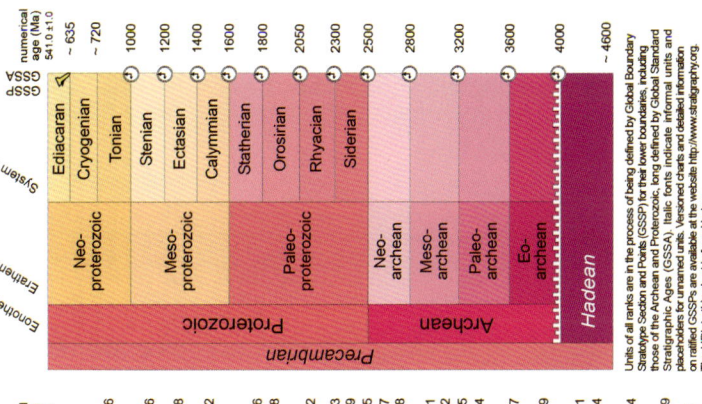

国际年代地层表 v 2018/08

国际地层委员会

www.stratigraphy.org

宇(宙)	界(代)	系(纪)	统(世)	阶(期)	年龄值(Ma)
					现今
显生宇	新生界	第四系	全新统 上/晚	梅加拉亚阶	0.0042
			中	诺斯格瑞比阶	0.0082
			下/早	格陵兰阶	0.0117
			更新统	上阶	0.126
				中阶	0.781
				卡拉布里雅阶	1.80
				杰拉阶	2.58
		新近系	上新统	皮亚琴察阶	3.600
				赞克勒阶	5.333
			中新统	墨西拿阶	7.246
				托尔托纳阶	11.63
				塞拉瓦莱阶	13.82
				兰盖阶	15.97
				波尔多阶	20.44
				阿基坦阶	23.03
		古近系	渐新统	夏特阶	27.82
				吕珀尔阶	33.9
			始新统	普利亚本阶	37.8
				巴顿阶	41.2
				卢泰特阶	47.8
				伊普里斯阶	56.0
			古新统	坦尼特阶	59.2
				塞兰特阶	61.6
				丹麦阶	66.0
	中生界	白垩系	上白垩统	马斯特里赫特阶	72.1 ±0.2
				坎潘阶	83.6 ±0.2
				圣通阶	86.3 ±0.5
				康尼亚克阶	89.8 ±0.3
				土伦阶	93.9
				塞诺曼阶	100.5
			下白垩统	阿尔布阶	~ 113.0
				阿普特阶	~ 125.0
				巴雷姆阶	~ 129.4
				欧特里夫阶	~ 132.9
				瓦兰今阶	~ 139.8
				贝里阿斯阶	~ 145.0

宇(宙)	界(代)	系(纪)	统(世)	阶(期)	年龄值(Ma)
					~ 145.0
显生宇	中生界	侏罗系	上侏罗统	提塘阶	152.1 ±0.9
				钦莫利阶	157.3 ±1.0
				牛津阶	163.5 ±1.0
			中侏罗统	卡洛夫阶	166.1 ±1.2
				巴通阶	168.3 ±1.3
				巴柔阶	170.3 ±1.4
				阿林阶	174.1 ±1.0
			下侏罗统	托阿尔阶	182.7 ±0.7
				普林斯巴阶	190.8 ±1.0
				辛涅缪尔阶	199.3 ±0.3
				赫塘阶	201.3 ±0.2
		三叠系	上三叠统	瑞替阶	~ 208.5
				诺利阶	~ 227
				卡尼阶	~ 237
			中三叠统	拉丁阶	~ 242
				安尼阶	247.2
			下三叠统	奥伦尼克阶	251.2
				印度阶	251.902 ±0.024
	古生界	二叠系	乐平统	长兴阶	254.14 ±0.07
				吴家坪阶	259.1 ±0.5
			瓜德鲁普统	卡匹敦阶	265.1 ±0.4
				沃德阶	268.8 ±0.5
				罗德阶	272.95±0.11
			乌拉尔统	空谷阶	283.5 ±0.6
				亚丁斯克阶	290.1 ±0.26
				萨克马尔阶	293.52 ±0.17
				阿瑟尔阶	298.9 ±0.15
		石炭系 宾夕法尼亚亚系	上	格舍尔阶	303.7 ±0.1
				卡西莫夫阶	307.0 ±0.1
			中	莫斯科阶	315.2 ±0.2
			下	巴什基尔阶	323.2 ±0.4
		密西西比亚系	上	谢尔普霍夫阶	330.9 ±0.2
			中	维宪阶	346.7 ±0.4
			下	杜内阶	358.9 ±0.4

宇(宙)	界(代)	系(纪)	统(世)	阶(期)	年龄值(Ma)
					358.9 ± 0.4
显生宇	古生界	泥盆系	上泥盆统	法门阶	372.2 ±1.6
				弗拉阶	382.7 ±1.6
			中泥盆统	吉维特阶	387.7 ±0.8
				艾菲尔阶	393.3 ±1.2
			下泥盆统	埃姆斯阶	407.6 ±2.6
				布拉格阶	410.8 ±2.8
				洛赫考夫阶	419.2 ±3.2
		志留系	普里道利统		423.0 ±2.3
			罗德洛统	卢德福特阶	425.6 ±0.9
				高斯特阶	427.4 ±0.5
			温洛克统	侯墨阶	430.5 ±0.7
				申伍德阶	433.4 ±0.8
			兰多维列统	特列奇阶	438.5 ±1.1
				埃隆阶	440.8 ±1.2
				鲁丹阶	443.8 ±1.5
		奥陶系	上奥陶统	赫南特阶	445.2 ±1.4
				凯迪阶	453.0 ±0.7
				桑比阶	458.4 ±0.9
			中奥陶统	达瑞威尔阶	467.3 ±1.1
				大坪阶	470.0 ±1.4
			下奥陶统	弗洛阶	477.7 ±1.4
				特马豆克阶	485.4 ±1.9
		寒武系	芙蓉统	第十阶	~ 489.5
				江山阶	~ 494
				排碧阶	~ 497
			苗岭统	古丈阶	~ 500.5
				鼓山阶	~ 504.5
				乌溜阶	~ 509
			第二统	第四阶	~ 514
				第三阶	~ 521
			纽芬兰统	第二阶	~ 529
				幸运阶	541.0 ±1.0

宇(宙)	界(代)	系(纪)	年龄值(Ma)
			~541.0±1.0
前寒武系 元古宇	新元古界	埃迪卡拉系	~ 635
		成冰系	~ 720
		拉伸系	1000
	中元古界	狭带系	1200
		延展系	1400
		盖层系	1600
	古元古界	固结系	1800
		造山系	2050
		层侵系	2300
		成铁系	2500
太古宇	新太古界		2800
	中太古界		3200
	古太古界		3600
	始太古界		4000
冥古宇			~ 4600

所有全球年代地层单位均由其底界的全球界线层型剖面和点位（GSSP）界定，包括长期由全球标准地层年龄（GSSA）界定的太古宇和元古宇各单位。图件及已批准的GSSP的详情参见国际地层委员会官网。本图件的网址见右下角。

年龄值仍在不断修订；显生宇和埃迪卡拉系的单位不能由年龄界定，而只能由GSSP界定。显生宇中没有确定GSSP或精确年龄值的单位，则标注了近似年龄值（~）。

已批准的亚统/亚世简写为上/晚、中、下/早；第四系、古近系上部、白垩系、三叠系、二叠系和前寒武系的年龄值由各分会提供；其他年龄值引自Gradstein等主编《地质年代表2012》。

各单位的颜色依据世界地质图委员会的色谱（http://www.ccgm.org）。

本图件由K.M. Cohen、D.A.T. Harper、P.L. Gibbard和樊隽轩绘制。

引用：Cohen, K.M., Finney, S.C., Gibbard, P.L. & Fan, J.-X. (2013; updated) The ICS International Chronostratigraphic Chart. Episodes 36: 199-204.

http://www.stratigraphy.org/ICSchart/ChronostratChart2018-08Chinese.pdf